Why Birds Matter

Why Birds Matter

Avian Ecological Function and Ecosystem Services

ÇAĞAN H. ŞEKERCIOĞLU,
DANIEL G. WENNY, AND
CHRISTOPHER J. WHELAN, EDITORS

THE UNIVERSITY OF CHICAGO PRESS CHICAGO AND LONDON

ÇAĞAN H. ŞEKERCIOĞLU is professor in the Department of Biology at the University of Utah and distinguished visiting fellow at Koç University of Istanbul.
DANIEL G. WENNY is senior landbird biologist at the San Francisco Bay Bird Observatory and visiting research scholar at the Museum of Vertebrate Zoology, University of California, Berkeley.
CHRISTOPHER J. WHELAN is visiting research associate professor in the Department of Biological Sciences at the University of Illinois at Chicago.

The University of Chicago Press, Chicago 60637
The University of Chicago Press, Ltd., London

Printed in the United States of America

25 24 23 22 21 20 19 18 17 16   1 2 3 4 5

ISBN-13: 978-0-226-38246-3 (cloth)
ISBN-13: 978-0-226-38263-0 (paper)
ISBN-13: 978-0-226-38277-7 (e-book)
DOI: 10.7208/chicago/9780226382777.001.0001

Library of Congress Cataloging-in-Publication Data

Names: Şekercioğlu, Çağan, editor. | Wenny, Daniel G., editor. |
Whelan, Christopher J. (Christopher John), 1958– editor.
Title: Why birds matter : avian ecological function and ecosystem services /
Çağan H. Şekercioğlu, Daniel G. Wenny, and Christopher J. Whelan, editors.
Desciption: Chicago ; London : The University of Chicago Press, 2016. |
Includes bibliographical references and index.
Identifiers: LCCN 2015049863 | ISBN 9780226382463 (cloth : alk. paper) |
ISBN 9780226382630 (pbk. : alk. paper) | ISBN 9780226382777 (e-book)
Subjects: LCSH: Birds—Ecology. | Birds—Environmental aspects.
Classification: LCC QL698.95 .W49 2016 | DDC 598—dc23 LC record
available at http://lccn.loc.gov/2015049863

♾ This paper meets the requirements of ANSI/NISO Z39.48-1992 (Permanence of Paper).

Contents

Foreword

"Birdwatching, huh? You mean you just *watch* them? What's the point?"

Millions of people watch birds, some casually, some devoting staggering amounts of time and resources to an avocation that can border on obsession. And yet I'll wager that nearly all birders have repeatedly faced some variant of the line of questioning above. No matter how essential an awareness and appreciation of birds may seem to those of us who already love them, lots of people—many of whom are in positions that involve making decisions with significant ecological impact—just don't get it. Why should anyone care about birds?

One of the core virtues of this book is that the editors, Çağan Şekercioğlu, Daniel G. Wenny, and Christopher J. Whelan, along with their contributors, pursue the question of the value of birds along a broad front. Even as they seek, with authority and rigor, to explicate the many ways in which birds and their activities add real, measurable value to human economies, they never lose sight of the value birds have in their own right.

A single flock of snow geese might, over the year, provide various people with food, clothing, recreation, an aesthetic thrill, and a deep, even spiritual sense of connection to the passing of seasons and time. That same flock might also provide a level of ecosystem disservice, perhaps overgrazing certain habitat areas. But apart from all these human-assigned, instrumental values, there is the intrinsic value of the geese themselves: sentient, social beings amazingly adapted to some truly challenging conditions. This book is large enough in scope and wide enough in outlook to embrace all these things.

If we are to successfully advocate for the conservation of the birds and the habitats that have done so much to sustain our own lives, we must adopt just such an outlook because we need as many arrows in our quivers as we can get. In one situation, the simple aesthetic beauty of birds and bird song may be persuasive; in others, dollars and cents may be the only language spoken. Most often, though, it will take a combination of approaches to carry the day and build consensus around bird conservation.

Why Birds Matter is a most welcome example of how scientific ornithology undergirds and enriches the pleasure we get from even informal, recreational birdwatching. Birders who read it will come away with a new understanding and appreciation of just what a contribution birds make to our world and to our lives. They'll also be better able to answer the persistent queries of skeptics, likely winning converts to bird appreciation and conservation in the process. And that matters a great deal indeed.

Jeffrey A. Gordon
President, American Birding Association
Delaware City, Delaware
August 2015

Preface

As we are the editors of this volume, it should go without saying that to us and to all the contributors, as well as to countless people around the world, birds matter. Each of us has a unique story for how we became passionate about birds.

Birds matter in many ways—maybe as many as there are people who care about them. In this volume, we examine how birds matter from the perspective of ecosystem services, those aspects of nature that benefit humans.

Despite the anthropocentric perspective of the ecosystem services documented here, we and this book's contributors know that birds matter in many ways that transcend this utilitarian perspective. Indeed, birds indirectly benefit humans by facilitating other biological processes or products that humans use. We will not lose sight of these other roles of birds, and we hope that other advocates for bird conservation and research do not lose sight of them either.

The contributions that birds and other taxa make to human well-being are profound; they deserve recognition by policy makers and by the public at large. Of course, if the public recognizes the value of birds and votes accordingly, policy makers will do so as well.

The need to inform the public about the contributions of birds to ecosystem services became abundantly clear when one of us (CJW) gave a public presentation on the subject in a rural area southwest of Chicago dominated by soy and corn agriculture. When we presented the results of a Dutch study showing that bird predation on apple-damaging insects increased the apple yield by 66%, two farmers in the audience asked: "Why isn't this information public?"

Of course the information is available to the public, but those of us who are most privy to it do a bad job of getting it out beyond the community of professionals. Hence, we designed this book to convey the current science in a way that an educated nonscientist could relate to and understand. For helping us realize such a book, we are grateful to the University of Chicago Press, which has supported us since the book's conception in 2007, and to our dedicated colleagues who have contributed chapters. We also thank in particular Christopher Chung and Evan White, both of whom helped steer the book through the review and publication process at the Press.

Despite the importance of birds to the functioning of ecosystems and their provision of ecosystem services to people, we continue to lose bird species, more of which are at great risk of extinction than ever. The most striking examples of the ecological consequences are the vultures of South Asia, particularly India. As a result of poisoning with the veterinary drug diclofenac, vulture populations in India crashed 100- to 1,000-fold in the last decade of the 20th century. Despised by many people, vultures provide critical sanitary services in India, and indeed around much of the world. The decline of vultures, and thus the decline of their sanitary services, in South Asia has led to irruptions of rats and feral dogs, many of which carry rabies, and rabies deaths in India have thus increased. Two economists have calculated that the disappearance of vultures has led to approximately 48,000 additional human deaths due to rabies, and has carried an economic cost of $34 billion (from 1992 to 2006). Furthermore, adherents of the Parsi religious sect have experienced a spiritual crisis because they can no longer leave their dead to the vultures in their traditional "sky burial"; they now have changed their method of disposing the bodies of their loved ones. The fact that vultures are by far the most threatened functional group of birds is an unfortunate but telling indication of how little humanity values birds' crucial ecosystem services.

We hope that in covering the gamut of birds' ecosystem functions and services, from pollination to ecosystem engineering, this book provides an informative and entertaining overview of birds' crucial and fascinating contributions to the functioning of our planet and to the well-being of humanity. Several chapters have extended tables and additional online content that can be viewed at www.press.uchicago.edu/sites/whybirdsmatter/.

Çağan H. Şekercioğlu, Daniel G. Wenny, Christopher J. Whelan

CHAPTER ONE

Bird Ecosystem Services

Economic Ornithology for the 21st Century

Christopher J. Whelan, Çağan H. Şekercioğlu, and Daniel G. Wenny

FIGURE 1.1. *Earthrise*

The iconic photograph *Earthrise*, taken by astronaut William Anders as the US space mission *Apollo 8* first orbited the moon, captured as never before the isolation of our blue planet in the vast blackness of space. The fragility, interconnectedness, and mutual dependencies of the myriad inhabitants of Spaceship Earth depicted in this stunning photograph propelled the environmental movement around the world.

In this volume, we explore one of the lasting legacies of *Earthrise* and the environmental movement it helped spark. Ecosystem services are those aspects of the earth that benefit humans (Şekercioğlu 2010). The history of ecosystem services has been explored in depth elsewhere (chapter 2; Daily 1997; Gómez-Baggethun et al. 2010). After the term was coined by Ehrlich and Mooney (1983), the number of scientific papers that address the subject increased slowly for a number of years, but soon exploded exponentially (fig. 1.2). Investigation of the ecosystem services provided by a variety of taxa is now well underway (fig. 1.3), though much work lies ahead. Here we provide an update on the state-of-the-art regarding ecosystem services provided by birds.

The Millennium Ecosystem Assessment (MA 2003) brought the concept of ecosystem services to the forefront of policy debate throughout the world (Gómez-Baggethun et al. 2010). The major objectives of the MA were to evaluate the potential consequences of ecosystem change from a broad perspective of human well-being, with an emphasis on ecosystem services. The MA (2003) identified four classes of ecosystem ser-

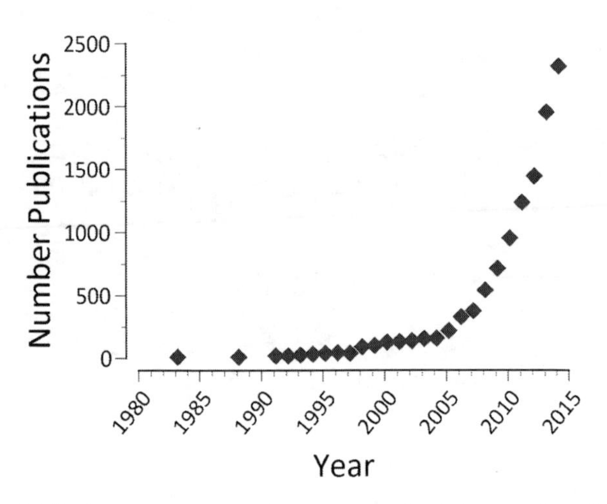

FIGURE 1.2. Publications using the term "ecosystem services," by year of publication; The numbers were generated by a search of ISI Web of Science through 2015, using the search term "ecosystem services." The term's use increased exponentially after the publication of the Millennium Ecosystem Assessment (2003), which brought the concept of ecosystem services to the forefront of the global policy debate.

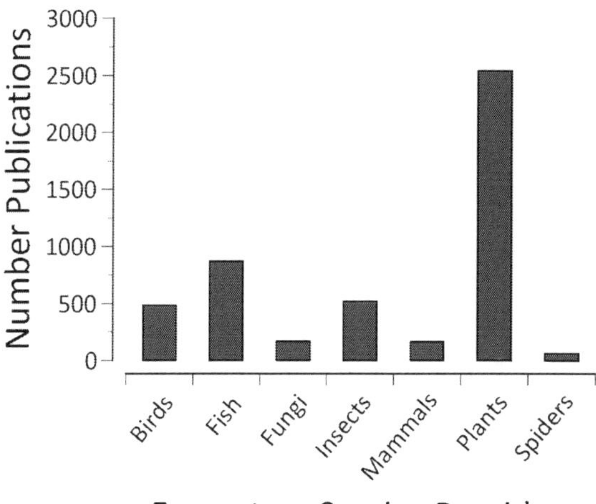

FIGURE 1.3. Publications using the term "ecosystem services" contingent upon taxonomic grouping. The numbers were generated by multiple searches of ISI Web of Science through 2015, using the search terms "ecosystem services birds," "ecosystem services fish," and so on, for each major taxonomic grouping.

vices: "**provisioning services** such as food, water, timber, and fiber; **regulating services** that affect climate, floods, disease, wastes, and water quality; **cultural services** that provide recreational, aesthetic, and spiritual benefits; and **supporting services** such as soil formation, photosynthesis, and nutrient cycling." We will use these four classes throughout this volume.

Two important offshoots of the MA include the Natural Capital Project (NatCap) and the Intergovernmental Platform on Biodiversity and Ecosystem Services (IPBES). NatCap, a partnership among the Nature Conservancy, the World Wide Fund (WWF) for Nature, and Stanford University, promotes scientifically rigorous approaches and tools to incorporate the value (natural capital) of ecosystem services into both public and private investment and development decisions. IPBES, founded in April 2012, aims to provide a forum for the scientific community, governments, and other stakeholders, for dialog and exchange of information centered on biodiversity and ecosystem services. IPBES is an independent intergovernmental body open to all member countries of the United Nations. Both NatCap and IPBES represent mechanisms by which the different types of values of nature, including economic value, can be discussed and accounted for in a wide range of policy-formation processes.

Birds and Ecosystem Services

Birds contribute the four types of ecosystem services recognized by the MA. Provisioning services are provided by both domesticated (poultry; Larson 2015) and nondomesticated species. Birds have long been important components of human diets (Moss and Bowers 2007), and many species, particularly waterfowl (Anatidae) and landfowl (Galliformes), are still today (Peres 2001; Peres and Palacios 2007). In developed countries, many birds are hunted for consumption and recreation (chapter 6; Bennett and Whitten 2003; Green and Elmberg 2014). In some developing countries, many species are hunted for subsistence. Bird feathers provide bedding, insulation, and ornamentation (Green and Elmberg 2014). As discussed by DeVault et al. (chapter 8), scavengers contribute regulating services, as efficient carcass consumption by obligate and facultative scavengers helps regulate human disease. Through their place in art, photography, religious custom, and bird-watching, birds contribute cultural services. Bird-watching, or birding, is one of the most popular outdoor recreational activities in the United States and around the world (see below). Numerous bird species contribute supporting services, as their foraging, seed dispersal, and pollination activities help maintain ecosystems that humans depend upon for recreation, natural resources, and solace throughout the world (Şekercioğlu 2006a; Wenny et al. 2011; Sodhi, et al. 2011). However, the decline of bird populations worldwide, especially those of more specialized species (Şekercioğlu 2011), means that birds' ecosystem services are also declining.

Birds possess a variety of characteristics making them particularly effective providers of many ecosystem services. Most birds fly, so most are highly mobile, with high mass-specific metabolic rates (hence, high metabolic demands). These characteristics allow birds to respond to irruptive or pulsed resources in ways generally not possible for other vertebrates. Their mobility also allows them to track resource abundance, vacating areas in which resources are no longer sufficient and moving to areas where they are. Because many species are migratory, birds link geographic areas separated by great distances over a variety of temporal scales. Different bird species exhibit a wide range of social structures during any given phase of the annual cycle. For example, many species are territorial when breeding, but others breed colonially. Finally, in various bird species, social structure changes dramatically between breeding and nonbreeding phases of the annual cycle. For many such species, breeding communities are composed of individual pairs of relatively low density, owing to intraspecific (and sometimes interspecific) territoriality. In the nonbreeding season, these species may form heterospecific flocks that can attain extremely high densities. Such differences in social structure may produce large differences in avian impact on the environment.

Importance of natural history

Many of the ecosystem services provided by birds arise through their ecological functions. Evaluating the value of these services requires sound knowledge of natural history—especially resource exploitation, habitat requirements, and interactions with other species (Şekercioğlu 2006a,b; Whelan et al. 2008; Wenny et al. 2011). Following the MA classification, most bird services are supporting and regulating services, and of these, most result from bird resource exploitation. Most of the regulating and supporting services arise via top-down effects of resource consumption (chapter 3). With more than 10,600 bird species on earth, birds consume a wide variety of resources in terrestrial, aquatic, and aerial environments. In many cases, the consumed resource is a pest of agricultural crops or forests. In other cases, bird resource consumption facilitates pollination (chapter 4), or movement and deposition of seeds, promoting successful plant reproduction in a large number of plant species (chapters 5, 6, and 7). When the plants are of economic or cultural significance, these services benefit humans. An extremely important bird regulating service is consumption of animal carcasses (chapter 8). Of particular significance is the global reach of these services, particularly insect control, seed dispersal, and scavenging. Through these services, birds have a large but as yet mostly unquantified impact on ecosystems. Developing methods to quantify bird impact in this functional sense, along with methods to monetize or place value on it, is critical. Clearly, the first step is to thoroughly understand the natural history involved. Key issues include:

How many species perform particular ecological functions?
How variable in space and time are they?
What ecological factors promote them?

Direct and Indirect Ecosystem Services

Some ecosystem services benefit humans directly, as when eider down is used as insulation in jackets, vests, and sleeping bags. Ecosystem services may also benefit humans indirectly, as when seed dispersal of plant species not used directly by humans as commodities nonetheless supports other plant and/or animal species that humans do depend on for utilitarian purposes. Cultural and provisioning services tend to fall into the direct service category, whereas regulating and supporting services tend to fall into the indirect category. As the following chapters describe in detail, the majority of ecosystem services provided by birds are indirect supporting services. Because of this, the ecological roles of birds are not usually included specifically in models assessing ecosystem services (e.g., Kareiva et al. 2011). Yet the examples throughout this book suggest that birds' ecological roles—and, therefore, their ecosystem services—are critical to the health of many ecosystems and to human well-being.

Cultural Services Provided by Birds

Humans and birds have a long history of interaction, dating back thousands of years (Podulka et al. 2004). Examples of this long history include the 16,500-year-old cave paintings in Lascaux, France, which clearly depict birds, and 3,000-year-old murals of ancient Egyptians with domesticated ducks and cranes. These two examples illustrate that while human-bird interaction was important enough for ancient peoples to document, the process of documentation transcends the original use of birds as commodities, becoming something of larger cultural significance. While the domestication of birds for food was no doubt important for the Egyptians, the murals depicting those scenes transform a product-driven ecosystem service with quantifiable economic value into an intangible social and cultural service. Fujita and Kameda (chapter 9) discuss a modern-day equivalent, when a cormorant colony husbanded for guano over 100 years was transformed, owing to the adoption of modern fertilizers, into a symbol and source of sentimental attachment by a community in Japan.

Birds often embody symbolic values and important roles in mythology and religion across many cultures, from ancient times to today (Groark 2010; Mazzariegos 2010). Particular bird species, owing to their majestic appearance, power of flight or sheer beauty (both visual and vocal), were considered symbols of deities, and today are used as mascots and even national symbols (e.g., the bald eagle [*Haliaeetus leucocephalus*], representing the United States of America). In Guatemala the national currency is called the quetzal, after the national bird, the resplendent quetzal (*Pharomachrus mocinno*). Birds have variously represented omens of hope, wisdom, and fear.

Birds and Recreation

Recreational services provided by birds can be quantified accurately. Prominent among them is bird-watching, or birding. As of 2008, some 81 million people in the United States enjoy watching birds, and this population is projected to increase to 108 million people by 2030 (White et al. 2014). In 2001 alone, US birders spent $32 million enjoying their hobby, creating $85 million in indirect economic impact and supporting nearly one million jobs (LaRouche 2001; see also Şekercioğlu 2002). The rise in popularity of bird-watching, especially in the past 50 years, has spawned field identification guides, an entire genre of books (Dunlap 2011) that has also expanded to cover other taxa. Similarly, the demand for bird-watching trips contributes to the current boom in ecotourism. An offshoot of recreational interest in birds is the rise of citizen science programs that use knowledgeable volunteers to help monitor bird populations on a large geographic scale (e.g., www.ebird.org; Abolafya et al. 2013). Bird-watching is an international industry employing guides in many countries, including developing countries, where it can be a significant source of income (Şekercioğlu 2002). Some of these endeavors contribute proceeds to the preservation of land for bird conservation, often of critically endangered birds (Stevens et al. 2013).

More than 37 million people visited the 97 million acres encompassing the National Wildlife Refuge System in the United States in 2006 (Carver and Caudill 2007). There, they engaged in nonconsumptive activities like observing and photographing wildlife, hiking, and environmental education, as well as consumptive activities like fishing and hunting. Eighty-two percent of total expenditures were spent on nonconsumptive activities, 12% were spent on fishing, and the remaining 6% were spent on hunting. Beyond the National Wildlife Refuge system, more than 87 million Americans (38% of the US population 16 years old or over) spent some $120 billion pursuing outdoor recreation of some kind. Of those 87 million Americans engaged in outdoor recreation, about 48 million spent that time birding, a 5% increase over 2001.

Falconry

The art or sport of falconry, the "sport of kings," dates to 2000 BCE in Mesopotamia, and it remains popular today in many countries. Traditionally, actual falcons (genus *Falco*) were the bird of choice. Today, species in various other genera (e.g., *Accipiter*, *Buteo*, *Circus*) are often used. Unfortunately, the great value placed on certain species (e.g., the peregrine falcon, *Falco peregrinus*) has over the years spawned an illicit trade for those species, which can fetch extraordinary prices on the black market (Wyatt 2011). For example, saker falcons (*Falco cherrug*) are globally endangered as a result of the falconry trade (Dixon et al. 2011).

Cage Birds

Hundreds of millions of people keep cage birds, a practice that is increasingly impacting bird populations worldwide, particularly in Asia (BirdLife International 2013a). Captive-bred birds in accredited zoos can raise awareness, educate the public, and contribute to bird conservation. A few conservationists have even succeeded in involving cage bird enthusiasts in bird conservation programs based on captive breeding, most famously exemplified by the Spix' Macaw (*Cyanopsitta spixi*; BirdLife International 2013b). However, the cage bird trade overall threatens wild bird populations worldwide (BirdLife International 2013a).

Birds in Art

Birdlife International (2008) reports that Herbert Friedmann (1946) investigated Renaissance religious paintings and found that the European goldfinch (*Carduelis carduelis*) appears frequently in them. The goldfinch is often held by an infant Jesus, apparently symbolizing a variety of religious themes such as the soul, resurrection, sacrifice, and death. Artists include Leonardo da Vinci (*Madonna Litta*, 1490–91), Raphael (*Solly Madonna*, 1502, and *Madonna of the Goldfinch*, 1506), Zurbarán (*Madonna and Child with the Infant St. John*, 1658) and Tiepolo (*Madonna of the Goldfinch*, 1760; Birdlife International 2008). Birds are often depicted or have inspired various musical compositions, such as Vivaldi's Flute Concerto in D ("The Goldfinch"), Oliver Messiaen's *Réveil des Oiseaux*, and Jerry Garcia's "Bird Song." On the other hand, birds are often portrayed inaccurately in movies and on television (Chisholm 2007). Examples include the numerous instances of red-tailed hawk (*Buteo jamaicensis*) vocalizations used to convey the fierceness of a bald eagle (*Haliaeetus leucocephalus*) or even of the mostly nonvocal turkey vulture (*Cathartes aura*).

Birds in Science

Birds have long played a critical role in the sciences (Konishi et al. 1989). From the age of Aristotle, who speculated on the appearance and disappearance over the annual cycle of certain species (now known as migration), birds have stimulated and challenged the human intellect. Study of birds has contributed to the fields of navigation (Griffin 1964; Wiltschko and Wiltschko 2013), aerodynamics (Kantha 2012), ecology (MacArthur 1972), evolutionary ecology (Lack 1947), neurobiology (Nottebohm 1984), and physiology (Karasov 1990), among numerous others. Birds were also the inspiration for the initial large-scale citizen science projects, the Christmas Bird Count and the North American Breeding Bird Survey (Altshuler et al. 2013), and they continue to inspire such projects today (e.g., ebird.org; Greenwood 2007). The data gathered by bird-watching has led to important conservation findings, such as projection of the effects of climate change on bird distributions (Abolafya et al. 2013; Wormworth and (Şekercioğlu 2011).

Bird Ecosystem Services: Economic Ornithology for the 21st Century?

The last two decades of the 19th century, an age of expansion and exploration of the North American continent by a growing United States, marked an era of great interest in how both native fauna and flora potentially impact the economic well-being of the country. Although not recognized formally as ecosystem services, the notion that humans may gain benefits from nature clearly was intellectually embraced by an enthusiastic cadre of students of natural history. From early studies on food habits of native fauna, for example, by Stephen Forbes of the Illinois Natural History Survey, to systematic expeditions scrupulously cataloging wildlife and flora in far reaches of the continent and around the world, the field of economic ornithology was embraced enthusiastically by naturalists at the end of the 18th century. The meteoric speed with which the burgeoning field took off, however, was matched by its meteoric demise. We believe there are lessons to be learned in this rise and fall for today's students of ecosystem services.

The Rise and Fall of Economic Ornithology

> The early New England settlers were troubled by some birds against which they declared war, and cheered by others to which they extended the offerings of friendship. —Clarence M. Weed and Ned Dearborn, 1903

As any fruit or vegetable gardener can attest, animal marauders, from rodents to songbirds, often greatly reduce the anticipated bounty. Such must have been true since the earliest days of agriculture. As a taxonomic group, birds consume almost every resource imaginable. Hence, from the perspective of human agriculture, birds can obviously compete with humans for many crops, both plant (e.g., cereal grains like wheat, rice, and oats; beans such as lentils) and animal (e.g., poultry). Until fairly recently in human history, birds and other wildlife, including many mammals, were largely viewed as pests that competed, literally in many cases, for the fruits of human labor.

A pest is any organism that decreases the fitness, population size, growth rate, or economic value of some resource important to humans. Casual observers may often detect a wide variety of animal species, including many bird species, in gardens and agricultural fields. Such observations may suggest that these animals are using our gardens or crops as foraging areas, and, moreover, that they are pests—they are "raiding" our crops. For such species to be pests, however, they must actually consume the crops, and their consumption must be of sufficient magnitude that it decreases the crops' value. This decrease could be quantified in a variety of ways: from the yield and/or the monetary value earned from harvest, or from the number of joules (or nutrients) delivered to potential consumers.

Pest control is a reduction in the effect of a pest that results in an increase in the fitness, population size, growth rate, or economic value of a resource important to humans. An insectivorous bird may thus decrease the population size of an insect pest (exerting top-down effects), yet not in sufficient magnitude to constitute pest control. In other words, to constitute pest control, the direct negative effect of a pest control agent on the pest must cascade into a positive indirect effect on the resource consumed by the pest to an extent that is economically important to humans.

BIRDS AS PESTS. From the perspective of food chains or food webs, the groups of birds most likely to be considered pests are those that consume resources valued by humans. Most frequently these resources will be agricultural or horticultural crops, such as cereal grains (wheat, oats, corn), legumes (beans), fruits (apples, berries), and nuts (walnuts, pecans). In some cases, birds may damage human structures or pose health risks through defecation (large roosts of blackbirds and European starlings, *Sturnus vulgaris*), nesting (starlings, woodpeckers), and excavating for food or nest cavities (woodpeckers). Increasingly important are the risks some large, flocking bird species (Canada geese, gulls, large birds) pose for colliding with and bringing down aircraft.

BIRDS AS PEST CONTROL AGENTS. Many studies have now examined the top-down effects of birds on invertebrates in a wide variety of natural and agro-ecosystems (chapter 3). Most, though not all, of these investigations have found that birds reduce the population density of invertebrates. Of these studies, many found that the top-down effects on invertebrates cascade to the level of the effects on plants (Whelan et al. 2008; Mäntylä et al. 2011). Less studied are the potential roles of birds of prey as pest control agents of small mammals such as rodents and rabbits, and granivorous birds as pest control agents of agricultural weeds (chapter 3; Whelan et al. 2008).

The Origin of Economic Ornithology in the United States

> In the wilderness, the Great Horned Owl exerts a restraining influence on both the game and the enemies of game. . . . But on the farm or the game preserve, it cannot be tolerated. —Edward Howe Forbush, 1907

Humans must have noticed animals, including birds, pilfering their crops and livestock from the earliest days of primitive agriculture. When they recognized that animals, including birds, could benefit their agricultural labors through pest control is unknown, and bird pest control services are not widely recognized even today. Kirk et al. (1996) report that the herding of ducks through rice paddies to remove pests is an old practice in China, but they do not indicate how long ago it originated. The 1762 importation of the common myna (*Acridotheres tristis*) to Mauritius from India to control the red locust (*Nomadocris septemfasciata*) is apparently the oldest documented introduction of a bird species for insect control (Coppel and Mertens 1977, in Kirk et al. 1996).

The most complete history of early developments leading to the establishment of economic ornithology as a discipline in the United States, where it reached its greatest prominence, is that of Weed and Dearborn (1903; see also Kronenberg 2014). Particularly relevant is their appendix IV, an annotated bibliography of notable publications dating to 1854, selections of which are presented in supplemental table S1.1 (www.press.uchicago.edu/sites/whybirdsmatter/). Two comprehensive early works investigated the food habits of birds in relation to agricultural production, and deserve special mention. In Illinois, Stephen A. Forbes (first director of the Illinois Natural History Survey) published a series of reports on the food habitats of birds as they relate to various agricultural interests (see supplemental table 1.1 for a few examples), culminating in his publication on the regulative effect of birds on insect population dynamics (Forbes 1883). In Wisconsin, F. H. King published a lengthy account of the "economic relations" of birds in that state (King 1882). In this work, King devotes considerable effort to defining the various ways in which birds provide humans a service, versus those in which they cause harm.

Interest in both the positive and the negative roles of birds spurred the US Congress to appropriate $5,000 to the US Department of Agriculture in the 19th century to establish its Section of Economic Ornithology. Originally directed by the physician and naturalist Dr. C. Hart Merriam, the mission of the section was to investigate the food, habits, and migrations of birds in relation to both insects and plants. In 1886 the section expanded to division status as the Division of Economic Ornithology and Mammalogy. It then grew quickly, becoming the Division of Biological Survey in 1896 and the Bureau of Biological Survey in 1905.

Early work of the Division of Economic Ornithology and Mammalogy concentrated on the relationship of birds and agriculture, with an emphasis on both positive and negative effects of birds on agricultural production. A goal of the division was to supply the growing agricultural industry with practical and useful information. Consequently, work over the first few years concentrated on food habits of birds and mammals believed to be injurious, beneficial, or possibly both to agriculture. A clear idea of the scope of what the division considered to fall under the heading of agriculture at that time is evident in the "prefatory letter" that C. Hart Merriam wrote in the first bulletin prepared by the division: "The term agriculture here must be understood in its broadest and most comprehensive sense as including the grain-growing industries, truck-gardening, fruit-growing, the cultivation of flowers and ornamental shrubs and vines, and even forestry" (Barrows 1889).

That first bulletin, published in 1889, was a 405-page treatise on the house sparrow (*Passer domesticus*) in North America by assistant ornithologist Walter Barrows. This work typifies characteristics of the many subsequent bulletins in the series in a number of respects. First, it is built upon a compilation of solicited observations resulting from the distribution of circulars of inquiry to agriculturists and naturalists around the country and other sources (such as members of the American Ornithologists' Union). Second, a large number of bird stomachs are examined, and their contents categorized. Finally, building upon the evidence thus assembled, recommendations are suggested for management of the species. In this particular instance, the evidence on diet and distribution reinforced a perception of the English or house sparrow as a pest species, and management recommendations focused on methods for reducing if not eliminating populations through means such as poisons and traps.

From 1885, when the division was first established, through 1929, when Cameron (1929) summarized the history and activities of the Bureau of Biological Survey, the bureau contributed a wide variety of publications, including

- annual reports (as part of the Annual Report of the Department of Agriculture);
- technical bulletins on subjects including bird migration, the relations of certain bird and mammal species to agriculture, and the natural histories of particular bird and mammal species;
- farmers' bulletins, reporting information of interest to farmers in nontechnical language, and the North American Fauna series, which included systematic treatises, accounts of natural history of particular geographical locations, and results of biological surveys of particular locales;
- miscellaneous reports on a variety of subjects, such as annotated lists of particular taxa collected during biological expeditions;
- and circulars—some distributed by mail, requesting submission of personal observations, and some reporting useful and practical information for farmers and others interested in birds and mammals.

Cameron (1929) provides an exhaustive compilation communicated by the editorial office of the Bureau of Biological Survey.

Under the leadership of Dr. Merriam, the Division of Economic Ornithology and Mammalogy quickly expanded its area of inquiry beyond the food habits of birds and mammals to faunal distributions. Division scientists conducted faunal surveys within the contiguous United States, particularly in the yet sparsely populated Western states, and also mounted expeditions to more distant locations such as Alaska, British Columbia, Mexico, and the West Indies. Over the succeeding years, the name and affiliations of the division changed, and today it exists as the Biological Survey Unit, part of the United States Geological Survey. Detailed histories can be found in Cameron (1929), Henderson and Preble (1935), and R. D. Fisher (http://www.pwrc.usgs.gov/history/bsphist2.htm).

Historical Case Studies from the Division of Economic Ornithology and Mammalogy

Many of the early bulletins focusing on economic ornithology were contributed by well-known figures in the history of American ornithology,

Walter D. Barrows, Foster E. L. Beal, Wells W. Cooke, Sylvester D. Judd, and W. L. McAtee among them. Here we examine in some detail three case studies published by Bureau scientists.

HAWKS AND OWLS OF THE UNITED STATES. A. K. Fisher (1893) examined the 73 species and subspecies of hawks and owls recognized at that time in the United States for bulletin 3 of the division. At the time, birds of prey were widely viewed as "enemies of the farmer," with some states even offering bounties to promote their destruction. Fisher examined the stomachs of more than 2,200 birds and found that, with the exception of six species, both hawks and owls largely consumed rodents and other small mammals, as well as insects. In his letter of transmittal, Dr. C. Hart Merriam wrote: "The result proves that a class of birds commonly looked upon as enemies to the farmer, and indiscriminately destroyed whenever occasion offers, really rank among his best friends, and with few exceptions should be preserved, and encouraged to take up their abode in the neighborhood of his home." This work may have helped change the widespread condemnation of hawks and owls amongst the general public at that time within the United States. Weeds and Dearborn (1903) report that following the publication of this bulletin, many states and localities repealed or stopped enforcing bounties on birds of prey, and some states even passed laws protecting them.

METHODS IN ECONOMIC ORNITHOLOGY. Determination of diet through a combination of controlled experimentation with captive birds and observational studies is exemplified in a publication in the *American Naturalist* by Sylvester D. Judd (1897). Judd notes that to "ascertain the food of any bird, and to determine its relation to agriculture" one typically relies on observations of foraging birds, a method often yielding fragmentary knowledge from which false conclusions are derived. Judd continues that the final arbiter, examination of stomach contents, demonstrates conclusively what food a bird has selected, but sheds no light on what available items it has rejected. He next demonstrates how field observation, combined with collection of observed birds and examination of gut contents, could be used to determine which available foods a given bird species, in this case gray catbirds (*Dumatella carolinensis*), will consume, and in what order of preference. Judd proceeds to relate experiments with captive catbirds in which a series of insects and other arthropods, as well as various fruiting species, were offered to them, and their acceptance or rejection noted. Through careful presentation of fruits, Judd found that mulberries were preferred to cherries and strawberries. He suggested that both cherries and strawberries, important agricultural crops, could be protected by planting mulberries close to wherever cherries and strawberries were cultivated. Judd concludes that through knowledge of what birds will eat and of what is available where they are foraging, "it will be possible to ascertain what the bird will eat, its preferences, and what it will refuse."

BIRDS OF A MARYLAND FARM. Judd (1902) obtained agreement to use the O. N. Bryan farm, along the south Maryland shore of the Potomac River approximately 24 km (15 miles) west of Washington, DC, to investigate the habitat use and food habits of birds over a seven-year period (July 1895 to July 1902). Judd noted that determining the food habits of a bird species typically relies on examination of the stomach contents of birds collected over much of its range. This method suffers from three large drawbacks. First, the results are necessarily composite and do not reveal specialization of diet attuned to local food availability. Second, there is no way to know what potential prey items are in fact not taken, which is particularly problematic when rejected prey are themselves agricultural pests. Third, it is impossible to evaluate the potential service to the farmer, since stomach contents do not reveal local infestations of particular crops. Judd contended that these methodological shortcomings can be overcome through a thorough accounting of food habits over an extended period of time at a single location.

Over the seven years of study, Judd determined the annual pulse of insect and other arthropod abundances in the various habitat types found on the Bryan farm and two adjacent farms that offered certain habitat features not found on the Bryan farm itself. Insects (arthropods) were classified as destructive or beneficial. Through direct field observation coupled with analysis of stomach contents, Judd also examined prey selection of both resident and migrant bird species found to use the farms. Judd concluded that a small number of species, including house sparrows, sharp-shinned (*Accipiter striatus*), Cooper's hawks (*A. cooperii*), yellow-bellied sapsuckers (*Sphyrapicus varius*), and American crows (*Corvus brachyrhynchos*), were detrimental and should be eliminated or reduced in numbers. The remaining species, including waterfowl, cuckoos, hummingbirds, woodpeckers, and songbirds, Judd deemed more beneficial than detrimental, owing to their consumption of folivorous insects and other arthropods (in the case of cuckoos, woodpeckers, vireos, thrushes, and warblers) or of weed seeds (in the case of quail and some songbirds).

Judd devoted considerable effort to documenting consumption by various bird species of the seeds of agricultural weeds. Judd reported that 20 different bird species found on the farm consumed seeds of 41 plant species considered to be agricultural weeds. He noted that seed consumption by the birds varied over the annual cycle, with the greatest occurring in the fall and winter. He also noted, however, that even in early spring many granivorous birds, such as song and chipping sparrows, sought out and consumed the developing seeds of newly growing plants—a process we would now classify as predispersal seed predation.

These three early publications illustrate the sorts of approaches used by members of the bureau when attempting to ascertain how various bird species likely affected agricultural production. It is noteworthy that the assessment of the relations of birds to agriculture appears not to be guided by preconceived notions. Based on their attempts to objectively assess actual food habits, bureau ornithologists classified particular bird species as having either positive or negative relations to human agriculture. In the words of Evenden (1995), the relations of birds to human agriculture were based on the "revealed truths" evident from analysis of stomach contents. It should also be noted that to a modern reader, conclusions regarding the "moral" status of birds with respect to their beneficial or injurious natures seems wholly out of place, if not outright bizarre.

Modern concepts, such as direct and indirect trophic interactions (see Abrams 1987; Brown 1994), are well appreciated, even if not referred to as such. For instance, Forbush (1907) wrote: "Let us now go back to the beginning of our chain of destruction. The Eagles, Hawks, Owls, and raccoons may indirectly allow an increase in the number of Robins by preventing too great an increase of the Crow. But Hawks and Owls also prey on the Robin, and, by dividing their attention between Robin and Crow, assist in keeping both birds to their normal numbers. Whenever Crows became rare, Robins as a consequence would become very numerous, were it not that the Hawks also eat Robins."

Criticism of methods and skepticism regarding the work of Biological Survey ornithologists developed in the early 20th century. Three important flaws contributed to the eventual decline of economic ornithology within the Biological Survey. First, when bureau scientists assessed diet through analysis of stomach contents, they mostly used the volumetric method to describe importance of the different components of the diet. In this method, relative bulk representation determines the importance of any component within the diet. The alternative numeric method, which ultimately gained favor in the larger ecological community, instead determined importance of items by frequency of occurrence. Second, a conclusion that a particular bird species (or group of species) was beneficial to agriculture was seldom followed up with practical advice about how farmers could enhance the positive bird effects through some practical manipulation. Third, and more important, the presence of a prey item (arthropod or rodent) in a bird's diet does not indicate the extent to which it may actually affect the prey's population dynamics. In the framework presented earlier, simple consumption of a putative pest organism by a bird does not demonstrate that the bird is exercising control of it. This line of argument led some entomologists to contend that the bureau's economic ornithologists overstated the importance of birds as agents of control. Interested readers should consult Evenden (1995) for a fascinating historical account of the rise and fall of economic ornithology within the USDA. Perhaps paradoxically, coincident with the fall of economic ornithology and the rise of synthetic organic pesticides around 1930, studies from the 1940s through the 1960s which examined birds in relation to agriculture emphasized their potential roles as pests rather than as agents of pest control.

Does Economic Ornithology Have a Role in Modern Agriculture?

Interest in the ecological roles of birds in the ecosystems they inhabit arose again in the 1970s. Although investigations of ecosystem energetics (Wiens 1973; Sturges et al. 1974; Holmes and Sturges 1975) suggested that birds may contribute little to overall ecosystem energy flux, exclosure studies demonstrated that bird predation significantly affected population densities (and, presumably, population dynamics) of arthropods, including some species considered pests (Askenmo et al. 1977; Solomon et al. 1977; Holmes et al. 1979).

Later studies determined that predation of insectivorous birds on herbivorous insects positively affected plant performance (Atlegrim 1989; Marquis and Whelan 1994). In contrast to the methods of economic ornithologists of the Bureau of Biological Survey, these studies explicitly examined the trophic interactions from the perspective of the plant, addressing whether bird predation on insect herbivores increased some component of plant performance or fitness. They also ushered in a renewed interest in birds as potential biological control agents in both natural systems and agroecosystems. Investigators today typically focus their data collection on direct or indirect assessment of a plant's performance, fitness and, in some circumstances, economic consequences for humans. For instance, Mols and Visser (2002) investigated the effect of bird control of herbivorous insects in Dutch apple orchards, and reported that increasing bird density through deployment of nest boxes led ultimately to a 50% reduction in damage per apple, and to an increase of about 60% in total apple crop yield. Koh (2008) attributed the prevention of a 9 to 26% fruit loss in oil palm (*Elaeis guineensis*) to bird pest control. Johnson et al. (2009) found that birds significantly reduced damage caused by the coffee berry borer beetle (*Hypothenemus hampei*), resulting in increased income from coffee yield from US$44 to $310 per hectare (also chapter 2).

Charges that the economic ornithologists of the Biological Survey overstated the case for birds as agents of biological control can be countered with charges that agriculturists often overstated the case for birds as agents of crop depredation. For instance, on the basis of energetics and population size estimates, Weatherhead et al. (1982) estimated that damage to regional corn crops based on subjective governmental surveys overestimated crop losses by an astronomical amount (more than 1,000 times). Their energetics estimates, moreover, were well within empirically based, replicated field estimates (Weatherhead et al., 0.41% of crop damaged; field estimates, 0.25 to 0.80% of crop damaged). Using essentially similar methods, Basili and Temple (1999) estimated comparable magnitudes of regional loss of sorghum (0.37%) and rice (0.73%) crops owing to dickcissel (*Spiza americana*) wintering flocks in Venezuela. This compared with a subjective impression among area farmers of an average loss of 25% of each crop. Subjective impressions and estimates of crop loss are rather easily exaggerated for several reasons, including the size of wintering bird flocks, their great conspicuousness, and their highly localized concentrations.

The need for reliable quantitative assessment of bird-caused losses of agricultural production is necessary to assure that damage prevention does not actually cost more than the reduction in crop yield. A good example is provided by Borkhataria et al. (2012), who used exclosures to estimate consumption by red-winged blackbirds (*Agelaius phoenicius*) both of the rice stinkbug (Hemiptera: Pentatomidae, *Oebalus* spp.), which is a pest of rice, and of the rice itself. They found no reduction in stinkbugs, but a tenfold increase in the percentage of rice panicles that were damaged by blackbirds. Despite the increased damage to rice panicles, they found no difference in the rice yield. The researchers concluded that blackbirds actually remove very few rice grains per panicle, which thus results in an essentially unmeasurable impact on total crop yield. Borkhataria et al. (2012) note that some field studies examining the efficacy of expensive bird repellents in rice fields found no effect on rice yield, in spite of studies demonstrating that the repellent strongly curtails bird foraging in the laboratory. They suggest that, rather than indicating a failure of the repellent in the field, these results may simply show that blackbirds do not significantly impact crop yield.

In light of modern approaches and renewed interest in bird ecosystem function, we suggest that economic ornithology deserves a second look. We conclude with a prospectus highlighting knowledge gaps with the aim of identifying promising avenues for applied ecological research.

Birds of Prey as Agents of Pest Control

Unfortunately, few studies since the early work of the Biological Survey have examined the potential of birds of prey to help control pests in agro-ecosystems. Much existing information suggests a strong potential. Many studies have documented the effects of birds of prey on the population dynamics of various rodent and other small mammal species in a variety of ecosystems around the world (Korpimäki and Krebs 1996). Many raptor species readily inhabit agricultural ecosystems (Williams et al. 2000). Several studies have found that deployment of hunting perches boosts populations and/or activity-density of various hawks and owls (Kay et al. 1994; Wolff et al. 1999; Sheffield et al. 2001), and in some cases decreases rodent population density (Kay et al. 1994). The existing evidence thus suggests (1) that birds of prey have potential to control pests in agro-ecosystems, and (2) that rather simple techniques, such as deployment of perches and nesting platforms or boxes, may facilitate the recruitment of those services (Kross et al. 2012).

As has been argued by Şekercioğlu (2006b), birds of prey may affect prey populations indirectly, by establishing a landscape of fear (Laundre et al. 2001). Prey under risk of predation curtail their use of habitats (Brown and Kotler 2004; Ale and Whelan 2008) and the intensity with which they crop their resources (Abramsky et al. 2002). A large-scale predator exclosure revealed a reduction in home range size and an increase in runways of the herbivorous *Octodon degus* between shrubs, both of which concentrated herbivore foraging intensity when predators were absent (Lagoes et al. 1995). Moreover, some bird species receive indirect "defense" by nesting in closer proximity to birds of prey (Haemig 2001; Ueta 2001;

Quinn et al. 2003; Halme et al. 2004). Taken together, the indirect effects of avian predators on prey and nonprey species—from direct mortality but also, importantly, from behavioral, ecological, and evolutionary influences induced by their presence—affect the prey's use of habitat and the intensity of their resource exploitation. Declines in avian predators will likely result in cascading effects in ecosystems as prey no longer die and are freed from fear of predation.

Granivorous Birds as Agents of Weed Seed Control

Since the studies of Judd (1901, 1902), almost no effort has been made to assess the role of granivorous birds in the abundance and distribution of herbaceous agricultural weeds. Again, the potential seems real (Holmes and Froud-Williams 2005; White et al. 2007). Numerous bird species around the globe are at least seasonal granivores, and many of them consume seeds of native forb and grass species, some of which are weeds in agro-ecosystems. The work of Barrows (1889) and Judd (1901, 1902) suggested that native avian granivores in the United States may preferentially consume the seeds of native herbs and forbs (unlike the nonnative house sparrow, which appears to prefer the seeds of commercial cereal grains). This is an area that deserves careful, quantitative research. Given that many of the passerine granivores are insectivorous during their breeding seasons and granivorous during their nonbreeding seasons, they potentially could serve as both insect and weed seed pest control agents. Various impact studies could be useful. For instance, some bird species, such as American goldfinches (*Spinus tristis*) in North America, often forage on seeds before dispersal, while others seem to forage on seeds after dispersal. What is the relative magnitude of these seed losses? Are they additive or compensatory? Can we manipulate habitat mosaics within agricultural landscapes in ways that encourage habitat use by granivorous birds while maintaining high agricultural yield and efficiency as demonstrated for granivorous insects (Landis et al. 2000)?

Measures to Counteract Bird Disservices

The quantification of avian ecosystem disservices deserves a level of rigor comparable to that demanded for quantifying their ecosystem services (chapter 12). When the disservice arises from apparent consumption of an agricultural crop, useful techniques include use of exclosures, application of energetics models coupled with population estimates, and analysis of stable isotopes to determine diet (e.g. Ferger et al. 2012). In past decades, bird species and populations judged to be responsible for crop damage were often the targets of lethal control methods including shooting, poisons, and traps. Today the emphasis is on nonlethal means such as repellents, diversionary crops, and sonic or mechanical deterrents (e.g., Avery et al. 1997; Avery et al. 2001; Tracey et al. 2007).

Conclusions

Keeping in mind the eventual demise of economic ornithology in the early 20th century, efforts to revitalize the field today must be rigorous and repeatable, with a focus on the tangible measurement of plant yield in managed systems and of plant fitness in natural systems (Whelan et al. 2008; Kronenberg 2014). Cost-benefit analysis may permit the evaluation of bird contribution to ecosystem services and disservices in light of available alternatives (e.g., bird predation versus pesticides). Methods that allow scaling up from experimental plots to entire ecosystems (Whelan et al. 2008; Rogers et al. 2012) are critical.

References

Abolafya, M., Onmus, O., Şekercioğlu, Ç.H., and Bilgin, R. 2013. Using citizen science data to model the distributions of common songbirds of Turkey under different global climatic change scenarios. *PLOS ONE* 8:e68037.

Abrams, P. 1987. Indirect interactions between species that share a predator: Varieties of indirect effects. In *Predation: Direct and Indirect Impacts on Aquatic Communities*, ed. W. C. Kerfoot and A. Sih, 38–54. University Press of New England, Hanover, NH.

Abramsky, Z., Rosenzweig, M. L., and Subach, A. 2002. The costs of apprehensive foraging. *Ecology* 83:1330–40.

Atlegrim, O. 1989. Exclusion of birds from bilberry stands: Impact on insect larval density and damage to the bilberry. *Oecologia* 79:136–139.

Altshuler, D. L. Cockle, K. L. and Boyle, W. A. 2013. North American ornithology in transition. *Biology Letters* 9:20120876.

Askenmo, C., Bromseen, A.V., von Ekman, J., and Jansson, C. 1977. Impact of some wintering birds on spider abundance in spruce. *Oikos* 28:90–94.

Avery, M. L., Humphrey, J. S., and Decker, D. G. 1997. Feeding deterrence of anthraquinone, anthracene, and anthrone to rice-eating birds. *Journal of Wildlife Management* 61:1359–65.

Avery, M. L., Tillman, E. A., and Laukert. C. C. 2001. Evaluation of chemical repellents for reducing crop damage by Dickcissels in Venezuela. *International Journal of Pest Management* 47:311–14.

Barrows, W. B. 1889. The English sparrow (*Passer domesticus*) in North America, especially in its relation to agriculture. *USDA Division of Economic Ornithology and Mammalogy Bulletin 1*. Government Printing Office, Washington.

Basili, G. D., and Temple, S. A. 1999. Dickcissels and crop damage in Venezuela: Defining the problem with ecological models. *Ecology* 9:732–39.

Bennett, J., and Whitten, S. 2003. Duck hunting and wetland conservation: Compromise or synergy? *Canadian Journal of Agricultural Economics* 51:161–73.

BirdLife International. 2008. *The Goldfinch in Renaissance Art*. Presented as part of the BirdLife State of the World's Birds website. Available from http://www.birdlife.org/datazone/sowb/casestudy/95. Accessed 22 August 2013.

———. 2013a. Overexploitation threatens many bird species. Presented as part of the BirdLife State of the World's Birds website. Available at http://www.birdlife.org/datazone/sowb/pressure/PRESS7.

———. 2013b. Species factsheet: *Cyanopsitta spixii*. Downloaded from http://www.birdlife.orgon 16/09/2013.

Borkhataria, R. R., Nuessly, G. S., Pearlstine, E., and Cherry, R. H. 2012. Effects of blackbirds (*Agelaius phoenicius*) on stink bug (Hemiptera: Pentatomidae) populations, damage, and yield in Florida rice. *Florida Entomologist* 95:43–149.

Brown, J. S. 1994. Restoration ecology: Living with the prime directive. In *Restoration of Endangered Species: Conceptual Issues, Planning and Implementation*, ed. M. L. Bowles and C. J. Whelan, 355–80. Cambridge University Press, Cambridge.

Cameron, J. 1929. *Bureau of Biological Survey: Its History, Activities and Organization*. John Hopkins University Press, Baltimore.

Carver, E., and Caudill, J. 2007. *Banking on Nature 2006: The Economic Benefits to Local Communities of National Wildlife Refuge Visitation*. Division of Economics, US Fish and Wildlife Service, Washington.

Chisholm, G. 2007. Movies don't give a hoot. *Washington Post*, Sept. 16, 2007.

Coppel, H. C., and Mertins, J. W. 1977. *Biological Insect Pest Suppression*. Springer-Verlag, Berlin.

Daily, G. C. 1997. *Nature's Services: Societal Dependence on Natural Ecosystems*. Island Press, Washington, DC.

Dixon, A., Batbayar, N., Purev-ochir, G., and Fox, N. 2011. Developing a sustainable harvest of saker falcons (*Falco cherrug*) for falconry in Mongolia. In *Gyrfalcons and Ptarmigan in a changing World*, ed. R. T. Watson, T. J. Cade, M. Fuller, G. Hunt, and E. Potapov, 363–72. Peregrine Fund, Boise, Idaho.

Dunlap, T. 2011. *In the Field, among the Feathered: A History of Birders and Their Guides*. Oxford University Press, New York.

Ehrlich, P. R., and Mooney, H. A. 1983. Extinction, substitution, and ecosystem services. *Bioscience* 33:248–54.

Evenden, M. D. 1995. The laborers of nature: Economic ornithology and the role of birds as agents of biological pest control in North American agriculture, ca. 1880–1930. *Forest and Conservation History* 39:172–83.

Ferger, S. W., Böhning-Gaese, K., Wilcke, W., Oelmann, Y. and Schleuning, M. 2012. Distinct carbon sources indicate strong differentiation between tropical forest and farmland bird communities. *Oecologia* 171:473–86.

Fisher, A. K. 1893. Hawks and owls of the United States in their relation to agriculture. *USDA Division of Ornithology and Mammalogy Bulletin 3*. Government Printing Office, Washington.

Fisher, R. D. http://www.pwrc.usgs.gov/history/bsphist2.htm. Accessed 30 October 2012.

Forbes, S. A. 1883. The regulative action of birds upon insect oscillations. *Bulletin of Illinois State Laboratory Natural History* 1:3–32.

Forbush, E. H. 1907. *Useful Birds and Their Protection*. Massachusetts State Board of Agriculture, Boston.

Friedmann, H. 1946. *The Symbolic Goldfinch: Its History and Significance in European Devotional Art*. Pantheon Books, Washington.

Griffin, D. R. 1964. *Bird Migration*. Doubleday, New York.

Gómez-Baggethun, E., de Groot, R., Lomas, P.L., and Montes, C. 2010. The history of ecosystem services in economic theory and practice: From early notions to markets and payment schemes. *Ecological Economics* 69:1209–18.

Green, A. J, and Elmberg, J. 2014. Ecosystem services provided by waterbirds. *Biological Reviews* 89:105–22.

Greenwood, J. J. D. 2007. Citizens, science and bird conservation. *Journal of Ornithology* 148:S77–S124.

Groark, K. P. 2010. The angel in the gourd: Ritual, therapeutic, and protective uses of tobacco (*Nicotiana tabacum*) among the Tzeltal and Tzotzil Maya of Chiapas, Mexico. *Journal of Ethnobiology* 30:5–30.

Haemig, P. D. 2001. Symbiotic nesting of birds with formidable animals: A review with applications to biodiversity conservation. *Biodiversity and Conservation* 10:527–40.

Halme, P., Häkkilä, M., and Koskela, E. 2004. Do breeding Ural owls *Strix uralensis* protect ground nests of birds? An experiment using dummy nests. *Wildlife Biology* 10:145–48.

Henderson, W. C., and Preble, E. A. 1935. 1885–Fiftieth anniversary notes–1935. *The Survey* 16:59–65.

Holmes, R. J., and Froud-Williams, R.J. 2005. Post-dispersal weed seed predation by avian and non-avian predators. *Agricultural Ecosystems and Environment* 105:23–27.

Holmes, R. T., and Sturges, F. W. 1975. Bird community dynamics and energetics in a northern hardwoods ecosystem. *Journal of Animal Ecology* 44:175–200.

Holmes, R. T., Schultz, J. C., and Nothnagle, P. 1979. Bird predation on forest insects: An exclosure experiment. *Science* 206:462–63.

Johnson M. D., Levy, N. J., Kellermann, J. L., and Robinson, D. E. 2009. Effects of shade and bird exclusion on arthropods and leaf damage on coffee farms in Jamaica's Blue Mountains. *Agroforestry Systems* 76:139–48.

Judd, S. D. 1897. Methods in economic ornithology with special reference to the catbird. *American Naturalist* 31:392–97.

———. 1901. The relation of sparrows to agriculture. In *USDA Division of Economic Ornithology and Mammalogy Bulletin 15*. Government Printing Office, Washington.

———. 1902. Birds of a Maryland farm: A local study of economic ornithology. In *USDA Division of Economic Ornithology and Mammalogy Bulletin 17*. Government Printing Office, Washington.

Kantha, L. 2012. *Migration on Wings: Aerodynamics and Energetics*. Springer, New York.

Karasov, W. H. 1990. Digestion in birds: Chemical and physiological determinants and ecological implications. *Studies in Avian Biology* 13:391–415.

Kareiva, P., Tallis, H., Ricketts, T. H, Daily, G. C., and Polasky, S., eds. 2011. *Natural Capital: Theory and Practice of Mapping Ecosystem Services.* Oxford University Press, Oxford, UK.

Kay, B. J., Twigg, L. E., Korn, T. J., and Nichol, H. I. 1994. The use of artificial perches to increase predation on house mice (*Mus domesticus*) by raptors *Wildlife Research* 21:95–106.

King, F. H. 1882. Economic relations of Wisconsin Birds. In *Geological Survey of Wisconsin, Vol 1*, 441–610.

Kirk, D. A., Evenden, M. D., and Mineau, P. 1996. Past and current attempts to evaluate the role of birds as predators of insect pests in temperate agriculture. *Current Ornithology* 13:175–269.

Koh, L. P. 2008. Birds defend oil palms from herbivorous insects. *Ecological Applications* 18:821–25.

Konishi, M., Emlen, S.T., Ricklefs, R.E., Wingfield, J.C. 1989. Contributions of bird studies to biology. *Science* 246:465–472.

Korpimäki, E., and Krebs, C. J. 1996. Predation and population cycles of small mammals. *BioScience* 46:754–64.

Kronenberg, J. 2014. What can the current debate on ecosystem services learn from the past? Lessons from economic ornithology. *Geoforum* 55:164–77.

Kross, S. M., Tylianakis, J. M., and Nelson, X. J. 2012. Effects of introducing threatened falcons into vineyards on abundance of Passeriformes and bird damage to grapes. *Conservation Biology* 26:142–49.

Lack, D. 1947. *Darwin's Finches*. Cambridge University Press, New York.

Landis, D. A., Wratten, S. D., Gurr, G. M. 2010. Habitat management to conserve natural enemies of arthropod pests in agriculture. *Annual Review of Entomology* 45:175–201.

LaRouche, G. P. 2001. *Birding in the United States: A demographic and economic analysis*. Report 2001-1, US Fish and Wildlife Service, Washington.

Larson, G. 2015. Rulers of the roost. How an unassuming bird changed the world as we know it. *Science* 347:1077.

Millennium Ecosystem Assessment. 2003. *Ecosystems and Human Well-being. A Framework for Assessment.* Island Press.

MacArthur, R. H. 1972. *Geographical Ecology*. Harper & Row, New York.
Mäntylä, E., Klemola, T., Laaksonen, T. 2011. Birds help plants: A meta-analysis of top-down trophic cascades caused by avian predators. *Oecologia* 165:143–51.
Marquis, R. J., and Whelan, C. J. 1994. Insectivorous birds increase growth of white oak through consumption of leaf-chewing insects. *Ecology* 75:2007–14.
Mols, C. M. M., and Visser, M. E. 2002. Great tits can reduce caterpillar damage in apple orchards. *Journal of Applied Ecology* 39:888–99.
Moss, M. L., and Bowers, P. M. 2007. Migratory bird harvest in northwestern Alaska: A zooarchaeological analysis of Ipiutak and Thule occupations from the Deering archaeological district. *Arctic Anthropology* 44:37–50.
Nottebohm, F. 1984. Birdsong as a model in which to study brain processes related to learning. *Condor* 86:227–36.
Peres, C. A. 2001. Synergistic effects of subsistence hunting and habitat fragmentation on Amazonian forest vertebrates. *Conservation Biology* 15:1490–1505.
Peres, C. A., and Palacios, E. 2007. Basin-wide effects of game harvest on vertebrate population densities in Amazonian forests: Implications for animal mediated seed dispersal. *Biotropica* 39:304–15.
Podulka, S., Eckhardt, M., and Otis, D. 2004. Birds and humans: A historical perspective. In *Handbook of Bird Biology*, ed. A. Podulka, R. Rohrbugh Jr., and R. Bonney, 1–42. Cornell Lab of Ornithology, Ithaca, NY.
Rogers, H., Hille Ris Lambers, J., Miller, R., Tewksbury, J. J. 2012. 'Natural experiment' demonstrates top-down control of spiders by birds on a landscape level. *PLoS ONE* 7:e43446.
Şekercioğlu, Ç. H. 2002. Impacts of birdwatching on human and avian communities. *Environmental Conservation* 29:282–89.
———. 2006a. Increasing awareness of avian ecological function. *Trends in Ecology and Evolution* 21:464–71.
———. 2006b. Ecological significance of bird populations. In *Handbook of the Birds of the World*, vol. 11, ed. J. del Hoyo, A. Elliott, and D. A. Christie, 15–51. Lynx Press and BirdLife International, Barcelona and Cambridge.
———. 2010. Ecosystem functions and services. In *Conservation Biology for All*, ed. N. S. Sodhi and P. R. Ehrlich, 45–72. Oxford University Press, Oxford, UK.
———. 2011. Funtional extinctions of bird pollinators cause plant declines. *Science* 331:1019–20.
Sheffield, L. M., Crait, J. R., Edge, W. D., and Wang, G. M. 2001. Response of American kestrels and gray-tailed voles to vegetation height and supplemental perches. *Canadian Journal of Zoology* 79:380–85.
Sodhi, N. S., Şekercioğlu, Ç. H., Robinson, S., and Barlow, J. 2011. *Conservation of Tropical Birds*. Wiley-Blackwell, Oxford.
Solomon, M. E., Glen, D. M., Kendall, D. A., and Milsom, N. F. 1977. Predation of overwintering larvae of codling moth (*Cydia pomonella* [l.]) by birds. *Journal of Applied Ecology* 13:341–53.

Steven, R., Castley, J. G., and Buckley, R. 2013. Tourism revenue as a conservation tool for threatened birds in protected areas. *PLoS ONE* 8:e62598.

Sturges, F. W., Holmes, R. T., and Likens, G. E. 1974. The role of birds in nutrient cycling in a northern hardwoods ecosystem. *Ecology* 55:149–55.

Tracey, J., Bomford, M., Hart, Q., Saunders, G., Sinclair, R. 2007. Managing bird damage to fruit and other horticultural crops. Bureau of Rural Sciences, Canberra.

Weatherhead, P. J., Tinker, S., and Greenwood, H. 1982. Indirect assessment of avian damage to agriculture. *Journal of Applied Ecology* 19:773–82.

Weed, C. M., and Dearborn, N. 1903. *Birds in Their Relations to Man: A Manual of Economic Ornithology for the United States and Canada.* J. B. Lippincott Company, Philadelphia and London.

Wenny, D. G., DeVault, T., Kelly, D., Johnson, M. D., Şekercioğlu, Ç. H., Tombak, D., and Whelan, C. J. 2011. The need to quantify ecosystem services provided by birds. *Auk* 128:1–14.

Whelan, C. J., and Marquis, R. J. 1995. Songbird ecosystem function and conservation. *Science* 268:1263.

Whelan, C. J., Wenny, D. G., and Marquis, R. J. 2008. Ecosystem services provided by birds. *Annals of the New York Academy of Sciences* 1134:25–60.

Whelan, C. J., Wenny, D. G., and Marquis, R. J. 2010. Policy implications of ecosystem services provided by birds. *Synesis* 1:11–20.

White, S. S., Renner, K. A., Menalled, F. D., Landis, D. A. 2007. Feeding preferences of weed seed predators and effect on weed emergence. *Weed Science* 55:606–12.

White, E. M., Bowker, J. M., Askew, A. E., Langner, L. L., Arnold, J. R., English, D. B. K. 2014. Federal outdoor recreation trends: Effects on Economic Opportunities National Center for Natural Resources Economic Research (NCNRER). NCNRER working paper number 1.

Wiens, J. A. 1973. Pattern and process in grassland bird communities. *Ecological Monographs* 43:237–70.

Williams, C. K., Applegate, R. D., Lutz, R. S., Rusch, D. H. 2000. A comparison of raptor densities and habitat use in Kansas cropland and rangeland ecosystems. *Journal of Raptor Research* 34:203–9.

Wiltschko, R., and Wiltschko, W. 2013. The magnetite-based receptors in the beak of birds and their role in avian navigation. *Journal of Comparative Physiology A: Neuroethology, Sensory, Neural, and Behavioral Physiology* 199:89–98.

Wolff, J. O., Fox, T., Skillen, R. R., and Wang, G. M. 1999. The effects of supplemental perch sites on avian predation and demography of vole populations. *Canadian Journal of Zoology* 77:535–41.

Wormworth, J., Şekercioğlu, Ç. H. 2011. *Winged Sentinels: Birds and Climate Change.* Cambridge University Press, New York.

Wyatt, T. 2011. The illegal trade of raptors in the Russian Federation. *Contemporary Justice Review* 14:103–23.

CHAPTER TWO

Why Birds Matter Economically

Values, Markets, and Policies

Matthew D. Johnson and Steven C. Hackett

The concept of ecosystem services[1] was originally conceived as a metaphor to increase the visibility of societal dependence on ecosystems in a language that reflects dominant political and economic views (Gómez-Baggethun and Ruiz-Pérez 2011; Norgaard 2010). As Costanza (2003) notes, the first formal efforts to bring ecologists and economists together was in 1981, when Ann-Mari Jansson organized a symposium in Saltsjöbaden, Sweden, funded by the Wallenberg foundation, entitled "Integrating Ecology and Economics" (Jansson 1984). In the last 20 years the concept of ecosystem service values has been mainstreamed in the field of ecological economics (Gómez-Baggethun et al. 2010). Ecological economists developed methods to value ecosystem services in monetary terms to foster understanding of the economic benefits of conservation (Costanza et al. 1997, 2014). This economic framework gained international policy prominence with the release of the Millennium Ecosystem Assessment (MA 2005) and the report *The Economics of Ecosystems and Biodiversity* (*TEEB*; Kumar 2010). The result has been increasing on-the-ground application of policy instruments rooted in ecosystem service values (e.g., Balmford et al. 2002; Tallis et al. 2008; Barbier et al. 2009). We review this history as it relates to the conservation of birds.

First, we distinguish ethics and values that give rise to the valuation of ecosystem services provided by birds. Next, we offer a brief primer of the environment in economics, with attention especially to the groundwork for the monetization of ecosystem services provisioned by animals. We then describe examples of the economic valuation of birds, drawing attention to advances and shortcomings in the practice. Lastly, we conclude with recommendations for future directions and a plea for a rigorous, innovative, and ethical approach to the interdisciplinary integration of economics and ornithology.

Value and Ethics

At its core, economics is concerned with the problem of allocating scarce resources among competing uses, and a good economy should allocate limited resources to reflect the value system of its society (Hackett 2011). Philosophers distinguish two values of nature: instrumental and intrinsic (Callicott 1986). The instrumental value of something is its utility as a means to some end. The intrinsic value of something is its inherent value as an end itself. The instrumental values of birds include not only their obvious economic "use value" as goods traded in conventional markets (e.g., food and feathers), they also extend to less obvious "nonuse values" as providers of ecosystem services that can, in some cases, be monetized (Sekercioglu 2006a). Newcomers to the field of environmental ethics sometimes wrongly ascribe the aesthetic value of birds—citing the beauty and grace of their plumages, songs, or behaviors—as intrinsic value, and often regard it as "higher" or "better" than use value (Nash 1973; Justus et al. 2008). But aesthetic value is nonetheless instrumental, a means for human fulfillment. The intrinsic value of a bird, of course, does not depend on its beauty. Instead, intrinsic values for birds have philosophical foundations in society's valuation of sentience, theology, and kinship (Callicott 2008).

The growth in the economic valuation of birds and other components of biodiversity has triggered a heated debate. We briefly review the controversy, clarify some common misconceptions, and point the reader to some additional literature (drawing from recent reviews by Gómez-Baggethun and Ruiz-Pérez 2011 and Luck et al. 2012). Competing viewpoints in this debate range from an outright rejection of utilitarian rationalization for conservation (e.g., McCauley 2006; Child 2009) to the endorsement of valuation and markets as the only viable solutions to current environmental problems, which are framed as market failures (e.g., Heal et al. 2005; Engel et al. 2008). In between, many conservation organizations now embrace valuation of ecosystem services as a practical short-term conservation tool to influence policy (Daily et al. 2009; de Groot et al. 2002; Justus et al. 2008; Gómez-Baggethun and Ruiz-Pérez 2011).

Understanding the debate requires clarifying three interrelated but well-differentiated stages of an economic argument for conservation: economic framing, monetization, and commodification (Gómez-Baggethun and Ruiz-Pérez 2011). Economic framing employs the metaphor of ecosystems as (natural) "capital," and ecosystem functions as "services," to communicate the value of biodiversity in a way that reflects dominant political and economic views (e.g., Natural Capital Project of the World Wildlife Fund, 2013). Monetization takes place when natural capital or ecosystem services are expressed as exchange values (e.g., in dollars), which has become increasingly common since the publication of the much-discussed paper by Costanza et al. (1997) that estimated the total worth of the earth's natural capital. Finally, commodification of ecosystem services refers to the inclusion of previously nonmarketed ecosystem functions into pricing systems and markets, including the creation of institutional structures for the sale and exchange of ecosystem services (Gómez-Baggethun and Ruiz-Pérez 2011).

Critics have raised concerns over each stage of an economic argument for conservation. The economic framing of birds (or of nature in general) implicitly endorses an anthropocentric perspective that prioritizes their instrumental value to human well-being. Some have argued that this may undermine or diminish their intrinsic value (Ludwig 2000; Sagoff 2011; Weidensaul 2013), which may be the most or only enduring justification for their conservation (Rosenzweig 2003; Callicott 2008). Adopting an economic language and metaphor to frame human-nature relationships may also imply substitutability—the notion that components of nature can be replaced by human-derived alternatives. For example, loss of the biological control of pests by birds can be compensated for, economically, by the use of pesticides. The monetary valuation of birds (or nature) is the logical conclusion of economic framing, and this stage has been a volatile source of ethical controversy (McCauley 2006). Indeed, Leopold stated, "The last word in ignorance is the man who says of an animal or plant: What good is it?" (Leopold 1966; p. 190). More recently, Kronenberg (2014) has argued that a narrow anthropocentric focus on identifying useful and harmful birds was the downfall of economic ornithology, which was popular from the late 1800s until the 1920s, and that proponents of ecosystem services must carefully rethink the way they argue for environmental conservation, or it may befall the same fate (chapter 1). Indeed, recent debate over the so-called new conservation science (Karieva and Marvier 2012; Doak et al. 2014) indicates that the conservation community is grappling with how to balance the various approaches to framing values and conservation.

Methods to monetize the value of wildlife raise ethical issues concerning the anthropocentric bias in how value is assigned. For example, charismatic species may attract more attention, and thus higher willingness to pay from the public, than do less well known species (Martín-López et al. 2008; Luck et al. 2012). Others have suggested that while monetization itself may be benign, it paves the way for commodification of nature, which may ultimately undermine conservation (Gómez-Baggethun and Ruiz-Pérez 2011). The commodification stage has received the most ire in the debate. Critiques range from those based on moral grounds (e.g., the idea that some things just ought not be for sale; McCauley 2006) to exposés of undesirable economic consequences (e.g., the argument that substitutability in markets ignores essential ecological realities; Spash 2008), sociocultural impact (e.g., loss of traditional cultural practices: Grieg-Gran et al. 2005), complexity blinding (Gómez-Baggethun and Ruiz-Pérez 2011), or human inequity (conversion of open-access services to commodities that can be accessed only by those with purchasing power, Corbera et al. 2007).

A root critique of the monetary valuation of birds rests on the alleged dependence of intrinsic and instrumental value. For example, Weidensaul (2014, p. 40) stated that valuation of services provided by birds "inevitably cheapens the very thing we're trying to protect." This is a conflation of intrinsic and instrumental value, which need not be at odds (Vucetich et al. 2015). Consider your plumber: recognizing the undeniably useful and valuable service a plumber provides in no way cheapens his or her intrinsic value as a human. Indeed, intrinsic human value and rights remain inviolate regardless of professional skill, as a plumber or otherwise. Meanwhile, failing to value a good plumber's services is foolish. Those of us who have a deep and abiding value for humans consider it ludicrous and immoral to base someone's intrinsic value conditionally on a practical one; those of us who make careful economic decisions consider it imprudent to ignore instrumental value. And so it should be for birds.

Criticism of any proposed commodification of nature must acknowledge that in reality the controversy is over where to draw the line in what should and should not be commodified. In fact, material elements of nature (food, fiber) have been traded and sold as commodities since the birth of markets. Theoreticians acknowledge that all economic products result from the transformation of raw material provided by nature (Farley 2012). Indeed, the ancient philosophers (*ex nihilo nihil fit*) and modern physicists (First Law of Thermodynamics) got it right: you can't make something from nothing.

A Brief Primer on the Environment in Economics

In classical economics, which dominated from the late 1600s to the late 1800s, natural capital maintained a core position in economic analysis. For example, land was considered a nonsubstitutable production input, which partly explains the emphasis on physical constraints to growth offered by some classical economists, including Malthus's famous essay on population growth (Gómez-Baggethun et al. 2010). However, the industrial revolution, technological advancements, and the rapid accumulation of capital triggered a series of changes in economic thought that diminished the distinct analytical treatment of nature. Economists gradually restricted their analysis to the sphere of exchange values. Hence, nonmonetized inputs and consequences became externalities,[2] and economists' interest in natural resources languished (Crocker 1999). This evolution gave rise to neoclassical economics, which together with Keynesian economics dominates mainstream economics today. Neoclassical economic theory began to elaborate how technological innovation could substitute for production inputs such as land and capital, eventually pushing concerns over resource scarcity to near oblivion (Georgescu-Roegen 1975). This notion is perhaps best summarized by the American economist Robert Solow, who stated, "If it is very easy to substitute other factors for natural resources, then there is in principle no 'problem.' The world can, in effect, get along without natural resources, so exhaustion is just an event, not a catastrophe" (Solow 1974, p. 11).

The environmental movements of the 1960s and afterward gave rise to new schools of thought in economics with alternative treatments of the environment. The field of environmental economics expanded the scope of neoclassical economics by developing methods to value and internalize economic impacts on the environment into decision-making—for example, through extended cost-benefit analyses (Turner et al. 1994; Hackett 2011). Environmental economists assert that pure neoclassical economics neglects the contributions and impacts of nature by restricting its scope to those goods and services that bear a price. Therefore, it is argued, the chronic underappreciation of nature is rooted in the fact that its value is not expressed in terms comparable with economic services and manufactured capital (Costanza et al. 1997). To better capture a comprehensive treatment of ecological inputs and consequences, economic value is often divided into use and nonuse values, each subsequently disaggregated in different value components that are added up to the so-called total economic value (Gómez-Baggethun et al. 2010; Hackett 2011). To quantify these different values, a range of valuation techniques has been developed (e.g., revealed preferences, and travel cost methods; see below).

Environmental economics operates mainly within the axiomatic framework of neoclassical economics; it merely expands it to internalize what are more conventionally considered externalities of nature. Beginning in the 1980s, the new field called ecological economics emerged, and it challenged former assumptions by conceptualizing the economic system as part of the ecosphere that coevolves with social and ecological systems, with which it exchanges energy, materials, and waste (Daly 1977; Norgaard 1994). While environmental economics and ecological economics differ in their qualitative framework, the two schools of thought overlap considerably in the use of techniques to measure sustainability, evaluate policies, and assist in decision making. However, some ecological economists remain critical of valuing ecosystem services because of "incommensurability"—the idea that different types of value cannot be meaningfully expressed in common units (i.e., money). Some advocate for deliberative and multicriteria-based decision processes (e.g., Munda 2004; Spash 1998). The distinction between environmental economics and ecological economics remains controversial, and readers may want to consult Turner (1999) for additional information.

Over the last 20 years, the rise of these new schools of economic thought, combined with a growing recognition of the inadequacy of traditional conservation, has thrust ecosystem services into the mainstream (Marvier and Kareiva 2014). Although the state of the environment would undoubtedly be worse without traditional conservation, it has so far failed to reverse or stabilize biodiversity and habitat loss (Armsworth et al. 2007; Gómez-Baggethun et al. 2011). Many believe that conservationists have been reluctant to mix economics and conservation, thereby failing to act on the economic and sociopolitical roots of environmental problems (Child 2009; Steffen et al. 2004; Gómez-Baggethun et al. 2011). The ecosystem service approach offers an alternative to move away from the logic of "conservation versus people" and toward a logic of "conservation for people" (Kareiva and Marvier 2007).

The expansion of economics to recognize ecosystem services has followed two main approaches. The first involves public intervention to correct the market failure of externalizing environmental costs by imposing state taxes and subsidies. Implementing this approach involves markets for ecosystem services that invoke the so-called polluter pays principle (Gómez-Baggethun et al. 2011). The second approach relies on private transactions to correct market failures, often in markets where ecosystem services can be bought and sold. Implementing this approach involves payments for ecosystem services that invoke the so-called steward earns principle (Gómez-Baggethun et al. 2011).

The polluter pays principle that underlies markets for ecosystem services (MES) reflects an alleged ethic of responsibility. Usually, taxes or other penalties ensure that the agent causing the environmental harm bears the economic cost that would otherwise be externalized and thus shouldered by society. For example, the US Clean Air Act of 1990 promoted cap and trade mechanisms for atmospheric pollutants such as sulphur dioxides (Stavins 1998; Hackett 2011). The US Clean Water Act also offers a market for ecosystem service, in this case more directly related to the conservation of birds and other wildlife. Mandatory mitigation of wetland losses has prompted the development of "wetland banking," a market-based system designed to incentivize the consolidation of many small wetland mitigation projects into larger, potentially more ecologically valuable sites. This innovative idea has since been expanded to other habitats, and is now more generally referred to as habitat banking or simply conservation banking. This federally regulated system allows a "conservation bank," usually a private entity, to restore resources (e.g., a wetland) to compensate for authorized impacts to similar resources at development sites. Conservation banks operate similarly to other financial institutions that describe transactions in terms of credits and debits. Credits represent the composite of ecological function at a habitat bank, while debits represent the loss of ecological function at a development site. Bank sponsors can sell mitigation credits to permittees who are required to compensate for jurisdictional impacts incurred at their development sites. We are not aware of a review of the effect of conservation banks on bird conservation per se, but in a review of 35 conservation banks in the United States, Fox and Nino-Murcia (2005) found they cumulatively covered almost 16,000 ha and housed more than 22 species listed under the US Endangered Species Act. However, there are serious ecological and implementation issues in using conservation banks for wildlife species, including the suitability of banking sites and the establishment of endowments necessary to properly maintain them over time (Bonnie and Wilcove 2008). Nonetheless, conservation banking is arguably an effective means to conserve habitat for species such as birds (and, collaterally, other co-occurring species), by providing a market mechanism for the funding, strategic selection, and perpetual maintenance of habitats for wildlife and other biodiversity.

While negative externalities are addressed by the polluter pays approach of markets for ecosystem services, positive externalities can be dealt with through the "steward earns principle" with payments for ecosystem services. Payments for ecosystem services (PES) are offered to landowners in exchange for managing lands to provide some ecological service. Payments are voluntary and conditioned transactions between at least one provider and one beneficiary of the service (Tacconi 2012). PES may at first seem like a radically new approach, but rudimentary forms have been in place for many decades. After the Dust Bowl, for example, the US government provided payments to farmers who adopted measures to guard against soil erosion. This notion was expanded in the 1950s to offer payments to protect farmlands from urban expansion (Jacobs 2008), and is now firmly rooted in the United States via the inclusion of the conservation title in the passage of the 1985 Food Security Act (aka the "Farm Bill") and its successors.[3] By many accounts, the conservation programs in the Farm Bill have had demonstrably positive effects on wildlife conservation (see review by Heard et al. 2000). For instance, the Farm Bill's Conservation Reserve Program (CRP) alone includes more than 7 million ha and has had a huge influence on grassland bird populations (Johnson 2000); research suggests that CRP lands in the Prairie Pothole Region contributed to a 30% improvement in waterfowl production (Reynolds 2000).

Though rudimentary forms have been in place for decades, the widespread expansion of PES as an integrated development and conservation scheme has advanced mainly since the mid 1990s (Gómez-Baggethun et al. 2011). Costa Rica was the first country to establish PES schemes at the national scale, by offering landowners US$45 per ha for following ecological protection requirements (Pagiola 2008). These large scale PESs paved the way for global implementation, now emerging from the Conference of the Parties of the Kyoto Protocol, and from the United Nations' Reducing Emissions from Deforestation and Forest Degradation programs (REDD and REDD+). These programs create a financial value for the carbon stored in forests, offering incentives for developing countries to reduce emissions from forested lands and invest in low-carbon paths to sustainable development, and they may yield financial flows that reach up to US$30 billion a year. This significant north-south flow of funds could reward a meaningful reduction of carbon emissions, and could also support pro-poor development, help conserve biodiversity, and secure vital ecosystem services (UN REDD 2013). More recently, Dinerstein et al. (2013) have offered a modification of REDD+ in the form of a "wildlife premium" to more directly ensure that REDD+ projects benefit wildlife. However, critiques of REDD+ note that the program ignores some costs and thereby exaggerates its cost-effectiveness (Fosci 2013), and several NGOs and indigenous organizations have raised concerns of equity (e.g., security of tenure and benefit sharing for local stakeholders) and environmental integrity (e.g., leakage, permanence, and inflated reference levels; Bohm and Dabhi 2009).

Birds and Economics

Some birds and bird products have conventional exchange values and are directly tradable in markets, such as in the legal trade of domestic birds and the poultry and down industries. We do not elaborate on the economics or conservation implications of these markets. Instead, we concern ourselves with the nonuse value of ecosystem services provided by birds, including supporting services (e.g., seed dispersal), regulating services (e.g., pest control), and cultural services (e.g., recreational experiences). The nonuse value of these services is much more difficult to measure, and the development of valuation techniques has captured the attention of environmental economists for the past couple of decades (Kopp and Smith 1993). The underlying concepts for monetizing nonuse values rely on measuring what a society would be willing to pay (WTP) for a service, or what it will be willing to accept (WTA) to forego that service (Farber et al. 2002; Hackett 2011).

For example, let us first consider the ecosystem service provided by an entire ecosystem, such as flood control provided by a wetland (this example follows one offered by Farber et al. 2002). This type of case is far more common in the literature than the type involving ecosystem services provisioned by specific organisms, such as birds. If damages from a future flood event are estimated at $1 million, and the wetland reduces the probability of flooding by 20%, then society receives $200,000 in services from the wetland. In principle, the owner of the wetland could capture this amount from society because it corresponds to society's WTP for the service. This assumes a mechanism to capture the value (and it also assumes that society is risk-neutral). Markets for ecosystem services, such as those developed for conservation banking (reviewed above) are one type of capture mechanism. These markets work well for private services (where landowners can deliver services to those making payments) and properly regulated public services (as is the case for conservation banking).

However, many services provided by nature, such as flood control, are inappropriate for market trading because they offer a public good available to many people. For these cases, economists have developed six major ecosystem service valuation techniques to infer the WTP or WTA (see supplemental table 2.1).

The value of an ecosystem service can be measured as the cost that would have been incurred in the absence of the service; this type of measurement is called the avoided cost method. For example, Markandya et al. (2008) valued the ecosystem services provided by vultures in Northern India. They found that a healthy vulture population avoided costs of elevated human disease and death because vultures remove carcasses, resulting in fewer dogs and fewer bites to humans from rabid dogs. Using link functions to associate vulture numbers to dogs, dog bites, and disease prevalence, and drawing from the literature to provide values for the loss of wages and loss of human life in India, the authors estimated that healthy vulture populations avoid human health costs of up to US$2.4 billion per year. Other applications of the avoided cost method may draw from the concept of substitutability, such as cases for which the value of bird-provided pest control or pollination could be estimated by calculating the costs of chemical and manual labor substitutes. As noted by McCauley (2006), using this approach to promote bird conservation may be tenuous, as advancing technology may render the substitutes more economically efficient than the cost of conservation. Indeed, agricultural technology, such as pesticides, accelerated the demise of the field of economic ornithology, which was widely practiced in the late 1800s and early 1900s (chapter 1; Evenden 1995; Whelan et al. 2008).

Similarly, the value of an ecosystem service can be measured as the costs to replace that service with human-made substitutes, an approach called the replacement cost method (Swinton et al. 2007). For example, Hougner et al. (2006) conducted an economic valuation of a seed dispersal service provided by Eurasian jays (*Garrulus glandarius*) in the Stockholm National Urban Park of Sweden (chapter 7). Based on the cost of replacing this seeding or planting service through human means, they estimated the replacement cost of a pair jays to be between US$4,900 and $22,500. However, this may not be an appropriate measurement of value (Barbier

1998; Bockstael et al. 2000), because people might not be willing to replace a service at the full replacement cost (Freeman 2003).

The value of an ecosystem service can also be measured as the enhancement of income associated with the service; this is called the factor income method. This method is commonly applied in agricultural settings by identifying the effect of birds on yields or costs. When birds enhance yield without altering cost, the increased yield directly translates into increased income (Swinton et al. 2007). For example, Johnson et al. (2010) compared saleable coffee berry production inside and outside bird exclosures to document the increase in yield from bird predation of the insect pest *Hypothenemus hampei* in Jamaica. They found that bird-provisioned pest control contributed about US$310 per hectare per year, a value that is approximately 10% of the per capita gross national income in Jamaica at the time. More generally, when an ecosystem service affects agricultural outputs and the need for various inputs, a production function can be used to value the ecosystem service. A production function relates the quantity of output (e.g., agricultural yields) to various levels and combinations of inputs, including natural (e.g., bird predation of insects) and human-provided (e.g., pesticides or fertilizers; Christiaans et al. 2007) inputs.

Some ecosystem services, especially cultural values such as recreational opportunities, require travel. In these cases, the value of an ecosystem service can be revealed by the cost of travel required to receive the service; this is called the travel cost method. Travel associated with birding, ecotourism, and bird hunting can be assessed with this technique. For example, Gürlük and Rehber (2008) used the travel cost method to quantify the recreational economic value of bird-watching in the Kuşcenneti National Park (KNP) at Lake Manyas—one of the Ramsar sites of Turkey and an important habitat for endangered birds. They found recreational value of the park to exceed US$103 million annually. Using a similar approach, Czajkowski et al. (2014) estimated that each person visiting Zywkowo, the best known "stork village" in Poland, paid a surplus of between US$60 and $120 for the opportunity. Given that Zywkowo receives 2,000 to 5,000 tourists annually, this represents a substantial willingness to pay for bird-based recreation. Observations of the relationship between people's recreation activity and their travel costs are used to estimate recreation demand functions. If the demand can also be related to levels of ecosystem provision, then changes in the ecosystem service will shift the demand functions and can be used to value changes in the value of the service. This approach has been used to estimate values associated with agricultural conservation programs that affect pheasant hunting (Hansen et al. 1999). Although the travel cost method is intuitive and has been used for decades, there are several challenges in applying and analyzing travel cost models, including choosing the dependent variable, choosing model specification, quantifying the value of time, and accounting for substitutes and multipurpose trips (Gürlük and Rehber 2008).

The value of a service can be measured by the change in value of a marketed good associated with the service; this is called the hedonic pricing method. In essence, hedonic approaches can measure values that get capitalized into the asset value of property (Swinton et al. 2007). Application of this method for valuing open space as a characteristic of residential properties is well established (Taylor 2003). A hedonic price function is estimated, often with regression, usually by characterizing the land, structure, neighborhood, and environment. The analysis can reveal the effect of environmental attributes on local real estate pricing after accounting for the other factors. For example, Neumann et al. (2009) conducted an analysis of real estate prices near the Great Meadows National Wildlife Refuge in Massachusetts, a popular place for bird-watching. They found that a property located 100 meters closer to the NWR than a neighboring property had a price premium of $984. This ecosystem service cannot be attributed solely to birds, of course, as open green space in general improves home prices; but the researchers noted that proximity to the refuge was valued more than proximity to agricultural land, cemeteries, and general conservation land. More generally, bird diversity positively influences human perception of natural areas (e.g., Fuller et al. 2007). Thus, an increase in the presence and diversity of birds in urban areas has the potential to provide a service that benefits humans directly.

Finally, in cases where none of the above methods is appropriate, it may be possible to monetize a service by surveying people to assess their willingness to pay. In these cases, the value of a service is measured by posing hypothetical scenarios that involve the valuation of alternatives based on preference that is explicitly stated or revealed by comparisons (Hackett 2011). For example, Clucas et al. (2014) used stated preferences to perform a cross-continental economic valuation of native urban songbirds, estimating the lower boundary to be about US$120 million/year in Seattle and US$70 million/year in Berlin. The contingent valuation method allows researchers to specify the exact scenario to be valued (Swinton et al. 2007). Unlike other methods, it is capable of measuring passive use values that people may hold regardless of whether or not they will directly use the ecosystem (Freeman 2003). This method has been applied especially for endangered wildlife species (Richard and Loomis 2009). For example, Reaves et al. (1999) showed that people were willing to pay between US$7.57 and $13.25 per person per year in South Carolina to restore the endangered red-cockaded woodpecker (*Picoides borealis*). Stevens et al. (1991) documented a figure of US$28.25 per person per year for bald eagles (*Haliaeetus leucocephalus*), and Bowker and Stoll (1988) found between US$21 and $42 for whooping cranes (*Grus americana*).

Future Directions

As the above review indicates, efforts to value ecosystem services by birds have been promoted by many in the last 20 years (Sekercioglu 2006b, Wenny et al. 2011). In theory, these efforts should help society and its institutions better recognize the value of birds, and this should increase investment in bird conservation while simultaneously advancing human well-being (Daily et al. 2009). To date, however, there are relatively few practices and policies that deliberately incorporate the ecosystem services delivered by birds into land-use decisions. Doing so will require ornithologists to work collaboratively and in interdisciplinary ways to advance our understanding not only of the ecological science underlying services provided by birds, but also of the social science linking those services to human values and policies. Here, we briefly review key steps and strategies so that ecosystem services can be harnessed for bird conservation and human well-being alike.

The ultimate goal of valuing ecosystem services provided by birds is to make better decisions about the use of land and other natural resources. Those decisions are part of a loop of information that must include acknowledgment of the instrumental values of birds (fig. 2.1). Navigating this loop requires key information and science to link each stage to the next. The ornithological and ecological sciences are essential for predicting how land use decisions affect birds and other elements of the ecosystem, and in turn how those changes affect the delivery of bird-provisioned ecosystem services. In this regard, ornithologists should focus additional future work on unraveling how habitat and landscape modification affects bird communities and populations, and how those changes affect ecologically functional groups (Díaz and Cabido 2001; Sekercioglu et al. 2004). In turn, economics and cultural sciences are necessary to translate those services

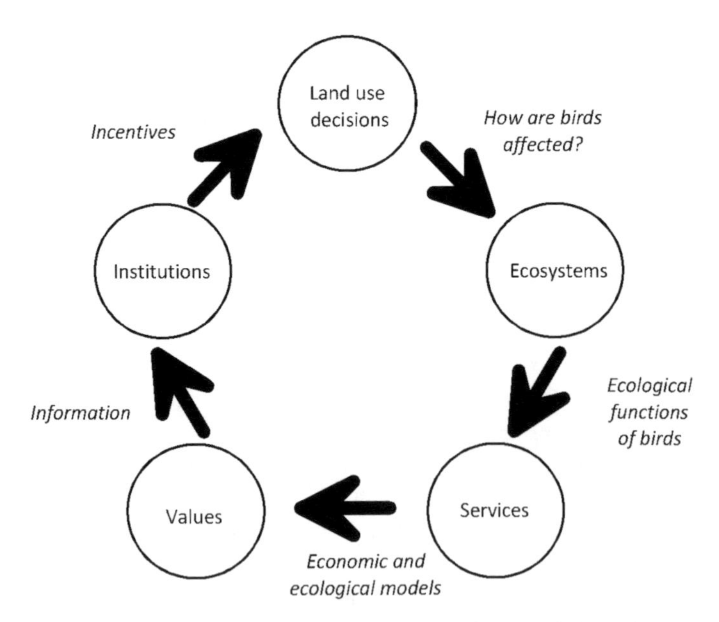

FIGURE 2.1. A framework for how ecosystem services provided by birds or other organisms can be integrated into land use decisions. Redrawn from Daily et al. 2009.

into values. The examples above show how ornithologists have drawn from the field of economics and, in the best cases, collaborated with natural resources economists to use valuation methods to quantify ecosystem services delivered by birds. As we have seen, these values are multidimensional (cultural, economic, intrinsic), so it will be important to characterize them as comprehensively and broadly as possible, and in ways that will be meaningful to many different audiences (Daily et al. 2009). New tools, such as the InVEST model (Nelson et al. 2009), can help land use managers integrate and balance multiple ecosystem services into proposed land use scenarios to, in the end, inform appropriate land use decisions.

The links between values, institutions, and ultimate land use decisions draw more from politics and social change than from science. Nonetheless, the scientists involved in the earlier links must remain engaged to inform the development of institutions and incentives. Influencing existing institutions and building new ones is a profound challenge to conservation science. The view of birds as economically essential can be cultivated by (1) promoting existing and proposing new incentives for bird and habitat conservation (e.g., Farm Bill conservation title) while fostering the recognition of the value of their services (Sekercioglu et al. 2004),

(2) evaluating existing policy and finance mechanisms with recognition of both their environmental and social consequences (Berkes et al. 2003), and (3) ensuring that essential stakeholders participate in the development of these new institutions and mechanisms (Cowling et al. 2008; Daily et al. 2009).

Finally, the link between institutions and decisions rests on human behavior. Societies that deliberatively value nature and resist an overemphasis on short-term economic gain will easily mainstream ecosystem services into decision-making (Diamond 2005; Daily et al. 2009). For this reason, perhaps the greatest benefit of documenting why birds matter economically is in advancing the ontological position that ecosystems and the species they hold not only are a matter of ethics and aesthetics, but are also essential for human subsistence and fulfillment (Gómez-Baggethun and de Groot 2010). There are many devils in the details of even the most basic decisions in setting up markets and payments for ecosystem services; these include contract duration, size and frequency of payments, and monitoring of outcomes (Daily et al. 2009). Indeed, proponents of ecosystem service science must be mindful that the commodification of birds' services may reproduce neoclassical market logic and its underlying ideology and institutional structures (Gómez-Baggethun et al. 2010). Commodification of ecosystem services may in the long term be counterproductive for biodiversity conservation and equity of access to an ecosystem's benefits (Gómez-Baggethun and Ruiz-Pérez 2011; see also "Value and Ethics," above). Working with philosophers can help ecologists free themselves from the misconception that recognizing instrumental value requires monetization and a diminution of intrinsic value, and such collaboration can introduce new decision support tools for conservation that do not demand monetization (Justus et al. 2008).

To advance the conservation of birds, should we appeal to peoples' hearts, minds, or wallets? All three. Our view is that the recognition that birds matter economically is a powerful tool for conservation and for improving human life. But it is just a tool—one that should be used not as a single decision-making criterion, but alongside recognition of the noneconomic value dimensions of nature (Justus et al. 2008; Gavin et al. 2015). So with one breath we should describe how, for example, black-throated blue warblers (*Setophaga caerulescens*) can deliver economically important ecosystem services by controlling insect pests of coffee (Kellermann et al. 2008), and thus advocate for policies that attract birds and improve human livelihood, while with the next breath we expound on the bird's lilting song, its awe-inspiring migration, and its place on this earth as our feathered kin. We can remind ourselves that, in Thoreau's words, a blue bird "carries the sky on his back" (Thoreau 1952).

Notes

1. Ecosystem services are the conditions and processes through which nature sustains and fulfills human life (Daily 1997).

2. In economics, an externality is a cost or benefit that results from an activity or transaction and that flows to members of society other than the buyer, seller, or owner (Hackett 2011).

3. At the time of publication, the current farm bill is the Agricultural Act of 2014.

References

Armsworth, P. R., Chan, K., Chan, M. A., Daily G. C., Kremen C., Ricketts T.H., and Sanjayan, M. A. 2007. Ecosystem-service science and the way forward for conservation. *Conservation Biology* 21:1383–84.

Balmford, A., Bruner, A., Cooper, P., Costanza, R., Farber, S., et al. 2002. Economic reasons for conserving wild nature. *Science* 297:950–53.

Barbier, E. 1998. The economics of soil erosion: Theory, methodology, and examples. In *The Economics of Environment and Development: Selected Essays*, ed. E. Barbier, 281–307. Edward Elgar Publishing, Cheltenham, UK.

Barbier, E. B., Baumgärtner, S., Chopra, K., Costello, C., Duraiappah A., et al. 2009. The valuation of ecosystem services. In *Biodiversity, Ecosystem Functioning, and Human Wellbeing: An Ecological and Economic Perspective*, ed. S. Naeem, D. Bunker, A. Hector, M. Loreau, and C. Perrings, 248–62. Oxford University Press: Oxford, UK.

Berkes, F., Colding, J., and Folke, C. 2003. *Navigating Social-Ecological Systems*. Cambridge University Press: Cambridge.

Bockstael, N. E., Freeman, A. M., Kopp, R. J., Portney, P. R., and Smith, V. K. 2000. On measuring economic values for nature. *Environmental Science and Technology* 34:1384–89.

Bohm, S., and Dabhi, S. 2009. *Upsetting the Offset: The Political Economy of Carbon Markets*. MPG Books Group: London, UK.

Bonnie, R., and Wilcove, D. S. 2008. Ecological consideration. In *Conservation and Biodiversity Banking: A Guide to Setting Up and Running Biodiversity Credit Trading Systems*, ed. N. Carroll, J. Fox, and R. Bayon, 53–68. Earthscan Publishing, London.

Bowker, J. M., and Stoll, J. R. 1988. Use of dichotomous choice nonmarket methods to value the whooping crane resource. *American Journal of Agricultural Economics* 70:20–28.

Callicott, J. B. 1986. On the intrinsic value of non-human species. In *The Preservation of Species*, ed. B. Norton, 138–72. Princeton University Press, Princeton, NJ.

———. 2008. Valuing wildlife. In *The Animal Ethics Reader, 2nd edition*, ed. S. J. Armstrong and R. G. Botzler, 439–43. Routledge Publishers, London.

Child, M.F. 2009. The Thoreau ideal as unifying thread in the conservation movement. *Conservation Biology* 23:241–43.

Christiaans, T., Eichner, T., and Rudiger, P. 2007. Optimal pest control in agriculture. *Journal of Economic Dynamics and Control* 31:3965–85.

Clucas, B., Rabotyagov, S., and Marzluff, J. M. 2014. How much is that birdie in my backyard? A cross-continental economic valuation of native urban songbirds. *Urban Ecosystems* 18:251–66.

Corbera, E., Kosoy, N., and Martínez-Tuna, M. 2007. The equity implications of marketing ecosystem services in protected areas and rural communities: Case studies from Meso-America. *Global Environmental Change* 17:365–80.

Costanza, R. 2003. The early history of ecological economics and the International Society for Ecological Economics (ISEE). Internet Encyclopedia of Ecological Economics, International Society for Ecological Economics, http://isecoeco.org/.

Costanza, R., d'Arge, R., de Groot, R., Farber, S., Grasso, M., Hannon, B., Limburg, K., Naeem, S., O'Neill, R. V., Paruelo, J., Raskin, G. R., Sutton, P. and van der Belt, M. 1997. The value of the world's ecosystem services and natural capital. *Nature* 387:253–60.

Costanza, R., de Groot, R., Sutton, P., van der Ploeg, S., Anderson, S. J., Kubiszewski, I., Farber, S., and Turner, R. K. 2014. Changes in the global value of ecosystem services. *Global Environmental Change* 26:152–58.

Cowling, R., Egoh, B., Knight, A. T., et al. 2008. An operational model for mainstreaming ecosystem services for implementation. *Proc. National Academy Sciences* 105:9483–88.

Crocker, T. D. 1999. A short history of environmental and resource economics. In *Handbook of Environmental and Natural Resource Economics*, ed. J. van der Bergh. J. Edward Elgar Publishers, Northampton, MA.

Czajkowski, M., Giergiczny, M., Kronenberg, J., and Tryjanowski, P. 2014. The economic recreational value of a white stork nesting colony: A case of "stork village" in Poland. *Tourism Management* 40:352–60.

Daily, G. C. 1997. *Nature's Services: Societal Dependence on Natural Ecosystems.* Island Press, Washington.

Daily, G. C., Polasky, S., Goldstein, J., Kareiva, P. M., Mooney, H. A., Pejchar, L., et al. 2009. Ecosystem services in decision making: Time to deliver. *Frontiers in Ecology and the Environment* 7:21–28.

Daly, H. 1977. *Steady-State Economics: The Political Economy of Bio-physical Equilibrium and Moral Growth.* W. H. Freeman and Co., San Francisco.

De Groot, R. S., Wilson, M., and Boumans, R. 2002. A typology for the description, classification and valuation of ecosystem functions, goods and services. *Ecological Economics* 41:393–408.

Diamond, J. 2005. *Collapse: How Societies Choose to Fail or Succeed*. Penguin Publishers, New York.

Díaz, S., and Cabido, M. 2001. Vive la différence: Plant functional diversity matters to ecosystem processes. *Trends in Ecology and Evolution* 16:646–55.

Dinerstein, E., Varma, K., Wikramanayake, E., Powell, G., Lumpkin, S., Naidoo, R., Korchinsky, M., Del Valle, M., Lohani, S., Seidensticker, J., Joldersma, D., Lovejoy, T., and Kushlin, A. 2013. Enhancing conservation, ecosystem services, and local livelihoods through a wildlife premium mechanism. *Conservation Biology* 27:14–23.

Doak. D. F., Bakker, V., Goldstein, B. E., and Hale B. 2014. What is the future of conservation? *Trends in Ecology and Evolution* 29:77–81.

Engel, S., Pagiola, S., and Wunder, S. 2008. Designing payments for environmental services in theory and practice: An overview of the issue. *Ecological Economics* 65:663–74.

Evenden, M. D. 1995. The laborers of nature: Economic ornithology and the role of birds as agents of biological pest control in North American agriculture, ca. 1880–1930. *Forest and Conservation History*, 39:172–83.

Faber, S. C., Costanza, R., and Wilson, M. A. 2002. Economic and ecological concepts for valuing ecosystem services. *Ecological Economics* 41:375–92.

Farley, J. 2012. Ecosystem services: The economics debate. *Ecosystem Services* 1:40–49.

Fosci, M. 2013. Balance sheet in the REDD +: Are global estimates measuring the wrong costs? *Ecological Economics* 89:196–200.

Freeman, A. M. 2003. *The Measurement of Environmental and Resource Values: Theory and Methods*. Resources for the Future, Washington.

Fuller, R. A., Irvine, K. N., Devine-Wright, P., Warren, P. H., and Gaston, K. J. 2007. Psychological benefits of greenspace increase with biodiversity. *Biology Letters* 3:390–94.

Gavin, M. C., McCarter, J., Mead, A., Berkes, F., Stepp, J. R., Peterson, D., and Tang, R. 2015. Defining biocultural approaches to conservation. *Trends in Ecology and Evolution* 30:140–45.

Georgescu-Roegen, N. 1971. *The Entropy Law and the Economic Process*. Harvard University Press, London.

Gómez-Baggethun, E. and de Groot, R. 2010. Natural capital and ecosystem services: The ecological foundation of human society. In *Ecosystem Services: Issues in Environmental Science and Technology*, ed. R. E. Hester and R. M. Harrison, 118–45. Royal Society of Chemistry, Cambridge, UK.

Gómez-Baggethun, E, de Groot, R., Lomas, P., and Montes, C. 2010. The history of ecosystem services in economic theory and practice: From early notions to markets and payment schemes. *Ecological Economics* 6:1209–18.

Gómez-Baggethun, E. and Ruiz-Pérez, M. 2011. Economic valuation and the commodification of ecosystem services. *Progress in Physical Geography* 35: 613–28.

Grieg-Gran, M., Porras, I., and Wunder, S. 2005. How can market mechanisms for forest environmental services help the poor? Preliminary lessons from Latin America. *World Development* 33:1511–27.

Gürlük, S., and Rehber, E. 2008. A travel cost study to estimate recreational value for a bird refuge at Lake Manyas, Turkey. *Journal of Environmental Management* 88:1350–60.

Hackett, S. C. 2011. *Environmental and Natural Resources Economics: Theory, Policy, and the Sustainable Society, 4th ed.* M. E. Sharpe, Armonk, NY.

Hansen, L., Feather, P., and Shank, D. 1999. Valuation of agriculture's multisite environmental impacts: An application to pheasant hunting. *Agricultural and Resource Economics Review* 28:199–207.

Heal, G. M., Barbier, E. E., Boyle, K. J., Covich, A. P., Gloss, S. P., Hershner, C. H.,

Hoehn, J. P., Pringle,C. M., Polasky, S., Segerson, K., and Shrader-Frechette, K. 2005. *Valuing Ecosystems Services: Toward Better Environmental Decision-Making*. National Research Council, Washington.

Heard, L. P., et al. 2000. *A Comprehensive Review of Farm Bill Contributions to Wildlife Conservation, 1985–2000*. USDA Natural Resources Conservation Service, Wildlife Habitat Management Institute, Technical Report USDA/NRCS/WHMI-2000.

Hougner, C., Colding, J., and Söderqvist, T. 2006. Economic valuation of a seed dispersal service in the Stockholm National Urban Park, Sweden. *Ecological Economics* 59:364–74.

Jacobs, H. M. 2008. *Designing Pro-Poor Rewards for Ecosystem Services: Lessons from the United States*? United States Agency for International Development, Tenure Brief, no. 8. Land Tenure Center, Madison, WI.

Jansson, A. M., 1984. *Integration of Economy and Ecology: An Outlook for the Eighties*. University of Stockholm Press, Stockholm.

Johnson, D. H. 2000. Grassland bird use of Conservation Reserve Program fields. In *A Comprehensive Review of Farm Bill Contributions to Wildlife Conservation, 1985–2000*, ed. L.P. Heard, 19–34. USDA Natural Resources Conservation Service, Wildlife Habitat Management Institute, Technical Report USDA/NRCS/WHMI-2000.

Johnson, M. D., Kellermann, J, and Stercho, A. M. 2010. Pest control services by birds in shade and sun coffee in Jamaica. *Animal Conservation* 13:140–47.

Justus, J., Colyvan, M., Regan H., and Maguire, L. 2008. Buying into conservation: Intrinsic versus instrumental value. *Trends in Ecology and Evolution* 24: 187–91.

Kareiva, P., and Marvier, M. 2007. Conservation for the people. *Scientific American* 297:50–57.

———. 2012. What is conservation science? *Bioscience* 62:962–69.

Kellermann, J. L., Johnson, M. D., Stercho, A. M., and Hackett, S. 2008. Ecological and economic services provided by birds on Jamaican Blue Mountain coffee farms. *Conservation Biology* 22:1177–85.

Kopp, R. J., and Smith, V. K. 1993. *Valuing Natural Assets: The Economics of Natural Resource Damage Assessment*. Resources for the Future, Washington.

Kronenberg, J. 2014. What can the current debate on ecosystem services learn from the past? Lessons from economic ornithology. *Geoforum* 55:164–77.

Kumar P., ed. 2010. *The Economics of Ecosystems and Biodiversity: Ecological and Economic Foundations*. Earthscan Publishing, London.

Leopold, A. 1966. *A Sand County Almanac*. Ballatine Books, New York.

Luck G. W., Chan K. M. A., Eser U., Gómez-Baggethun E., Matzdorf B., Norton B., and Potschin, M.B. 2012. Ethical considerations in on-the-ground applications of the ecosystem services concept. *BioScience* 62:1020–1929.

Ludwig, D. 2000. Limitations of economic valuation of ecosystems. *Ecosystems* 3:31–35.

Markandya, A., Taylor, T., Longo, A, Murty, M. N., Murty, S., and Dhavalad, K. 2008. Counting the cost of vulture decline: An appraisal of the human health and other benefits of vultures in India. *Ecological Economics* 67:194–204.

Martín-López, B., Montes, C., Benayas, J. 2008. Economic valuation of biodiversity conservation: The meaning of numbers. *Conservation Biology* 22:624–35.

Marvier, M., and Kareiva, P. 2014. The evidence and values underlying "new conservation." *Trends in Ecology and Evolution* 29:131–32.

McCauley, D. J. 2006. Selling out on nature. *Nature* 443:27–28.

Millennium Ecosystem Assessment, 2003. *Ecosystems and Human Well-Being: A Framework for Assessment*. Island Press: Covelo, CA.

Munda, G. 2004. Social Multi-Criteria Evaluation (SMCE): Methodological foundations and operational consequences. *European Journal of Operational Research*: 662. Natural Capital Project, WWF 2013.

Nelson, E., Mendoza, G., Regetz, J., Polasky, S., Tallis, H., Cameron, D. R., Chan, K. M. A., Daily, G. C., Goldstein, J., Kareiva, P. M., Lonsdorf, E., Naidoo, R., Ricketts, T. H., and Shaw, M. R. 2009. Modeling multiple ecosystem services, biodiversity conservation, commodity production, and tradeoffs at landscape scales. *Frontiers in Ecology and the Environment* 7:4–11.

Neumann, B. C., Boyle, K. J., and Bell, K. P. 2009. Property price effects of a national wildlife refuge: Great Meadows National Wildlife Refuge in Massachusetts. *Land Use Policy* 26:1011–19.

Norgaard, R. 2010. Ecosystem services: From eye-opening metaphor to complexity blinder. *Ecological Economics* 6:1219–27.

Norgaard, R. B. 1994. *Development Betrayed: The End of Progress and a Coevolutionary Revisioning of the Future*. Routledge Press, New York.

Pagiola, S. 2008. Payments for environmental services in Costa Rica. *Ecological Economics* 65:712–24.

Reynolds, R. E. 2000. Waterfowl responses to the Conservation Reserve Program in the Northern Great Plains. In *A Comprehensive Review of Farm Bill Contributions to Wildlife Conservation, 1985–2000*, ed. L. P. Heard, 35–44. USDA

Natural Resources Conservation Service, Wildlife Habitat Management Institute, Technical Report USDA/NRCS/WHMI-2000.

Rosenzweig, M. 2003. *Win-Win Ecology: How Earth's Species Can Survive in the Midst of Human Enterprise.* Oxford University Press, Oxford.

Sagoff, M. 2011. The quantification and valuation of ecosystem services. *Ecological Economics* 70:497–502.

Şekercioğlu, Ç. H. 2006a. Ecological significance of bird populations. In *Handbook of the Birds of the World*, vol. 11, ed. J. del Hoyo, A. Elliott, and D. A. Christie, 15–51. Lynx Press and BirdLife International, Barcelona and Cambridge.

———. 2006b. Increasing awareness of avian ecological function. *Trends in Ecology and Evolution* 21:464–71.

Şekercioğlu, Ç. H., Daily G. C., and Ehrlich, P. R. 2004. Ecosystem consequences of bird declines. *Proceedings of the National Academy of Sciences of the United States of America* 101:18042–47.

Solow, R. M. 1974. The economics of resources or the resources of economics. *American Economic Review* 64:1–14.

Spash, C. 2008. How much is that ecosystem in the window? The one with the biodiverse trail. *Environmental Values* 17:259–84.

Stavins, R. N., 1998. What can we learn from the Grand Policy Experiment? Lessons from SO2 allowance trading. *Journal of Economic Perspectives* 12:69–88.

Steffen, W., Sanderson, A., Tyson, P. D., Jäger. J., Matson, P. A., Moore, B., Olffield, F., Richardson, K. Schellenhuber, J. J., Turner, II, B. L., Wasson, R. J. 2004. *Global Change and the Earth System: A Planet under Pressure.* Springer, Heidelberg.

Stevens, T. H., Echeverria, J., Glass, R. J., Hager, T., and More, T. A. 1991. Measuring the existence value of wildlife: What do CVM estimates really show? *Land Economics* 67:390–400.

Swinton, S., Lupia, F., Robertson, G. P., and Hamilton, S. K. 2007. Ecosystem services and agriculture: Cultivating agricultural ecosystems for diverse benefits. *Ecological Economics* 64:245–52.

Tacconi, L. 2012. Redefining payments for environmental services. *Ecological Economics* 73:29–36.

Tallis H., Kareiva, P., Marvier, M., and Chang, A. 2008. An ecosystem services framework to support both practical conservation and economic development. *Proceedings of the National Academy of Sciences* 105:9457–64.

Taylor, L. O. 2003. The hedonic method. In *A Primer on Nonmarket Valuation*, ed. P. C. Champ, K. J. Boyle, and T. C. Brown, 331–93. Kluwer Academic Publishers, Dordrecht.

Thoreau, H. D. 1852. *The Journals of Henry David Thoreau, 1837–1861.* Ed. D. Searls. NYRB Classics, New York.

Turner, R. K. 1999. Environmental and ecological economics perspectives. In *Handbook of Environmental and Resource Economics*, ed. J. van der Bergh, 1001–33. Edward Elgar Press, Northampton, MA.

Turner, R. K., Pearce, D., and Bateman, I. 1994. *Environmental economics: An elementary introduction*. Harvester, Wheatsheaf, Hemel Hempstead.

UN REDD. 2013. The United Nations Collaborative Programme on Reducing Emissions from Deforestation and Forest Degradation in Developing Countries. http://www.un-redd.org/aboutredd/tabid/582/default.aspx; accessed 21 May 2013.

Vucetich, J. A., Bruskotter, J. T., and Nelson, M. P. 2015. Evaluating whether nature's intrinsic value is an axiom of or anathema to conservation. *Conservation Biology* 29:321–32.

Weidensaul, S. 2013. Beyond measure. *Audubon Magazine* 115 (March–April): 39–41.

Wenny, D., DeVault, T. L., Johnson, M. D., Kelly, D., Şekercioğlu, Ç. H., Tomback, D. F., and Whelan, C. J. 2011. The need to quantify ecosystem services provided by birds. *Auk* 128:1–14.

Whelan, C. J., Wenny, D. G., and Marquis, R. J. 2008. Ecosystem services provided by birds. *Annals of the New York Academy of Sciences* 1134:25–60.

CHAPTER THREE

Trophic Interaction Networks and Ecosystem Services

Christopher J. Whelan, Diana F. Tomback, Dave Kelly, and Matthew D. Johnson

Thanks to their power of flight, birds live on all continents and even on the most remote oceanic islands. Birds, as a group, function in every trophic category except primary producer and decomposer (although some birds are detritivores). They engage in local trophic interaction networks everywhere they occur. They forage in terrestrial, aquatic, and aerial environments. Through trophic interactions, they participate in antagonistic, mutualistic, and commensal interactions with other inhabitants of their ecosystems. In short, birds play many functional roles in their ecosystems, such as those of carnivore and herbivore, but also those of seed disperser, pollinator, scavenger, and ecosystem engineer.

Borrowing from Darwin, Terborgh et al. (2010) noted that species in ecosystems are engaged in tangled webs of interactions. Even simple food webs encompass a great number of direct and indirect interactions (see below). Using Holt's (1997) concept of the community module—a subset of the entire food web—the basic community module critical for pest control is a three-species food chain: a single species in the third trophic level (e.g., insectivorous bird) consumes a species of the second trophic level (caterpillar), which in turn consumes all or part of a species of the first trophic level (oak tree). The feeding interactions here comprise a trophic cascade, whereby the insectivore reduces the number of herbivores, which increases the biomass of the plant (fig. 3.1E). From the plant's perspective, the enemy (the bird) of its enemy (the caterpillar) is its friend: the bird indirectly benefits the plant by consuming the caterpillar.

A more complex community module consists of a trophic interaction network, in which multiple species at multiple trophic levels are linked. A complete food web consists of many such modules, each represented by an interaction chain or an interaction network. Through their participation in trophic interaction chains and networks, birds provide many ecosystem functions. When those functions benefit humans, birds provide ecosystem services. In fact, birds provide services in each of the four categories of ecosystem services identified by the Millennium Ecosystem Assessment (2003; see Whelan et al. 2008; Wenny et al. 2011).

Types of Trophic Interactions

Direct and Indirect

Interactions between or among species may be classified in the broad sense as direct or indirect. Direct interactions are those in which a given species

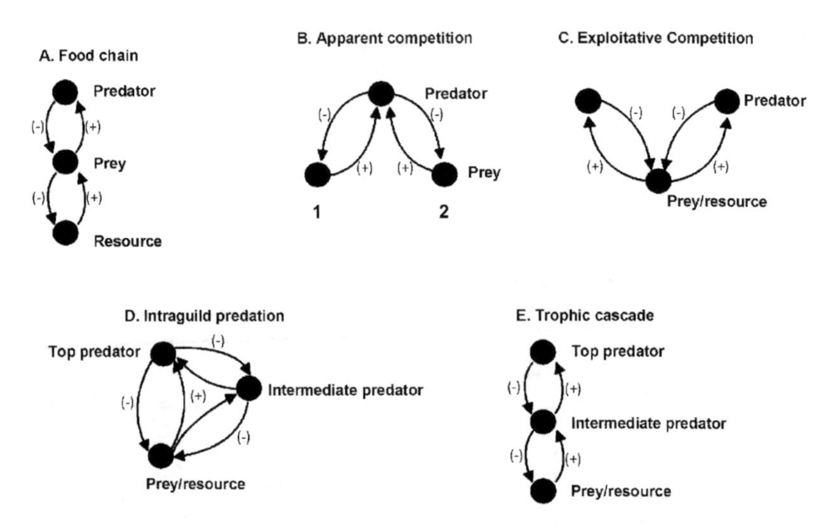

FIGURE 3.1. Examples of trophic interactions (following Holt 2009). Each example represents a community module, a simple subweb drawn from a potentially more complex community food web. Direct interactions are those between species linked by arrows (e.g., A. food chain: secondary consumer, primary consumer, and resource). Indirect interactions are those in which two species are linked via an intermediate species (e.g., B. apparent competition: prey 1 and prey 2 are linked indirectly by a shared predator). In intraguild predation (D.), a top predator both competes with and preys upon an intermediate predator. A trophic cascade (E.) is a food chain in which a top predator suppresses an intermediate predator, thus releasing the prey of the intermediate predator from suppression by the intermediate predator.

influences the behavior, population density, and reproductive success, among other characteristics, of a second species. Examples of direct interactions include predation, herbivory, and pollination. Indirect interactions are those in which species A interacts directly with species B, in turn influencing how species B interacts with species C; indirect effects (in this case between A and C) are the products of two or more direct effects. Examples include exploitative resource competition, apparent competition, and trophic cascades (fig. 3.1).

Top-Down, Bottom-Up

Ecological interactions are also classified by the direction in which their effects flow. With bottom-up interactions, or donor control, effects originate at one trophic level and move up the interaction network. Examples include pulsed inputs to ecosystems from mast crops (e.g., beech nuts, acorns) and insect irruptions (e.g., periodical cicadas) in terrestrial ecosystems, and algal blooms in aquatic systems. With top-down interactions, effects originate at one trophic level and move down the food web. Examples include the direct effects of predation and herbivory, and the indirect effects of trophic cascades.

Ecology of Trophic Interactions and Ecosystem Services

Trophic interactions are feeding relationships: one species consumes another. Trophic interactions that benefit humans represent ecosystem services. For instance, birds of prey may consume rodents, which in turn consume a valuable seed or nut crop. Delivery of ecosystem services through trophic interactions results from both direct and indirect effects and both top-down and bottom-up effects. Many of the ecosystem services provided by birds directly or indirectly result from their foraging interactions (Whelan et al. 2008). Through foraging, birds transfer energy and nutrients both within and among ecosystems (chapter 9), and thus contribute to ecosystem function and resilience (Lundberg and Moberg 2003). In this chapter we examine a variety of ecological contexts in which ecosystem services delivered by birds are the consequence of, and are impacted by, trophic interactions.

The impact of trophic interactions, whether direct or indirect, on other members of a community depends upon the scale at which resources are detected and, thus, the spatial scale of the functional response of the consumer species (Morgan et al. 1997). Consumers may respond to resources numerically, through population growth or aggregation, or functionally, through changes in foraging behavior (Solomon 1949). For a predator to deliver pest control services, for instance, it must be able to detect and respond to the spatial scale of heterogeneity in variation in pest densities (Schmidt and Whelan 1998). Under these conditions, predators will be able to respond to resources in a density-dependent manner, a condition necessary for a regulating impact on prey populations.

Trophic Cascades

A trophic cascade is an indirect trophic interaction in which a top predator benefits the prey or resource, owing to its consumption of an intermediate predator (fig. 3.1E). Most typically in trophic cascades with birds as the apex predators, insectivorous birds induce the top-down control of herbivorous insects, thereby benefiting the plants that would otherwise be consumed by the herbivores (Şekercioğlu 2006). Birds of prey potentially act as the apex predator in trophic cascades by consuming granivorous or herbivorous rodents, thus benefiting plants whose seeds or foliage would otherwise be consumed. Birds delivering ecosystem services via trophic cascades often serve as pest control agents.

Birds as Pest Control Agents

We broadly define a pest as any organism that decreases fitness, population size, growth rate, or economic value of any resource important to humans. Examples of pests abound: fungal pathogens destroy valuable crops and timber; herbivorous insects consume crops, and arthropods vector disease, to name a few. A biological pest control agent, therefore, is an organism that reduces the effect of a pest species on one or more resources, thus increasing the abundance, growth rate, or economic value of that resource for humans. Birds serve as pest control agents through their consumption of the pest, and pest control arises solely through cascading top-down trophic interactions. In the terminology of the Millennium Ecosystem Assessment (2003), pest control services may be classified as supporting services (Whelan et al. 2008).

Şekercioğlu et al. (2004) and Wenny et al. (2011) classified birds by trophic status, ecosystem services, and vulnerability to extinction. Pest control services potentially can be delivered by species in each of the trophic levels, with the likely exception of nectarivory, but data demonstrating such services are largely unavailable for most species. The trophic categories with the strongest evidence for pest control services are consumers of terrestrial invertebrates and scavengers. Of the approximately 10,000 species of birds in the world, about 5,706 terrestrial bird species consume invertebrates (Şekercioğlu et al. 2004). Pest control services by members of this trophic category have been documented in both natural and agro-ecosystems (chapter 1; see below). Obligate scavengers (36 species) control pests indirectly by ridding the environment of carrion (Buechley and Şekercioğlu 2016). This important service limits the spread of disease organisms and competing species like rodents and feral dogs that vector diseases to humans. As discussed in depth by Devault et al. (chapter 8), the value of these services were dramatically demonstrated in south Asia following a rapid and massive loss of four obligate scavenging vulture species (Oaks et al. 2004). The loss of scavengers enabled rodent and feral dog populations to increase, which in turn spread disease to humans, their domestic pets and livestock, and, likely, other species.

Natural Ecosystems

Trophic cascades in community modules with birds as the apex predator have been examined in a wide variety of natural ecosystems around the world, including grasslands and boreal, temperate, and tropical forests. The exclosure experiment of Holmes et al. (1979) provided the inspiration for later studies, though Holmes et al. (1979) did not examine the full cascade. Holmes et al. (1979) demonstrated that insectivorous forest birds depress abundances of lepidopteran larvae in forest understory vegetation when those densities are at endemic (nonirruptive) densities. The largest impacts on insect numbers coincided with nestling and fledgling periods of the nest cycle. Because their focus was on the direct effect of birds on insects, Holmes et al. (1979) did not examine the indirect effect of bird predation on herbivory levels or subsequent plant productivity. This study and other early exclosure experiments (Askenmo et al. 1977; Solomon et al. 1977; Joern 1986; Fowler et al. 1991) presented compelling evidence that birds can depress abundance of at least some arthropod prey, in some systems at some times. But they assessed only the predators (birds) and their prey (arthropods, especially herbivorous insects), and not the consequences for vegetation. Insect pest predators need not be insect pest control agents, because reductions in pests may not translate into greater plant productivity (chapter 1).

Subsequent studies expanded to examine all three trophic levels: insectivorous birds, arthropods, and plants. Atlegrim (1989) documented the effect of birds on herbivory: leaf damage to bilberry (*Vaccinium myrtillus*) increased significantly in the absence of birds. Marquis and Whelan (1994) found that excluding birds from sapling white oaks (*Quercus alba*) significantly increased both the density of leaf-damaging insects and leaf damage, which in turn decreased production of new biomass in the subsequent growing season. Marquis and Whelan (1994) included an insecticide treatment. Both birds and insecticide reduced arthropod abundance and leaf damage, and had about equal benefits for subsequent plant biomass production.

Top-down effects of insectivorous birds, including trophic cascades, have now been examined in many natural environments, including northern hardwood forest (Holmes et al. 1979; Strong et al. 2000), mixed grass prairie (Fowler et al. 1991), arid grassland (Bock et al. 1992), temperate oak forest (Marquis and Whelan 1994; Murakami and Nakano 2000; Lichtenburg and Lichtenburg 2002; Böhm et al. 2011), tropical forest (Van Bael et al. 2003), ponderosa pine forest (Mooney 2007), and hybrid cottonwoods (Bridgeland et al. 2010). The majority of these studies demonstrated, minimally, top-down effects of birds on arthropods. Many of them also found that bird predation on insects benefited plants (Whelan et al. 2008; Mäntylä et al. 2011; Wenny et al. 2011; Maas et al. 2015).

Accidental introductions of insects, such as that of the emerald ash borer (*Agrilus planipennis*) to North America, create opportunities to examine the responses of insectivorous birds to spreading and sometimes irruptive novel prey. Woodpeckers and other bark foragers prey upon emerald ash borer (Cappaert et al. 2005b; Duan et al. 2010), and their predation rates have been correlated with emerald ash borer density (Lindell et al. 2008). Some woodpecker species and white-breasted nuthatches (*Sitta carolinensis*) increased in density in regions infested with emerald ash borers (Koenig et al. 2013). Various bark-foraging species increased their use of ash trees in relation to the degree of infestation (Flower et al. 2013), which is indicative of a density-dependent response that could potentially contribute to population control. These results collectively suggest that bark-foraging birds may help slow the spread of this lethal pest in North America. Ecologists should be poised to take advantage of such "natural experiments," as they provide opportunities to examine ecological function at relevant spatial and temporal scales not attainable in manipulative experiments (Whelan et al. 2008; Rogers et al. 2012).

Agro-Ecosystems

The potential role of birds as agents of biological control was investigated by the economic ornithologists of the US Biological Survey from the late 1800s into the early 1900s (chapter 1). These studies, based on field observations and examination of stomach contents, implicated birds as effective pest control agents. Interest in this ecosystem function of birds waned with the advent of chemical insecticides and criticism of the methods employed by economic ornithologists. New investigations over the last decade confirm that, in some situations, birds do serve as effective pest control agents in agro-ecosystems. Many of these studies, like those in natural ecosystems, employ exclusion cages to reveal bird effects from their absence or "subtraction."

Moreover, some investigators employ "addition" manipulations, in which nest boxes are added to increase the density of birds inhabiting the study areas. For instance, Jedlicka et al. (2011) used nest boxes to attract western bluebirds (*Sialia mexicana*) to California (US) vineyards. The nest boxes greatly increased the abundance activity of bluebirds, and thereby increased the removal of larvae deployed in the field as bioassays of avian predation. Total avian abundance increased twofold before fledging, and 2.6-fold after fledging. Mols and Visser (2002) used a combination of nest boxes, to increase the density of great tits (*Parus major*), and exclusion cages to measure the tits' top-down effects in apple orchards, reporting that increased tit density decreased leaf and apple damage, and increased the apple yield by 66%. Bird control of insect pests has been documented in a variety of agricultural systems, including those of corn (Tremblay et al. 2001), apples (Mols and Visser 2002, 2007), broccoli (Hooks et al. 2003), kale (Ndang'ang'a et al. 2013), cacao (Van Bael et al. 2007; Maas et al. 2013), coffee (Kellerman et al. 2008; Johnson et al. 2009, 2010), oil palm (Koh 2008); and grapes (Jedlicka et al. 2011).

Ndang'ang'a et al. (2013) quantified bird diversity and foraging behavior in a Kenyan agroecosystem and found that most species foraged from the ground, consuming primarily seeds, fruits, and flowers of weed species. Ndang'ang'a et al. (2013) also observed two abundant aerial insectivores. The assemblage of observed bird species, in combination with their foraging behaviors, suggests a potential for beneficial pest control services by birds in this area. However, insectivorous birds decline in abundance over time in tropical agricultural ecosystems, in comparison to their rate of decline in tropical forests and agroforests (chapter 11; Şekercioğlu 2012).

Retaining native tree species and forest patches in agricultural areas may maintain higher numbers of insectivorous birds (Skreekar et al. 2013) and reduce crop damage.

A number of studies indicate that deployment of either nest boxes or hunting perches within agricultural systems may attract raptors in agro-ecosystems. In most cases, the increased density and activity of raptors resulted in decreased population sizes of rodent agricultural pests and, in some cases, decreased damage from those pests.

As reported by Smal et al. (1990), field trials investigating the use of barn owls (*Tyto alba*) as biological control agents of rodents (predominantly Malayan field rats, *Rattus tiomanicus*) in oil palm plantations in peninsular Malaysia began in 1986, at least in part owing to the evolution of resistance to the rodenticide warfarin. Smal et al. (1990) developed computer simulation models indicating that higher owl densities could reduce rat numbers, resulting in economically acceptable damage levels. This could be accomplished with biological control alone, or as part of an integrated pest management program (IPM) that reduces the use of rodenticides. Duckett (1991) recounted the history of natural spread of the barn owl in peninsular Malaysia following development of the oil palm industry, as well as the results of a nest box provisioning program aimed at enhancing barn owl density. This program proved biologically and economically successful, with the collateral benefit of population increases in mammalian predators (common palm civet, *Paradoxurus hermaphroditus*; leopard cat, *Prionailurus bengalensis*; feral house cat, *Felis catus*) which had previously declined owing to unintended consumption of warfarin-laced baits.

More recent work indicates that owl predators are effective biological control agents of rats in maize (Kenya: Ojwang and Oguge 2003), rice (Malaysia: Hafidzi and Na'iM 2003), alfalfa (Israel: Motro 2011), various field crops (wheat, sweet corn, alfalfa, clover, vetch, and oats), and date plantations (Israel: Meyrom et al. 2009), and also of rodents in semiurban (South Africa: Meyer 2008) and urban (Israel: Charter et al. 2007) environments. Nest boxes for barn owls were deployed successfully in Chile to control rodents, the reservoir for hantavirus syndome (Muñoz-Pedreros et al. 2010). Investigations of strategies to recruit raptors for biological control of vertebrate pests should be a priority in applied ecological research worldwide. In a twist on the use of raptors in agroecosystems (Kross et al. 2012), where birds can be both the top predator and the pests, New Zealand falcons (*Falco novaeseelandiae*) introduced into vineyards decreased the abundance of three species of introduced (Eurasian blackbird, *Turdus merula*; song thrush, *Turdus philomelos*; and starling, *Sturnus vulgaris*) and one native (silvereye, *Zosterops lateralis*) pest bird species, thus reducing grape losses by 95% in comparison to those in vineyards with no falcons.

Cascade Strength

Birds exert top-down trophic cascades in some but not all systems. What underlies this variability? Terborgh, Estes, and Holt (2010) examined ecological theory to investigate factors or relationships producing variability in the magnitude of trophic cascades. They found that trophic cascades will be of greater magnitude (1) in systems with high plant productivity, (2) when intense predation at higher trophic levels is coupled with strong density dependence at those levels, (3) with little intraguild predation and interference, and, (4) with greater predator niche complementarity.

Empirical studies are largely consistent with these expectations. Marquis and Whelan (1994) found cascading effects of bird predation on leaf-chewing insects on biomass production of white oak (*Quercus alba*) in relatively high-productivity oak forest in Missouri. Strong et al. (2000) found that bird predation decreased Lepidoptera abundance and mean size, but did not lead to a significant increase in biomass production of sugar maple (*Acer saccharum*) in less productive northern hardwood forest in New Hampshire. Similarly, Van Bael et al. (2003) found that bird invertebrate consumption decreased leaf herbivory in the more productive canopies of three Neotropical forest species than in the less productive understory. Mooney et al. (2010) found greater cascading effects, due to density dependence, on high-quality trees that enhance caterpillar growth more than did low-quality trees (though they found cascading effects on the latter trees as well). Van Bael et al. (2008) found greater cascading effects from birds in the canopy trees of tropical agroforests than from birds in the understory crop trees, and greater cascading effects when bird diversity was greatest owing to the presence of migratory species. These studies confirm expectations regarding productivity, density dependence, predator diversity, and niche complementarity.

In a meta-analysis of 114 empirical studies from aquatic and terrestrial ecosystems, Borer et al. (2005) reported that cascade strength was not well explained by ecosystem productivity, but was related to taxonomy of the herbivore (invertebrate) and the top predator (mammal or bird). These conditions apply to systems with insectivorous birds and herbivorous arthropods, which are now fairly widely studied (reviewed by Şekercioğlu 2006; Whelan et al. 2008; Wenny et al. 2011; and in this volume). Many of these investigations found birds effective at inducing strong cascading effects.

Bottom-Up Interaction Chains

Loss of Plants That Are Keystone Mutualists

Delivery of ecosystem services often depends on energy flux from primary producer to bird consumer. Consider the case history of whitebark pine (*Pinus albicaulis*) and Clark's nutcracker (*Nucifraga columbiana*), both discussed in detail by Tomback (chapter 7). The nutcracker is the primary seed disperser for whitebark pine. In this system, an introduced pathogen (*Cronartium ribicola*), which causes blister rust in five-needle white pines, and a natural episodic pest, the mountain pine beetle (*Dendroctonus ponderosae*)—or both—disrupt the seed dispersal services for the pine by the nutcracker in a lethal combination. Both the fungal pathogen and the insect pest may be considered predatory organisms, operating within a community module from high- to mid-level trophic levels down. The disease alone acts more slowly, but peak outbreaks of the beetle kill high proportions of mature cone-bearing trees (Tomback and Achuff 2010; Logan et al. 2010). Both mortality factors drastically reduce whitebark pine cone production. Following reduced cone production, nutcrackers seek higher rates of food rewards elsewhere, altering their use of whitebark pine communities and disrupting the regeneration cycle for whitebark pine. The likelihood of seed dispersal by nutcrackers consequently plummets, as does the likelihood of forest regeneration (McKinney et al. 2009; Barringer et al. 2012). Whitebark pine serves many important ecosystem functions as a foundation and keystone species, including the provision of ecosystem services to humans (chapter 7). These will decline as whitebark pine forests decline.

With growing globalization and increasingly rapid spread of exotic disease, other foundation and keystone species are at risk. For example, oak trees in Europe are succumbing to a previously unknown bacterial pathogen that causes the syndrome referred to as acute oak decline (Brady et al. 2010). In coastal California and Oregon, sudden oak death, caused by a fungal pathogen, has rapidly killed oaks and tanoaks as well as other plants (Rizzo and Garbelotto 2003). The acorns of oaks and tanoaks are important food sources for jays, which are important seed dispersal mutualists (chapter 7). The loss of these trees affects forest biodiversity and community structure, and disrupts the seed dispersal services of birds.

Intermediate Trophic Position

Predators may disrupt delivery of pollination (chapter 4) and seed dispersal (chapters 5, 6, 7) services by birds. Such disruption is commonly observed on oceanic islands, where endemic birds frequently evolved with no mammalian predators, but human colonists deliberately or inadvertently introduced a diversity of mammalian predators. Introduced mammals are often devastatingly successful predators on adults and young of native birds, reducing populations, sometimes to local or global extinction. Even when not driven to extinction, species may become functionally extinct (Şekercioğlu et al. 2004).

As reviewed by Innes et al. (2010), New Zealand now hosts 33 introduced mammal species, including devastating bird predators like the Pacific rat (*Rattus exulans*), ship rat (*Rattus rattus*), brushtail possum (*Trichosurus vulpecula*), and stoat (*Mustela erminea*). The disruption of ecosystem service delivery, including seed dispersal and pollination, has been particularly well studied in New Zealand (chapter 4). More importantly, New Zealand ecologists have investigated restoring those services.

As on many oceanic islands, birds play important functional roles as seed dispersal agents (for 59% of all tree species, and about 12% of all flora) and pollination agents (30% of trees, 4.5% of total flora) in New Zealand (chapters 5 and 6; Kelly et al. 2010). Delivery of these dispersal and pollination services is disrupted by introduced mammals, many assuming the role of apex predators in trophic interaction chains in which native New Zealand birds now occupy an intermediate trophic position. Predation has reduced some native bird species to functional extinction and reduced the density of others. Such declines end or limit the delivery of ecosystem services. A diversity of introduced mammals in New Zealand combine to prey on native bird species representing different communities, life history characteristics, body sizes, and ecological functions. For instance, stoats prey on many bird species, from yellowhead (*Mohoua ochrocephala*) to blue duck (*Hymenolaimus malacorhynchos*); three species of rat (ship rat; Pacific rat; and Norway rat, *R. norvegicus*) prey on various smaller adult songbirds, eggs of many species, and nestlings of larger species; brushtail possums prey upon larger species like kaka (*Nestor meridionalis*) and kokako (*Callaeas cinerea*; see Innes et al. 2010).

On the New Zealand mainland, Kelly et al. (2005) experimentally removed stoats from a 400-hectare Broken River site while using the 300-hectare Cheeseman site as a nontreatment area to test whether conservation management can restore bird pollination services for a native mistletoe (*Peraxilla tetrapetala*). Stoat removal rapidly increased bellbird reproductive success and an 85% increase in local densities, but Kelly et al. (2005) detected no significant increase in mistletoe pollination. These results suggest a dismal outcome for other desired projects, such as the restoring of bird pollination of *Rhabdothamnus* on the New Zealand mainland to levels quantified on New Zealand islands. An obvious lesson from this work, and from much else in conservation, is that preserving natural systems is much easier and less expensive than restoring or recreating them once they have been degraded or destroyed through human mismanagement and exploitation.

A dramatic example of loss of bird services (top-down trophic effects) results from the virtually complete extirpation of native forest birds from the Pacific island of Guam, the most southern island in the Mariana Island chain. Inadvertent introduction of the brown tree snake (*Boiga irregularis*) around World War II led to the annihilation of land birds and reductions in many mammals and lizards resident on Guam (Savidge 1987; Wiles et al. 2003; Mortenson et al. 2008). The nonnative snake assumed the role of apex predator while changing the position of insectivorous birds to that of intermediate predator (see fig. 3.1). The consequences of losing the Guam avifauna are (1) a precipitous decline in animal vectors of seed dispersal (Caves et al. 2013), and (2) a tremendous increase in the density of spiders (Rogers et al. 2012). Caves et al. (2013) found widespread seed dispersal by birds on Saipan, another island in the Marianas chain, where the native bird community is still intact, in contrast to the situation on Guam. Similarly, Rogers et al. (2011) found that spider density on Guam was up to 40 times greater than that on Saipan, an island farther to the north which lacks the tree snake. Moreover, to determine whether the loss of birds cascades down to affect plants, Rogers et al. (2011) compared seedling survival on Guam with that on the islands of Saipan, Tinian, and Rota, which all have relatively intact avifauna. For five or six plant species, seedling survival on Guam was equivalent or greater than on islands with birds. This suggests that the increased spider population on Guam in the absence of birds may control insect herbivores. Given that the extirpation of native birds on Guam occurred between 1945 and 1985, and that the first anecdotal reports of high spider densities were in the 1990s, spiders likely respond to bird loss quickly. Indeed, a meta-analysis of bird exclosure studies showed an increase in spiders after bird exclusion in 75% of the tests (Gunnarsson 2008), thus suggesting that spiders may frequently respond to bird declines or losses. However, the full effects of insectivorous birds is likely to be context-dependent, and too few landscape-level studies exist to make general predictions.

Human-Related Impact on Trophic Interaction Networks and Ecosystem Services

Many bird species known or likely to deliver ecosystem services are under risk of decline and extinction (Şekercioğlu et al. 2004). As discussed by Şekercioğlu and Buechley (chapter 11), human modification of habitats often changes the composition of bird communities, thus impacting the delivery of ecosystem services (see also Ferger et al. 2012). Maintaining services delivered by birds requires preservation of habitats and resources that support the bird species themselves (Whelan et al. 2008; Whelan et al. 2010). Conservation measures that generally enhance avian populations concomitantly strengthen their delivery of ecosystem services (Jedlicka et al. 2012; Barbaro et al. 2013). Increased understanding of the relationships among species richness, ecological function, and ecosystem service delivery will help reveal important consequences for the persistence of ecosystem services in the face of human impact (Philpott et al. 2009). As habitats are disturbed and climates change, species are not lost randomly; agricultural expansion and intensification selectively purge species with a distinct set of traits (Tscharntke et al. 2008). The same functional traits that confer species persistence may simultaneously affect service provision (Zavaleta and Hulvey 2004; Larsen et al. 2005). Dietary generalists survive in highly modified landscapes better than specialists (Lindell et al. 2004; Tscharntke et al. 2008), and they are less extinction-prone in general (Boyles and Storm 2007; Şekercioğlu 2011). Field data and models also indicate that generalists can exert stronger top-down control on their prey than can specialists (Symondson et al. 2002; Bianco Faria et al. 2008). Therefore, dietary generalism may dampen the adverse effects of land use intensification on the ecosystem services provided by avian trophic interactions. However, the conservation of ecosystem services may also hinge on retaining functionally unique species (Zavaleta and Hulvey 2004)—for example, species that consume a specific insect that is avoided by other insectivores. Because different birds have their own suite of preferred prey and their own foraging niches, the pest control ecosystem service provided by a bird assemblage may be noticeably changed by the functional extinction of a subset of the birds, even if all the common generalist bird species persist. The relationship between dietary specialization and functional uniqueness is uncertain. Understanding how both of those things change with the degree of competition among bird species should be a priority for diet and community research.

Climate Change

Global climate change affects birds around the world (Möller et al. 2010; Wormworth and Şekercioğlu 2011; Şekercioğlu et al. 2012), and, potentially, their delivery of ecosystem services. Global climate change causes shifts in the timing of ecological processes (Bradley et al. 1999; Ellwood et al. 2013), and abundance and distributions of numerous organisms, both animal and plant (Parmesan and Yohe 2003; Wormworth and Şekercioğlu 2011; Şekercioğlu et al. 2012). Changes in abundance and distribution of species are linked to the emergence of disease. The timing of nesting and migration of some bird species, in particular, has already changed (Dunn and Winkler 1999, 2010; Mills 2005; Kobori et al. 2011), and may reduce the ability of insectivorous birds to control populations of plant-eating insects that can influence the productivity of natural and agricultural systems.

Delivery of many ecosystem services may be threatened by global climate change, while others may be enhanced. Birds may even potentially render some ecosystems resilient to some consequences of global climate change. As demonstrated by the sentinel pest experiment of Jedlicka et al. (2011), birds may control agricultural pests that arrive in new areas in response to climate change or from accidental introductions. Indeed, the work of Koenig et al. (2013) and of Flower et al. (2013), as discussed above, indicates that a variety of bark-foraging bird species prey upon the introduced and expanding emerald ash borer in precisely such a manner. Although the expected negative consequences of global climate change often receive greater public attention, some of the changes may be beneficial. A study projecting changes in the range of species suggests that northern Europe may see an increase in diversity of species that provide ecosystem services following global warming, while southern Europe may see a decrease in those same species (Civantos et al. 2012).

Şekercioğlu et al. (2012) investigated the potential effects of climate change for tropical species. They concluded that species living in montane areas, those with no corridors to higher elevations, those living in coastal forests, and those with restricted geographical ranges are most vulnerable to population decline and extinction. Şekercioğlu et al. (2012) suggest that the establishment of new protected areas, or the enhancement of existing areas, must consider future climate change. This includes the development of area networks with extensive topographical diversity, wide elevational ranges, and high connectivity. These networks should be integrated into human-dominated landscapes to mesh with conservation priorities while simultaneously facilitating the delivery of ecosystem services (Tscharntke et al. 2005; Whelan et al. 2010; Woltz et al. 2012).

Şekercioğlu et al. (2012) argue that particular suites of bird species are particularly vulnerable, owing to geography and evolutionary history. For instance, some tropical mountain species living at particular elevations may have restricted ranges because of specialized habitat requirements and/or species interactions. Other species may occupy areas at high risk of increasing global temperatures, but have no ready access to higher elevations to mitigate rising temperatures (e.g., species in the central Amazon basin, far from the Andes). Coastal forest bird species and species with highly restricted geographic ranges (e.g., island species and many endemic species) are especially vulnerable. Some bird species may be especially susceptible to increased seasonality of annual rainfall (both increased and decreased), as such change may affect the abundance and/or timing of resources required for successful reproduction. Many species will also be vulnerable to extreme weather events such as heat waves, cold spells, and tropical cyclones. Birds that experience limited temperature variation and have low basal metabolic rates will be most prone to the physiological effects of warming temperatures and heat waves. Şekercioğlu et al. (2012) conclude by emphasizing the importance of using "various methods to estimate the economic value of ecosystem services delivered by birds and other animals."

Species will not respond to climatic changes uniformly or predictably. Some species may tolerate climate change, and even benefit from it (Civantos et al. 2012), while other species decline. The ecosystem services delivered by the former species will be persistent in the face of climate change, while those of the latter may be reduced, disrupted, and ultimately lost. However, these responses are likely to be complex, since bird species which persist in an area could change their diet or behaviors in response to the loss of other species from the area, as noted above.

Conclusions and Future Directions

Birds deliver important ecosystem services through a number of complex trophic interactions, with birds sometimes driving the interaction and sometimes under pressure from other trophic levels. These ecosystem services include pest control, seed dispersal, pollination, and scavenging services. Research is needed now to address at least three aspects of ecosystem services:

1. How Can We Mitigate against Human-Caused Disruption of These Services?

We need to identify traits that make some species good providers of ecosystem services, and to determine whether those traits make them more or less susceptible to anthropogenic disturbance such as habitat loss and climate change. With this knowledge, we may be able to target particularly important providers of ecosystem services and improve the conservation of the ecosystems or unique habitats and resources they require for persistence. Establishment and enhancement of networks of protected areas facilitate the ability of species to adjust their ranges in the face of climate change and thereby continue to provide the service. Any public and private actions that reduce human contribution to climate change will help conserve birds and their ecosystem services.

2. How Might We Facilitate and Enhance the Delivery of Ecosystem Services through Exploitation of Natural Interaction Networks?

Careful management of habitat availability, use of nest boxes and hunting perches, and control of invasive species that disrupt delivery of services may enhance birds' ability to deliver ecosystem services. This may even involve an otherwise uncommon species that has suffered from human disturbance, such as the New Zealand falcon, which provides conservation as well as economic benefits in vineyards. Research is needed to elucidate how we can manage or manipulate human-dominated environments, particularly urban and agricultural environments, in ways that promote high bird abundance and diversity and, hence, their ecosystem services. For instance, Jones and Sieving (2006) demonstrated that insect-consuming birds attracted to fields of organic vegetable crops with intercropped sunflower (*Helianthus annuus*) resulted in reduction of important crop pests, and, importantly, had no negative consequences for the crops themselves.

3. Can We Better Elucidate the Value of Ecosystem Services Provisioned by Trophic Interactions Involving Birds?

Both economists and ecologists have made conceptual advances in identifying the ethics and values leading to the valuation of ecosystem services (chapter 2), including those provided by birds. Chapter 2 points to various methods for estimating the economic value of ecosystem services delivered by birds and other animals. In the case of trophic cascades involving birds, techniques such as the avoided cost, replacement cost, and factor income methods may allow researchers to estimate the economic value of bird-provisioned services. In addition, because birds are mobile agents of ecosystem services, future research should be aimed at understanding how habitat and landscape composition may affect bird movements and the spatial delivery of their services. For example, Jirinec et al. (2011) found that pest-eating warblers commuted from diurnal foraging home ranges in coffee farms to nocturnal roosting sites in surrounding forests, a behavior which could link the delivery of pest control on farms to the preservation of forest in the landscape and vice versa. While estimating the value of birds' trophic interactions, researchers should be mindful that the commodification of birds' services may reproduce some of the pitfalls of neoclassical market economics (Gómez-Baggethun et al. 2010). The primary purpose for assigning economic value to the ecosystem services provided by birds must be to argue that conservation of birds is not only a matter of ethics and aesthetics, but is also essential for ecosystem function and human livelihood.

Acknowledgments

We thank Joel Brown and his lab group for helpful suggestions on an early draft of this chapter.

References

Askenmo, C., Bromssen, A., von Ekman, J., and Jansson, C. 1977. Impact of some wintering birds on spider abundance in spruce. *Oikos* 28:90–94.

Atlegrim, O. 1989. Exclusion of birds from bilberry stands: Impact on insect larval density and damage to the bilberry. *Oecologia* 79:136–39.

Barbaro, L., Dulaurent, A-M., Payet, K., Blache, S., Vetillard, F., and Battisti, A. 2013. Winter bird numerical responses to a key defoliator in mountain pine forests. *Forest Ecology and Management* 296:90–97.

Barringer, L. E., Tomback, D. F., Wunder, M. B., and McKinney, S. T. 2012. Whitebark pine stand condition, tree abundance, and cone production as predictors of visitation by Clark's nutcracker (*Nucifraga columbiana*). *PLoS ONE* 7(5):e37663.

Bianco Faria, L., Umbanhowar, J. and McCann, K. S. 2008. The long-term and transient implications of multiple predators in biocontrol. *Theoretical Ecology* 1:45–53.

Bock, C. E., Bock, J. H., and Grant, M. C. 1992. Effects of bird predation on grasshopper densities in an Arizona grassland. *Ecology* 73:1706–17.

Böhm S. M., Wells, K., Kalko, E. K. V. 2011. Top-down control of herbivory by birds and bats in the canopy of temperate broad-leaved oaks (*Quercus robur*). *PLoS ONE* 6:e17857.

Borer, E. T., Seabloom, E. W., Shurin, J. B., Anderson, K. E., Blanchette, C. A., Broitman, B., Cooper, S. D. and Halpern, B. S. 2005. What determines the strength of a trophic cascade? *Ecology* 86:528–37.

Boyles, J. G., and Storm, J. J. 2007. The perils of picky eating: Dietary breadth is related to extinction risk in insectivorous bats. *PLoS ONE* 2:e672.

Bradley N. L., Leopold, A. C., Ross, J., and Huffaker, W. 1999. Phenological changes reflect climate change in Wisconsin. *Proceedings of the National Academy of Sciences* 96:9701–4.

Brady, C., Denman, S., Kirk, S., Venter, S., Rodríguez-Palenzuela, and Coutinho, T. 2010. Description of *Gibbsiella quercinecans* gen. nov., sp. nov., associated with acute oak decline. *Systematic and Applied Microbiology* 33:444–50.

Buechley, E. R., Şekercioğlu, Ç. H. 2016a. The avian scavenger crisis: Looming extinctions, trophic cascades, and loss of critical ecosystem functions. *Biological Conservation*. In press.

Cappaert, D., McCullough, C. G., and Poland, T.M. 2005. The upside of the emerald ash borer catastrophe: A feast for woodpeckers. In *Emerald Ash Borer Research and Technology Development Meeting,* ed. V. Mastro and R. Reardon, 69–70. US Department of Agriculture Forest Service, Washington.

Caves, E. M., Jennings, S. B., Hille Ris Lambers, J., Tewksbury. J. J., and Rogers, H. S. 2013. Natural experiment demonstrates that bird loss leads to cessation of dispersal of native seeds from intact to degraded forests. *PLoS ONE* 8:e65618.

Charter, M., Izhaki, I., Shapira, L. and Leshem, Y. 2007. Diets of urban breeding barn owls (*Tyto alba*) in Tel Aviv, Israel. *Wilson Journal of Ornithology* 119:484–85.

Civantos, E., Thuiller, W., Maiorano, L., Guisan, A., and Araújo, M. B. 2012. Potential impacts of climate change on ecosystem services in Europe: The case of pest control by vertebrates. *BioScience* 62:658–66.

Duan, J. J., Ulyshen, M. D., Bauer, L. S., Gould, J. and Van Driesche, R. 2010. Measuring the impact of biotic factors on populations of immature emerald ash borers (Coleoptera: Buprestidae). *Environmental Entomology* 39:1513–22.

Duckett, J. E. 1991. Management of the barn owl (*Tyto alba javanica*) as a predator of rats in oil palm (*Elaeis quineensis*) plantations in Malaysia. *Birds of Prey Bulletin* 4:11–23.

Dunn, P. O., and Winkler, D. W. 1999. Climate change has affected the breeding date of tree swallows throughout North America. *Proceedings of the Royal Society of London Biology B* 266:2487–90.

———. 2010. Effects of climate change on timing of breeding and reproductive success in birds. In *Effects of Climate Change on Birds*, ed. A. P. Moller, W. Fiedler, and P. Berthold, 113–28. Oxford University Press, Oxford.

Ellwood, E. R., Temple, S. A., Primack, R. B., Bradley, N. L., and Davis, C. C. 2013. Record-breaking early flowering in the eastern United States. *PLoS ONE* 8:e53788.

Ferger, S. W., Böhning-Gaese, K., Wilcke, W., Oelmann, Y., and Schleuning, M. 2013. Distinct carbon sources indicate strong differentiation between tropical forest and farmland bird communities. *Oecologia* 171:473–86.

Flower, C. E, Long, L., Knight, K. S., Rebbeck, J., Brown, J. S., Gonzalez-Meler, M. A., and Whelan, C. J. 2013. Native bark-foraging birds preferentially forage in infected ash (*Fraxinus* spp.) trees and prove effective predators of the invasive emerald ash borer (*Agrilus planipennis* Fairmaire). *Forest Ecology and Management* 313:300–306.

Fowler, A. C., Knight, R. L., George, T. L., and McEwen, L. C. 1991. Effects of avian predation on grasshopper populations in North Dakota grasslands. *Ecology* 72:1775–81.

Hafidzi, M. N., and Na'iM, M. 2003. The use of the barn owl, *Tyto alba*, to suppress rat damage in rice fields in Malaysia. In *Rats, Mice and People: Rodent Biology and Management*, ed. G. R. Singleton, L. A. Hinds, C. J. Krebs, and D. M. Spratt, 274–77. Australian Centre for International Agricultural Research, Canberra.

Holmes, R, T., Schultz, J. C., and Nothnagle P. 1979. Bird predation on forest insects: An exclosure experiment. *Science* 206:462–63.

Holt, R. D. 1997. Community modules. In *Multitrophic Interactions in Terrestrial Ecosystems*, ed. A. C. Gange and V. K. Brown, 333–49. 36th symposium of the British Ecological Society. Blackwell Science, Oxford.

———. 2009. Predation and community organization. In *The Princeton Guide to Ecology*, ed. S.A. Levin, 274–81. Princeton University Press, Princeton, NJ.

Hooks, C. R., Pandey R. R., and Johnson, M. W. 2003. Impact of avian and arthropod predation on lepidopteran caterpillar densities and plant productivity in an ephemeral agroecosystem. *Ecological Entomology* 28:522–32.

Innes, J., Kelly, D., Overton, J. M., and Gillies, C. 2010. Predation and other factors currently limiting New Zealand forest birds. *New Zealand Journal of Ecology* 34:86–114.

Jedlicka, J. A., Greenberg, R., and Letourneau, D. K. 2011. Avian conservation practices strengthen ecosystem services in California vineyards. *PLoS ONE* 6:e27347.

Joern, A. 1986. Experimental-study of avian predation of coexisting grasshopper populations (Orthoptera, Acrididae) in a sandhills grassland. *Oikos* 46:243–49.

Johnson, M. D., Levy, N. J., Kellermann, J. L., and Robinson, D. E. 2009. Effects of shade and bird exclusion on arthropods and leaf damage on coffee farms in Jamaica's Blue Mountains. *Agroforestry Systems* 76:139–48.

Johnson, M. D., Kellermann, J. L., and Stercho, A. M. 2010. Pest reduction services by birds in shade and sun coffee in Jamaica. *Animal Conservation* 13:140–47.

Jones, G. A., and Sieving, K. E. 2006. Intercropping sunflower in organic vegetables to augment bird predators of arthropods. *Agriculture, Ecosystems & Environment* 117:171–77.

Kellermann, J. L., Johnson, M. D. Stercho, A. M., and Hackett. S. C. 2008. Ecological and economic services provided by birds on Jamaican Blue Mountain coffee farms. *Conservation Biology* 22:1177–85.

Kelly, D., Brindle, C., Ladley, J. J., Robertson, A. W., Maddigan, F. W., Butler, J., Ward-Smith, T., Murphy, D. J., and Sessions, L. A. 2005. Can stoat (*Mustela erminea*) trapping increase bellbird (*Anthornis melanura*) populations and benefit mistletoe (*Peraxilla tetrapetala*) pollination? *New Zealand Journal of Ecology* 29:69–82.

Kelly, D., Ladley, J. J., Roberston, A., Anderson, S. H. , Wotton, D. M., and Wiser, S. K. 2010. Mutualisms with the wreckage of an avifauna: The status of bird pollination and fruit dispersal in New Zealand. *New Zealand Journal of Ecology* 34:66–85.

Kobori, H., Kamamoto, T., Nomura, H., Oka, K., and Primack, R. 2011. The effects of climate change on the phenology of winter birds in Yokohama, Japan. *Ecological Research* 27:173–80.

Koenig, W. D., Liebhold, A. M., Bonter, D. N., Hochachka, W. M., Dickinson, J. L. 2013. Effects of the emerald ash borer invasion on four species of birds. *Biological Invasions* 15:2095–2103.

Koh, L. P. 2008. Birds defend oil palms from herbivorous insects. *Ecological Applications* 18:821–25.

Kross, S. M., Tylianakis, J. M., and Nelson, X. J. 2012. Effects of introducing threatened falcons into vineyards on abundance of Passeriformes and bird damage to grapes. *Conservation Biology* 6:142–49.

Lindell, C. A., McCullough, D. G., Cappaert, D., Apostolou, N. M., Roth, M. B. 2008. Factors influencing woodpecker predation on emerald ash borer. *American Midland Naturalist* 159:434–44.

Logan, J. A., MacFarlane, W. W., and Willcox, L. 2010. Whitebark pine vulnerability to climate-driver mountain pine beetle disturbance in the Greater Yellowstone Ecosystem. *Ecological Applications* 20:895–902.

Lundberg, J., and Moberg, F. 2003. Mobile link organisms and ecosystem functioning: implications for ecosystem resilience and management. *Ecosystems* 6:87–98.

Maas, B., Clough, Y., and Tscharntke, T. 2013. Bats and birds increase crop yield in tropical agroforestry landscapes. *Ecology Letters* 16:1480–87.

Maas, B., Karp, D. S., Bumrungsri, S., Darras, K., Gonthier, D., Huang, C. -C., Lindell, C. A., Maine, J. J., Mestre, L., Michel, N. L., Morrison, E. B., Perfecto, I., Philpott, S. M., Şekercioğlu, Ç. H., Silva R. M., Taylor, P., Tscharntke, T., Van Bael, S. A., Whelan, C. J. and Williams-Guillén, K. 2015. Bird and bat predation services in tropical forests and agroforestry landscapes. *Biological Reviews*. Article first published online 23 July 2015. DOI: 10.1111/brv.12211.

Mäntylä, E., Klemola, T., Laaksonen, T. 2011. Birds help plants: A meta-analysis of top-down trophic cascades caused by avian predators. *Oecologia* 165:143–51.

McKinney, S. T., Fiedler, C. E, and Tomback, D. F. 2009. Invasive pathogen threatens bird-pine mutualism: Implications for sustaining a high-elevation ecosystem. *Ecological Applications* 19:597–607.

Meyer, S. 2008. *Owl as a Control Agent for Rat Populations in Semi-Urban Habitats*. Master's thesis, University of the Witwatersrand, Johannesburg.

Meyrom, K., Motro, Y., Leshem, Y., Aviel, S., Izhaki,I., Argyle, F. and Charter, M. 2009. Nest-box use by the Barn Owl *Tyto alba* in a biological pest control program in the Beit She'an valley, Israel. *Ardea* 97:463–67.

Millennium Ecosystem Assessment. 2003. *Ecosystems and Human Well-Being: A Framework for Assessment*. Island Press, Washington.

Mills, A.M. 2005. Changes in the timing of spring and autumn migration in North American migrant passerines during a period of global warming. *Ibis* 147:259–69.

Möller, A. P., Fiedler, W., and Berthold, P. 2010. *Effects of Climate Change on Birds*. Oxford University Press, Oxford.

Mols, C. M. M., and Visser, M. E. 2002. Great tits can reduce caterpillar damage in apple orchards. *Journal of Applied Ecology* 39:888–99.

———. 2007. Great tits (*Parus major*) reduce caterpillar damage in commercial apple orchards. *PLoS ONE* 2:e202.

Mooney, K. A. 2007. Tritrophic effects of birds and ants on a canopy food web, tree growth, and phytochemistry. *Ecology* 88:2005–14.

Mooney, K. A., Halitschke, R. Kessler, A. and Agrawal, A. A. 2010. Evolutionary trade-offs in plants mediate the strength of trophic cascades. *Science* 327:1642–44.

Morgan, R. A., Brown, J. S., and Thorson, J. M. 1997. The effect of spatial scale on the functional response of fox squirrels. *Ecology* 78:1087–97.

Motro, Y. 2011. Economic evaluation of biological rodent control using barn owls *Tyto alba* in alfalfa. 8th European Vertebrate Pest Management Conference, 79–80. DOI: 10.5073/jka.2011.432.040.

Muñoz-Pedreros, A., Gil, C., Yáñez, J. and Rau, J. R. 2010. Raptor habitat management and its implication on the biological control of the Hantavirus. *European Journal of Wildlife Research* 56:703–15.

Murakami, M., and Nakano, S. 2000. Species-specific bird functions in a forest-canopy food web. *Proceeding of the Royal Society of London B*. 267:1597–1601.

Ndang'ang'a, P. K., Njoroge, J. B. M., Ngamau, K., Kariuki, W., Atkinson, P. W., and Vickery, J. 2013. Avian foraging behaviour in relation to provision of ecosystem services in a highland East African agroecosystem. *Bird Study* 60:156–68.

Ndang'ang'a, P. K., Njoroge, J. B. M., and Vickery, J. M. 2013. Quantifying the contribution of birds to the control of arthropod pests on kale, *Brassica oleracea* acephala, a key crop in East African highland farmland, *International Journal of Pest Management* 59:211–16.

Oaks, J. L. et al. 2004. Diclofenac residues as the cause of vulture population decline in Pakistan. *Nature* 427:630–33.

Ojwang, D. O., and Oguge, N. O. 2003. Testing a biological control programme for rodent management in a maize cropping system in Kenya. In *Rats, Mice and People: Rodent Biology and Management*, ed. G. R. Singleton, L. A. Hinds, C. J. Krebs and D. M. Spratt, 251–53. Australian Centre for International Agricultural Research, Canberra.

Parmesan, C., and Yohe, G. 2003. A globally coherent fingerprint of climate change impacts across natural systems. *Nature* 421:37–42.

Philpott, S. M., Soong, O., Lowenstein, J. H., Pulido, A.L., Lopez, D. T., Flynn, D. F. B., and DeClerck, F. 2009. Functional richness and ecosystem services: Bird predation on arthropods in tropical agroecosystems. *Ecological Applications* 19:1858–67.

Rizzo, D. M., and Garbelotto, M. 2003. Sudden oak death: Endangering California and Oregon forest ecosystems. *Frontiers in Ecology and the Environment* 1:197–204.

Rogers, H., Lambers, J. H. R., Miller, R., Tewksbury, J. J. 2011. "Natural experiment" demonstrates top-down control of spiders by birds on a landscape level. *PLoS ONE* 7:e43446.

Savidge, J. A. 1987. Extinction of an island forest avifauna by an introduced snake. *Ecology* 68:660–68.

Schmidt, K. A., and Whelan, C. J. 1998. Nest predation on woodland songbirds: When is nest predation density dependent? *Oikos* 87:65–74.

Şekercioğlu, Ç. H. 2006. Ecological significance of bird populations. In *Handbook of the Birds of the World*, ed. J. del Hoyo, A. Elliott, and D. A. Christie, volume 11, 15–51. Lynx Edicions, Barcelona.

———. 2011. Functional extinctions of bird pollinators cause plant declines. *Science* 331:1019–20.

———. 2012. Bird functional diversity in tropical forests, agroforests and open agricultural areas. *Journal of Ornithology* 153:S153–61.

Şekercioğlu, Ç. H., Daily, G. C. and Ehrlich, P. R. 2004. Ecosystem consequences of bird declines. *Proceedings of the National Academy of Sciences* 101:18042–47.

Şekercioğlu, Ç. H., Primack, R., Wormworth, J. 2012. Effects of climate change on tropical birds. *Biological Conservation* 148:1–18.

Smal, C. M., Halim, A. H. Din Amiruddin, M. 1990. *Predictive Modelling of Rat Populations in Relation to Use of Rodenticides or Predators for Control.* No. 25. Palm Oil Research Institute Malaysia, Kuala Lumpur.

Solomon, M. E. 1949. The natural control of animal populations. *Journal of Animal Ecology* 18:1–35.

Strong, A. M., Sherry, T. W., and Holmes, R. T. 2000. Bird predation on herbivorous insects: Indirect effects on sugar maple saplings. *Oecologia* 125:370–79.

Symondson, W. O. C., K. D. Sunderland, and M. H. Greenstone. 2002. Can generalist predators be effective as biocontrol agents? *Annual Review of Entomology* 47:561–94.

Terborgh, J., Holt R. D., and Estes, J. A. 2010. Trophic cascades: What they are, how they work, and why they matter. In *Trophic Cascades: Predators, Prey and the Changing Dynamics of Nature*, ed. J. Terborgh and J. A. Estes, 1–18. Island Press, Washington.

Tomback, D. F., and P. Achuff. 2010. Blister rust and western forest biodiversity: Ecology, values, and outlook for white pines. *Forest Pathology* 40:186–225.

Tremblay, A., Mineau, P., Stewart, R.K. 2001. Effects of bird predation on some pest insect populations in corn. *Agriculture, Ecosystems and Environment* 83: 143–52.

Tscharntke,T., Klein, A. M., Kruess, A., Ingolf, S.-D. and Thies, C. 2005. Landscape perspectives on agricultural intensification and biodiversity: Ecosystem service management. *Ecology Letters* 8:857–74.

Tscharntke, T., Şekercioğlu, Ç. H., Dietsch, T. V., Sodhi, N. S., Hoehn, P., and Tylianakis, T. M. 2008. Landscape constraints on functional diversity of birds and insects in tropical agroecosystems. *Ecology* 89:944–51.

Van Bael, S. A., Bichier, P., Greenberg, R. 2007. Bird predation on insects reduces damage to the foliage of cocoa trees (*Theobroma cacao*) in western Panama. *Journal of Tropical Ecology* 23:715–19.

Van Bael, S. A., and Brawn, J. D. 2005. The direct and indirect effects of insectivory by birds in two contrasting Neotropical forests. *Oecologia* 143:106–16.

Van Bael, S. A., Brawn, J. D., and Robinson, S. K. 2003. Birds defend trees from herbivores in a Neotropical forest canopy. *Proceedings of the National Academy of Sciences* 100:8304–07.

Van Bael, S. A., Philpott, S. M., Greenberg, R., Bichier, P., Barber, N. A., Mooney, K. A., and Gruner, D. S. 2008. Birds as predators in tropical agroforestry systems. *Ecology* 89:928–34.

Wenny, D. G., DeVault, T. L., Johnson, M. D., Kelly, D., Şekercioğlu, C. H., Tomback, D. F., and Whelan, C. J. 2011. The need to quantify ecosystem services provided by birds. *Auk* 128:1–14.

Whelan, C. J., Wenny, D. G., and Marquis, R. J. 2008. Ecosystem services provided by birds. *Annals of the New York Academy of Sciences* 1134:25–60.

Wiles, G. J., Bart, J., Beck, R. E. Jr., and Aguon, C. F. 2003. Impacts of the brown tree snake: Patterns of decline and species persistence in Guam's avifauna. *Conservation Biology* 17:1350–60.

Wormworth, J., Şekercioğlu, Ç. H. 2011. *Winged Sentinels: Birds and Climate Change*. Cambridge University Press, New York.

Zavaleta, E. S., and Hulvey. K. B. 2004. Realistic species losses disproportionately reduce grassland resistance to biological invaders. *Science* 306:1175–77.

CHAPTER FOUR

Pollination by Birds

A Functional Evaluation

Sandra H. Anderson, Dave Kelly, Alastair W. Robertson, and Jenny J. Ladley

Birds provide ecosystem services by facilitating plant reproduction as both pollinators and seed dispersers, but to date bird pollination has been considered relatively less important than dispersal (Corlett 2007), and the status of bird pollinators relatively more secure than bird seed dispersers (Şekercioğlu et al. 2004). However, the last 20 years have seen a gradual shift in ideas as less conspicuous bird-pollination links have become more evident and more important than first assumed. In this chapter, we build on several recent global reviews, and extend those by drawing lessons from studies of the actual (rather than expected) contribution of birds to pollination. This includes measures of pollination failure that result from bird declines. Our results confirm that bird pollination is less common than insect pollination, but they show that birds are unexpectedly important for effective reproduction, and surprisingly hard to replace. These studies are mostly on native species, so the benefits (ecosystem services) largely accrue to biodiversity rather than economic production, with a few interesting exceptions.

In its simplest form, the shift we stress in this chapter is from a morphological to a functional assessment of bird pollination. The former classifies the importance of bird pollination by whether the birds and the plants "appear" to have an important relationship. That evaluation depends heavily on the concept of floral syndromes, whereby bird-pollinated flowers are expected to have ornithophilous characters and bird visitors are expected to be nectarivore specialists from mainly tropical families (fig. 4.1). In contrast, the functional approach measures the importance of bird pollination by the level of fruit set achieved when birds visit, and the decrease in fruit set observed when bird visits are prevented (by experimental manipulation, or bird population declines). Functional studies show that even unspecialized birds can be effective pollinators, and even non-ornithophilous flowers may be reliant on birds (fig. 4.2).

At a more detailed level, the shift in emphasis on bird pollination includes reexamining several other assumptions, as Kelly et al. (2010) did for the well-studied New Zealand flora. They showed that local ecologists (e.g., Godley 1979) had down-weighted observations of birds visiting flowers in two other ways apart from discounting visits to nonornithophilous flowers.

FIGURE 4.1. Examples of ornithophilous bird-pollinated flowers from New Zealand: (a) Tui (*Prosthemadera novaeseelandiae*, Meliphagidae) opening an explosive bud of red mistletoe (*Peraxilla colensoi*; photo by Alastair Robertson). (b) Kokako (*Callaeas cinerea*, Callaeidae) pollinating New Zealand flax (*Phormium tenax*); the orange on forehead is entirely flax pollen (photo © Simon Fordham/NaturePix). (c) Silvereye (*Zosterops lateralis*, Zosteropidae) robbing flower of kowhai (*Sophora microphylla*), which is longer than its beak (photo by Dave Kelly). (d) New Zealand bellbird (*Anthornis melanura*, Meliphagidae) pollinating puriri *Vitex lucens* (photo by Abe Borker).

FIGURE 4.2. Examples of birds visiting non-ornithophilous flowers in New Zealand. (a) Tui pollinating five-finger (*Pseudopanax arboreus*); most of the flowers are not yet open (photo by Abe Borker). (b) New Zealand bellbird pollinating kohekohe (*Dysoxylum spectabile*; photo by Abe Borker). (c) Stitchbird (*Notiomystis cincta*, Notiomystidae) pollinating the five-millimeter-wide flowers of *Muehlenbeckia complexa* (photo © Simon Fordham/NaturePix). (d) New Zealand kaka (*Nestor meridionalis*, Psittacidae) pollinating *Pittosporum umbellatum* (photo © Simon Fordham/NaturePix).

First, the observed bird visits were considered incidental to visits by insects, which often also visit large ornithophilous flowers and were assumed capable of effecting pollination. Second, bird visits were speculated to result mainly in geitonogamy (self-pollination within the same plant) because birds might remain in and defend single flowering plants in ways that insect pollinators do not.

All of these assumptions are reiterated outside New Zealand in the global literature. In their comprehensive review of bird pollination, Proctor et al. (1996) state that "there are no bird-pollinated flowers in Europe" (p. 225), but on the next page describe Eurasian blue tits (*Cyanistes caeruleus*) pollinating *Salix* species. Here "no bird-pollinated flowers" is apparently a morphological rather than functional classification, and suggests an assumption that insect visitation is probably effective while bird visitation is not.

Background: Numbers and Distribution of Bird Pollinators and Bird-Pollinated Plants

Here we briefly review existing knowledge on regional patterns of bird-pollinator diversity and importance, and update these from the recent literature. This draws on a number of useful reviews (Stiles 1981; Proctor et al. 1996; Şekercioğlu 2006; Fleming and Muchhala 2008; Whelan et al. 2008) which give more detail than we can cover here.

We first consider how many bird species pollinate flowers, and this immediately raises problems of what to include. Birds vary widely from being reliable and frequent to occasional floral visitors, and this results in varying estimates of the total numbers of bird species involved. Pellmyr (2002) reported that more than 2,000 bird species (20% of all birds) in 50 families visited flowers, although many of these do so infrequently, and their importance as pollinators is unknown. Other authors have focused on birds that visit flowers more regularly, and give a total number of around 900 to 920 species (Şekercioğlu 2006; Whelan et al. 2008). The most important bird families are shown in table 4.1, where the total number of species is higher than 900 mainly because of the inclusion of 295 species in "other" families which are not traditionally thought of as flower specialists (Stiles 1981). We consider below to what extent these other birds may be important pollinators.

Because many of the important bird families are regionally localized (table 4.1), the flower-visiting avifauna varies between continents. It is widely accepted that, apart from Australia, most bird pollination takes place in the tropics, though there is less agreement about whether bird pollination is unimportant in temperate regions. Ford argued that, rather than Australia being odd in having relatively high bird pollination, it was Europe that was unusual among temperate regions in having practically none (Ford 1985a). In functional terms, Fleming and Muchhala (2008) identified three main regions: the Americas (especially the Neotropics) dominated by small, hovering, specialized birds; Africa, with large, nonhovering specialist birds; and Southeast Asia and Australasia, which feature large, nonhovering generalists.

We next consider how many plants are visited by birds, a question which also suffers from problems of demarcation (often based on flower morphology and frequency of observed bird visits). Information is variously presented at the level of plant species, genera, or families. Renner and Ricklefs (1995) estimated that around 500 of the 13,500 genera of vascular plants are pollinated by birds, but information is incomplete. At the level of plant genera and families, hummingbirds visit 311 genera in 95 families

TABLE 4.1 **Major groups of flower-visiting birds. Species numbers are approximate and for less specialized groups (Icteridae, Others) are only those known or suspected to be highly nectarivarous. The honeyeaters and white-eyes include a range of species, some of which are not closely associated with flowers. Specialization refers to importance of nectar in the diet and morphological adaptations (e.g,. brush tongues, hovering flight). After Stiles 1981, Proctor et al. 1996, and Fleming and Muchala 2008.**

Family	Region	No of spp.	Specialization
Hummingbirds (Trochilidae)	Americas especially Neotropics	330	Very high
Sunbirds (Nectariniidae)	Africa, Asia	130	High
Hawaiian honeycreepers (Drepanidinae)	Hawaii	23	High
Honeycreepers (Thraupidae)	Tropical America	15	High
Sugarbirds (Promeropidae)	South Africa	2	High
Honeyeaters (Meliphagidae)	Australasia	176	High-moderate
White-eyes (Zosteropidae)	Asia, Australia	85	Moderate
Lories (subfam Loriinae)	Australia, SE Asia	55	Moderate
Flowerpeckers (Dicaeini)	Asia, Australasia	50	Moderate
Sunbird-asities (Philepittidae)	Madagascar	2	Moderate
American orioles (Icteridae)	Americas	23	Low
Others	Various	295	Low
TOTAL		1,186	

(56% of the families in the Neotropics), sunbirds visit 279 genera in 94 families (59% of families present), and honeyeaters visit 250 species in 40 genera from 25 families (17% of the local nonaquatic nongrass genera; Fleming and Muchhala 2008).

As a percentage of local floras, bird pollinated (or visited) plants are typically low; perhaps around 5% in most regions and up to 10% on islands (Kato and Kawakita 2004; Whelan et al. 2008). Şekercioğlu (2006) gives a figure of around 2.1 to 3.4% of the flora being bird-visited at sites in Costa Rica. In temperate New Zealand, birds were recorded visiting the flowers of 85 species, representing 5% of the total seed-plant flora (Kelly et al. 2010). However, the figure is higher in Australia, where 15% of plant species are bird-visited (Armstrong 1979; Keighery 1982; Johnson 2013). Bird pollination has been considered almost absent in temperate

Asia (Corlett 2004; Cronk and Ojeda 2008), but recent studies are revising upward the estimates of bird visitation. Nectar foraging is reported for at least forty bird species in China (Mackinnon et al. 2000), and the widespread Zosteropidae have been implicated in pollination throughout Asia (Yumoto 1987; Kondo et al. 1991; Ali and Ripley 1999) notably in winter-flowering trees endemic to China (Gu et al. 2010; Fang et al. 2012). If, as Fang et al. (2012) suggest, winter flowering is a response to resident passerines acting as pollinators at low temperatures, then this may be an important trait to consider when assessing the relationship between birds and flowers in temperate and montane areas (see also Castro and Robertson 1997).

Interestingly, the level of bird pollination varies among plant life forms and regions, including among the three functional regions identified by Fleming and Muchhala (2008). It is generally considered that most bird-pollinated plants are shrubs and epiphytes (e.g. Whelan et al. 2008), but this applies most strongly in the Neotropics (Fleming and Muchhala 2008), while visits to trees are more important in the Paleotropics and Australasia (Whelan et al. 2008). In lowland New Guinea, 22% of canopy trees had bird-visited flowers (Brown and Hopkins 1995). In New Zealand, tallies of all bird-visited plant species found that 68% (58 of 85) were trees, with the rest divided among vines (13%), shrubs (8%), herbs (6%) and mistletoes (5%; Kelly et al. 2010).

Assessing the Importance of Birds as Pollinators

Pollinator assessment has to date been based on three approaches: (1) recognition of floral syndromes in flowers or adaptations to nectarivory in visitors, (2) records of visitation to flowering plants and interpretation of foraging behavior, and (3) comparison of fruit and seed production under different visitor regimes. Underlying assumptions in these approaches may have led to a systematic underreporting in the literature of the functional importance of birds as pollinators. Here we outline the limitations of these traditional approaches and identify the kinds of information required to more accurately assess the contribution of birds to pollination.

Morphology

The simplest appraisal of the level of bird interaction is an assessment of whether flower morphology conforms to a bird-pollinated "ornithophilous" syndrome, and whether bird visitors display morphological adaptations for nectarivory. Accepted ornithophilous features in flowers include large size, large quantities of dilute nectar, red coloration, and lack of a landing platform or odor (Faegri and van der Pijl 1979). In birds, features recognized as adaptive to nectarivory include a hovering habit, long bill, and brush tongue suitable for the uptake of nectar (Stiles 1978). The presence of these features is commonly used as a first approximation of the role of bird pollination in a flora.

Although proponents of this "pollinators for blossom types" theory did not intend that it should be prescriptive, the concept has pervaded pollination ecology. The most recognized cases of bird pollination are those where flowers have features that are obviously attractive to birds, and birds are predictable and conspicuous visitors (e.g. *Brunsvigia litoralis* and sunbirds [Geerts and Pauw 2012], and bromeliads and hummingbirds [Varassin and Sazima 2012]). These bird-flower mutualisms are more evident in the tropics than in the temperate zone (Proctor et al. 1996). The appeal of morphological matching has led to an emphasis in the literature on relationships displaying clear trait convergence between plant and pollinator. While specialization may characterize certain well-studied systems, it is not typical of most bird-plant interactions, particularly in temperate areas. Non-ornithophilous plants made up 54% and 84% of the species visited by hummingbirds in two different parts of Brazil (Las-Casas, Azevedo Júnior, and Dias Filho 2012; Araújo, Sazima, and Oliveira 2013), and 67% of the plants visited by birds in New Zealand (Kelly et al. 2010). Other examples are numerous (Ford et al. 1979; Keighery 1982; Brown and Hopkins 1995; Rodríguez-Rodríguez and Valido 2008; Ortega-Olivencia et al. 2012; Turner et al. 2012). Conversely, flowers are frequently visited and pollinated by apparently nonspecialized birds (Kay 1985; Willis 2002; Kunitake et al. 2004; Valido et al. 2004; Arena et al. 2013). A community-based study by Ollerton et al. (2006) confirms that classical syndromes have poor predictive power, particularly for bird visitation to flowers.

Relying on morphological fit can therefore seriously underestimate the importance of bird visitation to flowering plants. A recent review of the role of bird pollination in New Zealand provides an example; while less than 1% of the flora is ornithophilous, birds visit a much higher proportion (5%; Kelly et al. 2010). The same may be true in other regions, so that current data on the number of bird-visited flowers are probably minimum estimates.

Predictions based on morphology can be improved by adjusting the characters used. Brown and Hopkins (1995) investigated the Papuan bird pollinator system and confirmed a lack of correlation between bill and flower morphology despite evidence for strong bird-plant associations. Although flowers in these systems differ from the classical bird syndrome, they possess other features that suit them well to the birds that visit them, such as bracts to accommodate perching birds, and aggregated inflorescences. Similarly, in New Zealand a revision of visitors and blossom classes according to access explained observed visitation to flowers better than morphological classifications (Newstrom and Robertson 2005). That pollinating birds may visit flowers for rewards other than nectar (Willis 2002; Agostini, Sazima, and Sazima 2006) also highlights the need to reconsider the importance of morphology in flower-visitor interactions. A wider context is required for evaluating particular features of pollination systems if the syndrome concept is to be more broadly applied (Brown and Hopkins 1995; Wenny et al. 2011).

Flower Visitation

The second approach to assessing the level of bird pollination is to record the frequency and foraging behavior of flower visitors in various taxa. Observational information identifies which plants are visited by birds. This is sometimes many more species than morphology would indicate.

For example, in Europe bird pollination was considered practically unknown (Cronk and Ojeda 2008). There are numerous records of unspecialized European passerines visiting exotic bird-syndrome flowers (Ford 1985b) as well as unspecialized native flowers (Ortega-Olivencia et al. 2012), but the lack of obvious adaptations has meant that bird pollination was discounted. However, these birds are now known to regularly pollinate flowering plants on migration routes (Cecere et al. 2011) and to be effective pollinators for native flowering plants in two genera previously considered to be insect-pollinated (Ortega-Olivencia et al. 2005, 2012). Ortega-Olivencia et al. (2012) noted the resemblance of these mixed bird-insect pollinator systems to some Australasian pollination systems, thus raising the possibility that pollination by generalist birds of unspecialized flowers may also be more widespread in temperate Europe.

Closer scrutiny of pollination systems elsewhere also reveals new records of unexpected bird pollination. In South Africa, birds rather than insects have been confirmed as pollinators of the entomophilous-flowered heath *Erica halicacaba* (Turner et al. 2012), and sunbirds as pollinators of the iris *Babiana avicularis* despite its small flower size (de Waal et al.

2012), while opportunistic birds are apparently responsible for most visits to high-altitude aloes which had previously been assumed to be pollinated by specialist nectarivores (Arena et al. 2013).

These observations illustrate the degree to which plant-pollinator interactions vary from those predicted by morphology, but this interpretation assumes that all potential floral visitors are still available. Widespread anthropogenic disturbance has altered the pollinator landscape, so care should be taken when inferring from studies based on flower visitation rates. History has shown that birds are among the first and worst casualties of habitat change (Steadman 1995; Cox and Elmqvist 2000; Loehle and Eschenbach 2012; Szabo et al. 2012), so that any current lack of visitation may be partly an artifact of population loss or decline. Nectarivores are the bird guild apparently least at risk (Şekercioğlu et al. 2004), but there are examples from many geographic areas of changes in bird density or distribution which affect pollination (Paton 2000; Lindberg and Olesen 2001; Elliott et al. 2012; Geerts et al. 2012; Pauw and Louw 2012; Şekercioğlu 2012). Few observational studies report either the historical community that coexisted with a flowering plant or the current background abundance of floral visitors as a context for interpreting observed visitor preferences. Visitor observations, then, only reflect what occurs with what is left of the pollinating fauna.

Some progress in reconstructing the pollinator community of a flora, when components are already missing, can be made using forensic and paleobiological methods to detect past interactions. Techniques include the use of "arks," such as island nature reserves, which persist as vestiges of a former ecosystem and retain higher bird densities (Anderson et al. 2006, 2011; Mortensen et al. 2008). In New Zealand such studies have shown that, in addition to the expected honeyeaters, parrots and wattled crows are also reliable flower visitors (figs 4.1 and 4.2; Ladley and Kelly 1995; Thorogood et al. 2007).

Paleoecology on museum specimens of Hawaiian honeycreepers revealed the pollen of extinct lobelioid flowers on their feathers, and allowed identification of a prior pollination link (Cox 1983). Similarly, the study of herbarium flower specimens exposed historical changes in bird-pollination mechanisms in New Zealand mistletoes (Ladley and Kelly 1995). The study of coprolites of rare or extinct birds using pollen, genetics, and carbon dating has also allowed long-defunct bird-plant interactions to be identified in the moa species (ratites) and the kakapo (*Strigops habroptilus*) a flightless parrot of New Zealand (Wood et al. 2012). This combination of paleobiology and neoecology to revisualize the missing components of mutualisms is an increasingly accessible tool that enables a fuller understanding of current flower-pollinator assemblages.

The low profile of bird visitors is reinforced by their low contribution to agricultural crop pollination relative to insects. As cultivated landscapes replace native forest, commercial pollinators such as the honeybee (*Apis mellifera*) expand into remaining forest and utilize floral resources. Obvious invertebrate flower visitation can exacerbate underreporting of bird visitation, especially where the original birds are reduced or missing. If insects were adequately replacing missing birds, this would not matter, but we show below that this is often not the case.

A final point is that visitation importance measured by the proportion of the total flora may mask a greater importance of bird visitation to certain sectors of the plant community. As noted above, birds visit only 5% of the total New Zealand flora but almost 30% of tree species (Kelly et al. 2010). When weighted by basal area, bird-visited flowers represented 37% of mean forest basal area, greater than the percentage of basal area with bird-dispersed fruit (31%: Kelly et al. 2010). For a temperate country these numbers are unexpectedly high, revealing the importance of bird pollination when adequate visitation data are available.

Fruit Set and Seed Set

The third and most useful approach to quantifying bird interactions is to record the production of fruit or seeds under various pollination scenarios. This experimental approach has the key benefit of being able to discriminate between flower visitors and true (effective) pollinators. In many instances, birds may be only one of several visitors to a flower, but visitors will differ in their ability to transfer pollen adequately.

The most important experiments measure pollen limitation by comparing fruit set on unmanipulated flowers and hand-pollinated flowers. Additional information can be obtained on the effectiveness of each visitor by protecting flowers before and after exposure to a visit and monitoring fruit- or seed-set (Olsen 1996; Padyšáková et al. 2013). Another approach uses treatments to measure the effect of particular groups of pollinators, like wire mesh cages around flowers to exclude birds but not insects (e.g.,Whelan and Burbidge 1980; Vaughton 1992; Robertson et al. 2005; Botes et al. 2009; Schmidt-Adam et al. 2009; Symes et al. 2009; Schmid et al. 2015). Similarly, fine mesh bags exclude all flower visitors and measure the plant's ability to self-pollinate. However, the effects of absent pollinators (usually vertebrates) cannot be evaluated with these methods. Absence of a pollinator will increase the pollen limitation index of the plant, but will not be attributable to a particular taxon unless comparative studies can be made that include sites where the original pollinator communities persist.

Finally, it is important to consider the quality of seeds as well as their quantity. This requires measurement of outbreeding rates and inbreeding depression, and can be done through genetic data (Scofield and Schultz 2006) or by field measurements of fitness (e.g., Robertson et al. 2011), as discussed below.

Evaluating the Ecological Importance of Bird Pollination

In this section, we show to what extent new information is changing our evaluation of the contribution of birds to the ecosystem service of pollination. We follow Bond (1994) by asking (1) whether declining bird pollinators are being replaced by other vertebrates or insects, (2) whether the decline or change in bird pollinators reduces seed production, (3) whether there are genetic (seed quality) impacts of bird declines, and (4) whether there are demographic consequences for the plant population.

The loss of some flower-visiting birds from a pollinator network because of either functional or actual extinction will have differing effects on plants. Exclusive relationships between birds and the flowers they visit are the most vulnerable and have received the most attention, but they comprise the minority of cases. Generalized relationships involving birds as well as other visitors to flowers are more resilient and more widespread, although relatively less well studied. However, the extent of ecosystem modification globally means that many interactions which were once resilient are now reduced to one or few partners (Fritts and Rodda 1998; Cox and Elmqvist 2000; Mortensen et al. 2008), thus giving the misleading appearance of specialization.

CAN THE ORIGINAL BIRD POLLINATOR BE REPLACED BY OTHER VERTEBRATES?

The longevity of vertebrate flower visitors such as birds, relative to other visitors, means that they are unlikely to rely on a single plant species for resources, although those plants may depend on birds for pollination. If birds were the exclusive visitors to a flowering plant because of either coevolution or extinction of other visitors, their loss would be expected to result in lack of visitation. Few such examples are known. Circumstantial evidence suggests that the decline and extinction of the bird-adapted explosive-flowered mistletoe (*Trilepidia adamsii*) endemic to New Zealand was partly due to the rapid reduction in bird densities throughout its range (Ladley and Kelly 1995), although habitat clearance was probably the largest contributor (Norton 1991), and the possibility of other flower visitors cannot be discounted. In most situations, visitation is more generalized and alternative visitors persist, or replacement visitors occur.

In a few cases, the loss of coevolved flower-visiting birds has led to their replacement by other bird visitors. In Hawaii, the introduced Japanese white-eye (*Zosterops japonicus*) has replaced extinct endemic nectar-feeding birds as a visitor to endemic *Freycinetia* and Myrtaceae flowers (Cox 1983), although there is evidence that the successful invasion of *Z. japonicus* has also had negative effects on the remaining native bird community (Freed and Cann 2009). On the New Zealand mainland, where endemic flower-visiting birds have been drastically reduced following the introduction of mammalian predators, the recently self-introduced silvereye (*Zosterops lateralis*) is a frequent replacement visitor to native flowering plants (Kelly et al. 2006).

A recent review suggests, however, that substitute pollinators tend to be less effective than the original species (Aslan et al. 2012). In Australia, Paton and Ford (1977) reported that silvereyes did not extend their tongues far beyond their short (10 mm) bills, and therefore probably cannot reach nectar at the base of long tubular corollas. This applies in New Zealand, where the silvereye's tongue is too short for it to pollinate the large flowers of *Fuchsia excorticata* (Robertson et al. 2008), *Sophora* species (fig. 4.1c), *Alseuosmia macrophylla* (Pattemore and Anderson 2013) and *Rhabdothamnus solandri* (Anderson et al. 2011) which they rob. In other situations, novel bird-pollination associations may largely contribute to weed spread (Hoffmann et al. 2011; Linnebjerg et al. 2010).

Other animals may replace birds as floral visitors, and thereby sustain pollination services. However, generalized pollination strategies also allow introduced generalists to displace natives, with an overall loss of diversity (Waser et al. 1996). Introduced ship rats (*Rattus rattus*), which maintain a low level of pollinator function in the absence of native birds in New Zealand, partly caused the original loss of bird pollinators (Pattemore and Wilcove 2012).

CAN THE ORIGINAL BIRD POLLINATOR BE REPLACED BY INSECTS? Floral visitors vary in their effectiveness as pollinators, and flowering plants vary in their dependence on pollinators to set seed. For these reasons, measures of flower visitation need to be paired with measures of seed set, in order to assess the relative contribution of bird pollinators to plant reproduction. In some cases, visitation is a true indicator of pollination service. Alcorn et al. (1961) showed no difference in seed production and viability between *Saguaro* cactus flowers visited by birds, bats, and honeybees. However, variation in foraging behavior between visitors means that visitation is often not a good proxy for pollination. Exclusion experiments show that in many plants with mixed visitation, birds contribute disproportionately to seed set relative to insect visitors (Waser 1978, 1979; Bertin 1982; Collins and Spice 1986; Ramsey 1988; Celebrezze and Paton 2004; Anderson et al. 2006). In some of these, birds are the only effective pollinator, and insect visitors contribute little (Craig 1989b; Hargreaves et al. 2004; Kunitake et al. 2004; Gu et al. 2010; Weston et al. 2012; Liu et al. 2013; Schmid et al. 2015). In other cases, birds are responsible for significantly higher fruit or seed set, but insect visitation provides additional resilience (Carpenter 1976; Robertson et al. 2005; Cecere et al. 2011; Schmid et al. 2011; Etcheverry et al. 2012). In plants with winter flowering seasons, insects may be effective pollinators during the warmer period, but the importance of birds increases as temperatures decrease (Vaughton 1992; Fang et al. 2012). For montane aloes, opportunistic avian nectarivores increase seed production sevenfold in comparison to insects (Arena et al. 2013).

Perhaps most surprisingly, the contribution by bird visitors to reproductive success is significant even for apparently generalized flowers that lack ornithophilous features (Anderson 2003; Ortega-Olivencia et al. 2012). For example, caging to exclude birds, but not insects, reduced fruit set by half (Anderson 2003), both in *Metrosideros excelsa* (Myrtaceae), which has open brush-blossoms that attract a range of insects, and in *Pseudopanax arboreus* (Araliaceae), which has tiny florets (fig. 4.2a). More than any other single fact, these fruit set experiments on "entomophilous" flowers prove that bird pollinators are important and not readily replaced even for plant species well outside the morphological range previously thought to indicate dependence on birds.

Where insects have replaced "lost" flower-visiting birds, the most frequent replacement is likely to be the honeybee (*Apis mellifera*). Honeybees are globally distributed as an agricultural pollinator, and have successfully invaded most native pollination networks. Exotic honeybee visitation to native flowers may signal the loss of either birds (Junker et al. 2010; Pattemore and Anderson 2013) or native insects (probably bees) that once visited. For many temperate plants, honeybees are now the most frequent floral visitors and can remove more than 80% of the floral resources, even from plants that are pollinated largely by birds (Paton 1993). This may be why honeybees sometimes detract from seed set (Hargreaves et al. 2010) and can lower pollination success by interfering with more effective bird pollinators (Collins et al. 1984; England et al. 2001; Hansen et al. 2002; Celebrezze and Paton 2004; Botes et al. 2009).

Despite this, in the absence of native bird pollinators, honeybees may occasionally provide beneficial net effects (Paton 2000; Junker et al. 2010). In changing landscapes it is more important to retain ecosystem services than historical faithfulness to species composition (Hobbs et al. 2009), and natural communities may have the resilience to accommodate some pollinator change (Butz Huryn 1997).

HOW STRONG IS POLLEN LIMITATION IN BIRD-POLLINATED PLANTS? Without information on which alternative pollinators are present and whether they are effective, the overall sensitivity of plant reproduction to pollination failure can be measured by the degree of pollen limitation. When Kelly et al. (2010) did this, they found that measurements of pollen limitation for bird-pollinated plants in New Zealand—using the pollen limitation index (PLI; Larson and Barrett 2000) or log odds ratios (Knight et al. 2005), both calculated by comparing natural fruit set to hand-pollinated fruit set—were significantly higher than for a compilation of 482 cases of pollen limitation globally. The global compilation included plants with all types of pollinator, but was numerically dominated by insect-pollinated plants, which raises the question of whether bird-pollinated plants are more likely to be pollen-limited than plants in general. The alternative hypothesis is that New Zealand bird-pollinated plants might be unusually pollen-limited because of extensive human impacts on the avifauna. New Zealand has the unenviable distinction of having the highest proportion of extinction-prone (extinct, threatened, or near-threatened) bird species of any region (Şekercioğlu et al. 2004).

To test this, we searched the literature and compiled all data available on pollen limitation experiments on plants identified by the authors as bird-pollinated, using Knight et al. (2005) and Web of Science (using the terms bird AND [pollination OR PLI OR hand pollination]) in March 2013. Where there were data for a single species at several sites or years, we averaged these to include one number per plant species. We compared other parts of the world ($n = 41$ species across three regions) to the New Zealand data in Kelly et al. (2010; $n = 11$) using one-way analysis of variance on the log odds ratio data; analysis using PLIs gave similar results. There were sufficient

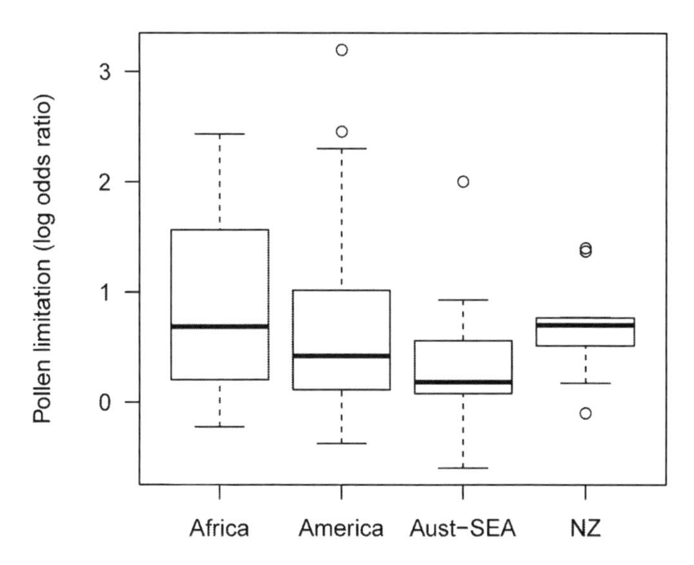

FIGURE 4.3. Degree of pollen limitation (log odds ratios) for bird-pollinated plant species in different geographic regions (Africa, n = 11; Americas, n = 20; Australia plus Asia, n = 10; New Zealand, n = 11). There was no significant difference among the regions ($F_{3,48} = 0.658$, $P = 0.58$).

species to compare the Americas (mostly hummingbird-pollinated), Africa, and Australia combined with Asia. There was no evidence that New Zealand plants were significantly more or less pollen-limited than bird-pollinated plants elsewhere (fig. 4.3). This leads us to reject the hypothesis that bird-pollinated plants are unusually pollen-limited in the New Zealand region, and it lends support to the idea that bird-pollinated plants globally, not just in New Zealand, currently tend to have higher levels of pollen limitation than insect-pollinated plants.

HOW STRONG IS THE DEMOGRAPHIC DEPENDENCE ON SEED FROM BIRD POLLINATION? There are three key issues in considering the demographic impact of changes to bird pollination on plants: (1) numerical effects through seed limitation, (2) genetic effects including inbreeding depression, and (3) the resulting probability of extinction.

(1) Seed limitation: Plant populations will only suffer demographic consequences of reduced seed set from disrupted bird visitation if germination is limited by available seed rather than available germination sites. Thus, it is necessary to consider alternative reproductive strategies available to plants, and to follow up evidence of pollen limitation (insufficient

pollen delivered to achieve optimal seed set) with tests for seed limitation (insufficient seed to fill all safe sites) in order to establish whether birds are essential to the process. Few studies do this, partly because the required experiments often must be run for long time periods (Ashman et al. 2004).

The handful of studies on bird-pollinated flowering plants that have combined pollen-augmentation and seed-sowing experiments show that seed is indeed limiting. Seedling counts increase with seed input, confirming that reduced densities of native bird pollinators are likely to reduce the density of adult plants in the next generation (Kelly et al. 2007; Price et al. 2008; Waser et al. 2010; Anderson et al. 2011). The long-term effect is less clear, since experimental studies show that density dependence can affect fecundity, thus complicating efforts to predict eventual population dynamics (Price et al. 2008). A similar outcome from parallel natural studies confirms the need for plant life history information before we can draw conclusions about population change (Waser et al. 2010).

A related question is whether reproduction by seed is essential at all for plant reproduction (Bond 1994). The ability of bird-visited flowering plants to use other means of reproduction is rarely considered, so it is assumed that seed is the only way by which the plants can reproduce. This assumption is true for some plants, for example the mistletoes; their requirement for seed dispersal to new hosts is obligate (Kelly, Ladley, and Robertson 2007). For others, such as understory shrubs, there may be some capacity for stems to 'sucker' away from the adult plant—the rare Californian endemic *Dirca occidentalis* is able to reproduce asexually from rhizomes (Graves and Schrader 2008), the South African iris *Babiana hirsuta* often produces multi-ramete clones (de Waal et al. 2012), and the New Zealand endemics *Rhabdothamnus solandri* and *Alseuosmia macrophylla* both extend over short distances using stolons (Anderson unpublished data)—although this creates concerns of low genetic diversity and severely truncated dispersal distances.

(2) Genetic consequences: The influence of bird visitation on reproductive efficiency and pollen flow varies with plant breeding system. Mobile bird visitors are particularly important for self-incompatible and dioecious flowering plants with generalized pollination systems (Vaughton 1992; Anderson 2003), and a lack of bird visitation is a likely reason for low seed set in these species. Although it has been argued that birds may promote geitonogamy because they can defend territories (Stiles 1981; Proctor et al.1996; Şekercioğlu 2006), this varies with bird behavior (such

as dominance hierarchies) and appears to be more applicable to hummingbirds in the Neotropics than in other parts of the world (Brown and Hopkins 1995). Several other features of birds mean that they tend to be more effective at outcrossing than insects; in particular, they have a large body surface and thus often carry large pollen loads, which increases pollen carryover (Price and Waser 1982; Robertson 1992); they can travel long distances between plants; and territorial aggression at high bird densities may drive subdominant birds away after short visits (enhancing cross-pollination) rather than preventing them from visiting at all (Brown and Hopkins 1995). Data for flowering plants with generalized pollination systems and mixed-mating strategies (i.e., the ability to produce a mix of outcrossed and selfed seed), show that birds are more likely to promote cross-pollination than insects (Richardson et al. 2000; Schmidt-Adam et al. 2000; Hingston and Potts 2005; Forrest et al. 2011), though one study found equal cross-pollination rates (Steenhuisen et al. 2012) and this may vary with human intervention (Richardson et al. 2000). This is also consistent with the observed lower levels of local genetic structuring in bird-pollinated species (see below).

The role of selfed seed deserves special consideration. The ability of some plants to produce selfed seed potentially provides reproductive assurance in the absence of bird pollinators. The actual mating strategy of a plant is influenced by pollinator and resource availability; plants may favor outcrossed pollen but, in the absence of pollinators, may invest in selfed seed according to resource availability (Craig 1989a; Becerra and Lloyd 1992). However, the value of selfed seed as reproductive assurance depends on its quality and fitness (Herlihy and Eckert 2002). Few studies have explored the relative fitness of selfed and outcrossed seed in plants with mixed-mating strategies. Theoretical considerations, combined with field genetic data on inbreeding rates, suggest that in long-lived plants (trees and shrubs) there is little evidence that selfed seedlings survive to reproduce, due to accumulated inbreeding depression that cannot be purged (Scofield and Schultz 2006). Therefore, inbreeding depression is likely to be significant, at least in woody bird-pollinated plants, with the selfed progeny representing "futile selfing" rather than providing reproductive assurance (Hardner and Potts 1997; Schmidt-Adam et al. 2000; Robertson et al. 2011; van Etten et al. 2015). In these cases, the higher outcrossing rates provided by birds assume even greater value.

Bird pollination also affects the spatial scale of gene flow between plants. The high mobility of bird pollinators relative to insects influences genetic structure in the plants they visit; bird-pollinated species or populations show significantly lower genetic differentiation and higher gene flow than related species or populations pollinated by insects (Hughes et al. 2007; Graves and Schrader 2008; Kramer et al. 2011). Although some territorial hummingbirds move only short distances, contributing to small neighborhood sizes in the plants they pollinate (Waser 1982; Parra et al. 1993), others maintain gene flow between populations separated by much greater distances (Graves and Schrader 2008), and even by mountain ranges (Kramer et al. 2011). Bird-mediated long-distance gene flow between plant populations has also been documented for honeyeaters (Byrne et al. 2007) and sunbirds (Hughes et al. 2007). Spatial memory and a long life span further enhance the ability of birds to provide high-quality pollen transfer between patchily distributed plants (Schuchmann 1999). As a result, bird pollinators enhance connectivity between vegetation fragments (Llorens et al. 2012) and their capacity for extensive long-distance pollination provides a mechanism for genetic rescue of isolated plant populations (Byrne et al. 2007).

Continuing anthropogenic disturbance of natural environments may erode this genetic rescue capacity. A study of bird movement between forest fragments in an agricultural landscape in Australia suggests that there is a limit to the ability of bird pollinators to salvage fragments of diminishing size; below a threshold number of plants, pollen immigration falls away and outcrossing stops (Byrne et al. 2007). Other studies of pollinator changes across an environmental gradient toward the geographical limit of a plant show that, while plants may be able to respond to climate change by shifting their ranges within a climatic envelope, their bird pollinators may not follow, and a shift to insect pollinators results in increased pollen limitation and eventually reproduction failure (Rovere et al. 2006; Chalcoff et al. 2012). Indeed, a modeling exercise on the bird assemblages of South African fynbos and grassland biomes under a climate change scenario predicts that by 2085, species richness will have reduced 30 to 40% on average in these habitats (Huntley and Barnard 2012). Three important bird pollinators—the Cape sugarbird (*Promerops cafer*), malachite sunbird (*Nectarinia famosa*), and orange-breasted sunbird (*Anthobaphes violacea*)—are all expected to show above-average contractions in their range. Worryingly, the malachite sunbird is the only remaining bird in the region with a bill long enough to probe the long tubes of a guild of seven plant species in the Cape Province that are otherwise robbed by shorter-billed sunbirds (Geerts and Pauw 2009), so a further contraction in the range of this bird species will endanger all seven members of the plant guild.

(3) Plant extinction: The ultimate demographic response to bird pollinator loss or functional extinction is plant extinction. Because of the long life span of many plants and the gradual nature of pollination declines, this may take a long time (Anderson et al. 2011). Unsurprisingly, the most compelling cases for this are in the Pacific Islands, where bird pollination is conspicuous and the ecological impact of recent human colonization has been extreme (Cox and Elmqvist 2000; Boyer 2008; Duncan et al. 2013). Studies from Hawaii report the extinction of one-third of the 52 endemic bird species present before European arrival, and the subsequent extinction of 31 species of Campanulaceae because of pollinator loss (Cox and Elmqvist 2000). These losses are self-reinforcing; pollinator numbers decline to a level that they no longer effectively serve plant populations, and plant populations become too small to support viable pollinator populations.

Contribution of Bird Pollination to Commercial Systems

Birds appear to play a relatively small role in the pollination of cultivated crops. Roubik (1995) and Nabhan and Buchmann (1997) classified the pollination requirements of 960 tropical crops where they had sufficient information on the pollen vectors. Of these, 52 (5.4%) were regularly visited by birds. However, in many of these cases, bees, bats or other visitors also visited the flowers, so the relative importance of birds is unclear. This is because there is usually only anecdotal information about flower visitors and no information on the effectiveness of each vector, which makes it difficult to evaluate the importance of birds. Ideally, the importance of birds in comparison to that of insects should be experimentally tested (as explained above).

However, there are examples of economically important plants in which there is reasonable evidence that birds are, if not the exclusive pollinator, the most effective flower visitors. As with many bird-plant alliances discussed in this chapter, these bird-pollination roles are often unexpected and involve birds or flower morphologies that fall outside the "ornithophily" paradigm.

The feijoa (*Acca sellowiana*) has an unusual reward system of succulent sweet petals (fig. 4.4a) which attract frugivorous birds that eat the petals and pollinate the self-incompatible nectarless flowers (Stewart and Craig 1989; Ducroquet and Hickel 1997). Stewart and Craig (1989) showed that in orchards in New Zealand, only large birds—common mynas (*Acridotheres tristis*) and Eurasian blackbirds (*Turdus merula*, fig. 4.4b)—routinely

FIGURE 4.4. Photos of feijoa (*Acca sellowiana*, Myrtaceae), showing (a) succulent petals (on the right flower) that are removed and eaten by birds (as seen on the left flower; photo by Dave Kelly), and (b) a blackbird (*Turdus merula*, Turdidae) eating the petals (photo by Alastair Robertson).

deposited pollen, since small silvereyes, honeybees, and other insects failed to contact the stigma regularly, and carried minimal pollen loads. Within the plant's native range in South America, wild plants were also visited by large frugivorous birds, and experimentally excluding birds with cages reduced fruit set considerably (Ducroquet and Hickel 1997). In contrast, in two orchards (in Japan and the United States) where no birds were seen, natural fruit set was practically zero. Honeybees are not only ineffective but probably detrimental, since they act as pollen thieves and reduce the pollen available for pollination by birds.

The winter-flowering self-incompatible loquat (*Eriobotrya japonica*), which is visited by birds and insects throughout its cultivated range, was studied in a plantation in its native China (Fang et al. 2012). As winter progressed, insect visitation, which had been high, slowed down, but two birds—light-vented bulbuls (*Pycnonotus sinensis*) and Japanese white-eyes—persisted. Excluding birds from visiting by enclosing flowers in cages significantly reduced seed set, and seed set on flowers that could only be accessed by insects was no different from seed set on flowers where all visitors were excluded. Thus, honeybees and other insects were ineffective pollinators, despite their high frequency early in the season.

The red silk cotton tree (*Bombax ceiba*) is an important tree in rural India used for oil and fiber (Bhattacharya and Mandal 2000; Raju et al. 2005). It requires outcrossing and attracts a wide range of birds and bats, which appear to be the main pollinators, with bees playing only a minor role. Fruit set is low, perhaps due to infrequent visits from the most effective of these pollinators.

The pollination requirements in seed orchards go beyond the need for high fruit and seed set. Seed quality is also an issue. Selfing can compromise the seed quality of self-compatible species due to inbreeding depression, so the best pollen vectors will regularly move compatible pollen between plants. Insects and honeybees in particular are considered undesirable pollinators for seed production in trees like *Eucalyptus* species, since they move largely within plants, resulting in geitonogamous self-pollination and poor quality seed (Paton and Ford 1977; Ford et al. 1979; Hopper and Moran 1981). Interestingly, this reverses the previously assumed pattern of higher geitonogamy by birds discussed above.

The partially self-compatible pulpwood species *Eucalyptus globulus* is frequently visited by honeybees as well as by birds, especially swift parrots (*Lathamus discolor*; Hingston et al. 2004). The parrot deposits more pollen in a single visit than honeybees do, and there is evidence that birds bring more outcross pollen, since the outcrossing rate was higher in the upper canopy of these trees where bird visitation was higher (Patterson et al. 2001; Hingston and Potts 2005). Moreover, genetic studies have shown that this plant species suffers severe inbreeding depression, and that the inbreeding coefficient decreases with plant age as selfed offspring are purged from the population (Hardner and Potts 1995; Hardner et al. 1998). Spotted gums (*Corymbia citriodora* subsp. *variegata*), which are visited by birds and bats as well as honeybees, show a similar requirement for outcross pollen and depend on vertebrate pollinators to provide this. The same pattern is replicated in silky oaks (*Grevillea robusta*), which are important farm trees in Africa (Kalinganire et al. 2001). Members of this genus require outcross pollen to avoid inbreeding depression; they receive only low-quality pollination service from honeybees and rely on birds and other vertebrates for optimal pollination (Vaughton 1996; Richardson et al. 2000; England et al. 2001; Whelan et al. 2009; Forrest et al. 2011).

Two points emerge from this analysis of commercial crop pollination, both of which are consistent with earlier sections on noncrop plants. First, it is likely that the extent of the role of birds in the pollination of crop plants has been underestimated, because the flowers often do not look ornithophilous, and birds other than the main nectarivores are often involved. For example, the role of birds in pollinating feijoa and loquat was overlooked in Roubik's (1995) survey. Second, birds have the potential to provide higher-quality pollination of mixed-mating tree crops than do insects such as honeybees. Though birds are often outnumbered by bees, rare visits from birds may be more important for seed production than their relative frequency suggests. A cost-benefit analysis for agricultural crops should include these less obvious benefits of enhancing bird habitats in agricultural landscapes (Triplett et al. 2012).

Conclusions

Recent work indicates that bird visitation to flowers is not confined to the tropics, shrubs, classic floral syndromes, or specialized nectarivores, as previously reported, but is more widespread and is often governed by a set of parameters different from those recognized to date. Bird pollination is less common than insect pollination, but it is perhaps two to five times more common than previously thought, and sometimes surprisingly hard for other animals (or alternative means of pollination) to effectively replace.

The scale of this revision may be surprising, but it is founded on three important factors that are not apparent unless they are specifically searched for. First, the absence of bird visits to non-ornithophilous flowers may be taken to support conventional floral syndromes, but could instead be due to reductions in the density or range of birds that previously visited such plants. Second, bird visits to non-ornithophilous flowers may appear to be incidental, but only manipulative experiments can test whether birds are actually important pollinators. If the visits are considered incidental, there is less motivation to conduct the manipulations. Third, the combination of better outcrossing provided by birds and strong inbreeding depression in some bird-pollinated trees means that evidence of seedlings does not equate to successful regeneration. Lower bird densities may have reduced the production of fit outcrossed seedlings, a hazardous situation that could be masked by abundant "futile selfing" resulting from insect or autonomous pollination.

We are aware that lessons learned from New Zealand might not initially seem relevant to the rest of the world. New Zealand is famously unusual for both its prehuman dominance by birds and its posthuman damage to the avifauna. But several lines of evidence indicate that the New Zealand situation should not be considered unique.

First, all the processes that contribute to the cryptic importance of bird pollination in New Zealand have been reported in other areas, as described above. There is no question about the existence of these processes in other regions—only uncertainty about their magnitude.

Second, New Zealand is famous for bird extinctions, but importantly, these extinctions spared the pollinating avifauna. Atkinson and Millener (1991) reviewed subfossil evidence and concluded that there had been no extinctions among the guild of nectarivores. The five endemic birds with brush tongues apparently specialized for nectarivory (the tui [*Prosthemadera novaeseeladiae*], New Zealand bellbird [*Anthornis melanura*], New Zealand kaka [*Nestor meridionalis*], North Island kokako [*Callaeas cinerea*], and stitchbird [*Notiomystis cincta*]) are still extant, although all have reduced ranges (Kelly et al. 2006). The colonization by silvereyes from Australia in 1856 added a sixth brush-tongued bird. Silvereyes, tui, and bellbirds are all currently widespread and common. Another six native and five introduced birds have been documented visiting flowers, thus giving a current total of 17 flower-visiting avian species. Although there have been human impacts on New Zealand's pollinating avifauna, the nectarivore guild is not in the parlous state currently seen in Pacific islands like Guam (Caves et al. 2013) and Hawaii (Smith et al. 1995). Human impact on New Zealand's pollinating avifauna is not obviously different from that on avifauna elsewhere.

Third, our analysis shows that levels of pollen limitation in bird-visited plants are similar in other regions to those seen in New Zealand, thus suggesting that bird-pollinated plants have a broadly similar level of pollination service impairment in all regions.

Fundamentally, this chapter is about calling attention to several poorly tested assumptions that need closer scrutiny. It has been generally assumed that insects will pollinate well if they go to flowers, whereas the efficacy of birds has only been accepted after seeing proof, especially on non-ornithophilous flowers. It was also assumed that birds are more likely to self-pollinate than insects, whereas the data now suggest that birds are less likely to self-pollinate. Finally, the idea that birds mainly cause self-pollination was proposed as a possible explanation for the putative rarity of bird pollination in trees (except in Australasia). Only when these assumptions are explicitly tested in different parts of the world will the full contribution of birds to ecosystem services through pollination be known.

References

Agostini, K., Sazima, M., and Sazima, I. 2006. Bird pollination of explosive flowers while foraging for nectar and caterpillars. *Biotropica* 38:674–78.

Alcorn, S. M., McGregor, S. E., and Olin, G. 1961. Pollination of saguaro cactus by doves, nectar-feeding bats, and honey bees. *Science* 133:1594–95.

Ali, S., and Ripley, S. D. 1999. *Handbook of the Birds of India and Pakistan*. 2nd ed. Oxford University Press, Delhi.

Anderson, S. H. 2003. The relative importance of birds and insects as pollinators of the New Zealand flora. *New Zealand Journal of Ecology* 27:83–94.

Anderson, S. H., Kelly, D., Ladley, J. J., Molloy, S., and Terry, J. 2011. Cascading effects of bird functional extinction reduce pollination and plant density. *Science* 331:1068–71.

Anderson, S. H., Kelly, D., Robertson, A. W., Ladley, J. J., and Innes, J. G. 2006. Birds as pollinators and dispersers: A case study from New Zealand. *Acta Zoologica Sinica* 52:112–15.

Araújo, F. P., Sazima, M., and Oliveira, P. E. 2013. The assembly of plants used as nectar sources by hummingbirds in a Cerrado area of Central Brazil. *Plant Systematics and Evolution* 299:1119–33.

Arena, G., Symes, C. T., and Witkowski, E. T. F. 2012. The birds and the seeds: Opportunistic avian nectarivores enhance reproduction in an endemic montane aloe. *Plant Ecology* 214:35–47.

Armstrong, J. A. 1979. Biotic pollination mechanisms in the Australian flora: A review. *New Zealand Journal of Botany* 17:467–508.

Ashman, T.-L., Knight, T. M., Steets, J. A., Amarasekare, P., Burd, M., Campbell, D. R., Dudash, M. R., Johnston, M. O., Mazer, S. J., Mitchell, R. J., Morgan, M. T., and Wilson, W. G. 2004. Pollen limitation of plant reproduction: Ecological and evolutionary causes and consequences. *Ecology* 85:2408–21.

Aslan, C. E., Zavaleta, E. S., Croll, D. O. N., and Tershy, B. 2012. Effects of native and non-native vertebrate mutualists on plants. *Conservation Biology* 26:778–89.

Atkinson, I. A. E., and Millener, P. R. 1991. An ornithological glimpse into New Zealand's pre-human past. In *Acta XX Congressus Internationalis Ornithologici*, ed. M. J. Williams, 129–92. New Zealand Ornithological Congress Trust Board, Wellington.

Becerra, J. X., and Lloyd, D. G. 1992. Competition-dependent abscission of self-pollinated flowers of *Phormium tenax* (Agavaceae): A second action of self-incompatibility at the whole flower level. *Evolution* 46:458–69.

Bertin, R. I. 1982. Floral biology, hummingbird pollination and fruit production of trumpet creeper (*Campsis radicans*, Bignoniaceae). *American Journal of Botany* 69:122–34.

Bhattacharya, A., and Mandal, S. 2000. Pollination biology in *Bombax ceiba* Linn. *Current Science* 79:1706–12.

Bond, W. J. 1994. Do mutualisms matter? Assessing the impact of pollinator and disperser disruption on plant extinction. *Philosophical Transactions of the Royal Society B: Biological Sciences* 344:83–90.

Botes, C., Johnson, S. D., and Cowling, R. M. 2009. The birds and the bees: Using selective exclusion to identify effective pollinators of African tree aloes. *International Journal of Plant Sciences* 170:151–56.

Boyer, A. G. 2008. Extinction patterns in the avifauna of the Hawaiian islands. *Diversity and Distributions* 14:509–17.

Brown, E. D., and Hopkins, M. J. G. 1995. A test of pollinator specificity and morphological convergence between nectarivorous birds and rainforest tree flowers in New Guinea. *Oecologia* 103:89–100.

Butz Huryn, V. M. 1997. Ecological impacts of introduced honey bees. *Quarterly Review of Biology* 72:275–97.

Byrne, M., Elliott, C. P., Yates, C., and Coates, D. J. 2007. Extensive pollen dispersal in a bird-pollinated shrub, *Calothamnus quadrifidus*, in a fragmented landscape. *Molecular Ecology* 16:1303–14.

Carpenter, F. L. 1976. Plant-pollinator interactions in Hawaii: Pollination energetics of *Metrosideros collina* (Myrtaceae). *Ecology* 57:1125–44.

Castro, I., and Robertson, A. W. 1997. Honeyeaters and the New Zealand forest flora: The utilisation and profitability of small flowers. *New Zealand Journal of Ecology* 21:169–79.

Caves, E. M., Jennings, S. B., HilleRisLambers, J., Tewksbury, J. J., and Rogers, H. S. 2013. Natural experiment demonstrates that bird loss leads to cessation of dispersal of native seeds from intact to degraded forests. *PLoS ONE* 8:e65618.

Cecere, J. G., Cornara, L., Mezzetta, S., Ferri, A., Spina, F., and Boitani, L. 2011. Pollen couriers across the Mediterranean: The case of migrating warblers. *Ardea* 99:33–42.

Cecere, J. G., Cornara, L., Spina, F., Imperio, S., and Boitani, L. 2011. Birds outnumber insects in visiting *Brassica* flowers on Ventotene Island (central Mediterranean). *Vie et Milieu: Life and Environment* 61:145–50.

Celebrezze, T., and Paton, D. C. 2004. Do introduced honeybees (*Apis mellifera*, Hymenoptera) provide full pollination service to bird-adapted Australian plants with small flowers? An experimental study of *Brachyloma ericoides* (Epacridaceae). *Austral Ecology* 29:129–36.

Chalcoff, V. R., Aizen, M. A., and Ezcurra, C. 2012. Erosion of a pollination mutualism along an environmental gradient in a south Andean treelet, *Embothrium coccineum* (Proteaceae). *Oikos* 121:471–80.

Collins, B. G., Newland, C., and Biffa, P. 1984. Nectar utilisation and pollination by Australian honeyeaters and insects visiting *Calothamnus quadrifidus* (Myrtaceae). *Australian Journal of Ecology* 9:353–65.

Collins, B. G., and Spice, J. 1986. Honeyeaters and the pollination biology of *Banksia prionotes* (Proteaceae). *Australian Journal of Botany* 34:175–85.

Corlett, R. T. 2004. Flower visitors and pollination in the Oriental (Indomalayan) Region. *Biological Reviews* 79:497–532.

———. 2007. Pollination or seed dispersal: Which should we worry about most? In *Seed Dispersal: Theory and Its Application in a Changing World*, ed. A. J. Dennis, E. W. Schupp, R. Green, and D. W. Westcott, 523–44. CABI Publishing, Wallingford, UK.

Cox, P. A. 1983. Extinction of the Hawaiian USA avifauna resulted in a change of pollinators for the Ieie *Freycinetia arborea*. *Oikos* 41:195–99.

Cox, P. A., and Elmqvist, T. 2000. Pollinator extinction in the Pacific Islands. *Conservation Biology* 14:1237–39.

Craig, J. L. 1989. A differential response to self pollination: Seed size in *Phormium*. *New Zealand Journal of Botany* 27:583–86.

———. 1989. Seed set in *Phormium*: Interactive effects of pollinator behavior, pollen carryover and pollen source. *Oecologia* 81:1–5.

Cronk, Q., and Ojeda, I. 2008. Bird-pollinated flowers in an evolutionary and molecular context. *Journal of Experimental Botany* 59:715–27.

De Waal, C., Anderson, B., and Barrett, S. C. H. 2012. The natural history of pollination and mating in bird-pollinated *Babiana* (Iridaceae). *Annals of Botany* 109:667–79.

Ducroquet, J. P. H. J., and Hickel, E. R. 1997. Birds as pollinators of Feijoa (*Acca sellowiana* Bera). *Acta Horticulturae* 452:37–40.

Duncan, R. P., Boyer, A. G., and Blackburn, T. M. 2013. Magnitude and variation of prehistoric bird extinctions in the Pacific. *Proceedings of the National Academy of Sciences* 110:6436–41.

Elliott, C. P., Lindenmayer, D. B., Cunningham, S. A., and Young, A. G. 2012. Landscape context affects honeyeater communities and their foraging behaviour in Australia: Implications for plant pollination. *Landscape Ecology* 27:393–404.

England, P. R., Beynon, F., Ayre, D. J., and Whelan, R. J. 2001. A molecular genetic assessment of mating-system variation in a naturally bird-pollinated shrub: Contributions from birds and introduced honeybees. *Conservation Biology* 15:1645–55.

Etcheverry, Á. V., Figueroa-Castro, D., Figueroa-Fleming, T., Alemán, M. M., Juárez, V. D., López-Spahr, D., Yáñez, C. N., and Gómez, C. A. 2012. Generalised pollination system of *Erythrina dominguezii* (Fabaceae: Papilionoideae) involving hummingbirds, passerines and bees. *Australian Journal of Botany* 60:484–94.

Faegri, K., and van der Pijl, L. 1979. *The Principles of Pollination Ecology*. 3rd ed. Pergamon Press, Oxford.

Fang, Q., Chen, Y.-Z., and Huang, S.-Q. 2012. Generalist passerine pollination of a winter-flowering fruit tree in central China. *Annals of Botany* 109:379–84.

Fleming, T. H., and Muchhala, N. 2008. Nectar-feeding bird and bat niches in two worlds: Pantropical comparisons of vertebrate pollination systems. *Journal of Biogeography* 35:764–80.

Ford, H. A. 1985. Nectar-feeding birds and bird pollination: Why are they so prevalent in Australia yet absent from Europe? *Proceedings of the Ecological Society of Australia* 14:153–58.

———. 1985. Nectarivory and pollination by birds in southern Australia and Europe. *Oikos* 44:127–31.

Forrest, C. N., Ottewell, K. M., Whelan, R. J., and Ayre, D. J. 2011. Tests for inbreeding and outbreeding depression and estimation of population differentiation in the bird-pollinated shrub *Grevillea mucronulata*. *Annals of Botany* 108: 185–95.

Freed, L. A., and Cann, R. L. 2009. Negative effects of an introduced bird species on growth and survival in a native bird community. *Current Biology* 19: 1736–40.

Fritts, T. H., and Rodda, G. H. 1998. The role of introduced species in the degradation of island ecosystems: A case history of Guam. *Annual Review of Ecology and Systematics* 29:113–40.

Geerts, S., Malherbe, S. D. T., and Pauw, A. 2012. Reduced flower visitation by nectar-feeding birds in response to fire in Cape fynbos vegetation, South Africa. *Journal of Ornithology* 153:297–301.

Geerts, S., and Pauw, A. 2009. Hyper-specialization for long-billed bird pollination in a guild of South African plants: The Malachite Sunbird pollination syndrome. *South African Journal of Botany* 75:699–706.

Godley, E. J. 1979. Flower biology in New Zealand. *New Zealand Journal of Botany* 17:441–66.

Graves, W. R., and Schrader, J. A. 2008. At the interface of phylogenetics and population genetics, the phylogeography of *Dirca occidentalis* (Thymelaeaceae). *American Journal of Botany* 95:1454–65.

Gu, L., Luo, Z., Zhang, D., and Renner, S. S. 2010. Passerine pollination of *Rhodoleia championii* (Hamamelidaceae) in subtropical China. *Biotropica* 42:336–41.

Hansen, D. M., Olesen, J. M., and Jones, C. G. 2002. Trees, birds and bees in Mauritius: Exploitative competition between introduced honey bees and endemic nectarivorous birds? *Journal of Biogeography* 29:721–34.

Hardner, C. M., and Potts, B. M. 1995. Inbreeding depression and changes in variation after selfing in *Eucalyptus globulus* ssp. *globulus*. *Silvae Genetica* 44:46–54.

———. 1997. Postdispersal selection following mixed mating in *Eucalyptus regnans*. *Evolution* 51:103–11.

Hardner, C. M., Potts, B. M., and Gore, P. L. 1998. The relationship between cross success and spatial proximity of *Eucalyptus globulus* ssp. *globulus* parents. *Evolution* 52:614–18.

Hargreaves, A. L., Harder, L. D., and Johnson, S. D. 2010. Native pollen thieves reduce the reproductive success of a hermaphroditic plant, *Aloe maculata*. *Ecology* 91:1693–1703.

Hargreaves, A. L., Johnson, S. D., and Nol, E. 2004. Do floral syndromes predict specialization in plant pollination systems? An experimental test in an "ornithophilous" African *Protea*. *Oecologia* 140:295–301.

Herlihy, C. R., and Eckert, C. G. 2002. Genetic cost of reproductive assurance in a self-fertilizing plant. *Nature* 416:320–23.

Hingston, A. B., and Potts, B. M. 2005. Pollinator activity can explain variation in outcrossing rates within individual trees. *Austral Ecology* 30:319–24.

Hingston, A. B., Potts, B. M., and McQuillan, P. B. 2004. The swift parrot, *Lathamus discolor* (Psittacidae), social bees (Apidae) and native insects as pollinators of *Eucalyptus globulus* ssp. *globulus* (Myrtaceae). *Australian Journal of Botany* 52:371–79.

Hobbs, R. J., Higgs, E., and Harris, J. A. 2009. Novel ecosystems: Implications for conservation and restoration. *Trends in Ecology and Evolution* 24:599–605.

Hoffmann, F., Daniel, F., Fortier, A., and Hoffmann-Tsay, S. S. 2011. Efficient avian pollination of *Strelitzia reginae* outside of South Africa. *South African Journal of Botany* 77:503–05.

Hopper, S. D., and Moran, G. F. 1981. Bird pollination and the mating system of *Eucalyptus stoatei*. *Australian Journal of Botany* 29:625–38.

Hughes, M., Möller, M., Edwards, T. J., Bellstedt, D. U., and De Villiers, M. 2007. The impact of pollination syndrome and habitat on gene flow: A comparative study of two *Streptocarpus* (Gesneriaceae) species. *American Journal of Botany* 94:1688–95.

Huntley, B., and Barnard, P. 2012. Potential impacts of climatic change on southern African birds of fynbos and grassland biodiversity hotspots. *Diversity and Distributions* 18:769–81.

Johnson, K. A. 2013. Are there pollination syndromes in the Australian epacrids (Ericaceae: Styphelioideae)? A novel statistical method to identify key floral traits per syndrome. *Annals of Botany* 112:141–49.

Junker, R. R., Bleil, R., Daehler, C. C., and Bluthgen, N. 2010. Intra-floral resource partitioning between endemic and invasive flower visitors: Consequences for pollinator effectiveness. *Ecological Entomology* 35:760–67.

Kalinganire, A., Harwood, C. E., Slee, M. U., and Simons, A. J. 2001. Pollination and fruit-set of *Grevillea robusta* in western Kenya. *Austral Ecology* 26:637–48.

Kato, M., and Kawakita, A. 2004. Plant-pollinator interactions in New Caledonia influenced by introduced honey bees. *American Journal of Botany* 91:1814–27.

Kay, Q. O. N. 1985. Nectar from willow catkins as a food source for Blue Tits. *Bird Study* 32:40–44.

Keighery, G. J. 1982. Bird pollinated plants in Western Australia. In *Pollination and Evolution*, ed. J. A. Armstrong, J. M. Powell, and A. J. Richards, 77–89. Royal Botanic Gardens, Sydney.

Kelly, D., Ladley, J. J., and Robertson, A. W. 2007. Is the pollen-limited mistletoe *Peraxilla tetrapetala* (Loranthaceae) also seed limited? *Austral Ecology* 32: 850–857.

Kelly, D., Ladley, J. J., Robertson, A. W., Anderson, S. H., Wotton, D. M., and Wiser, S. K. 2010. Mutualisms with the wreckage of an avifauna: The status of bird pollination and fruit-dispersal in New Zealand. *New Zealand Journal of Ecology* 34:66–85.

Kelly, D., Robertson, A. W., Ladley, J. J., Anderson, S. H., and McKenzie, R. J. 2006. The relative (un)importance of introduced animals as pollinators and dispersers of native plants. In *Biological Invasions in New Zealand*, ed. R. B. Allen and W. G. Lee, 227–45. Springer-Verlag, Heidelberg.

Knight, T. M., Steets, J. A., Vamosi, J. C., Mazer, S. J., Burd, M., Campbell, D. R., Dudash, M. R., Johnston, M. O., Mitchell, R. J., and Ashman, T.-L. 2005. Pollen limitation of plant reproduction: Pattern and process. *Annual Review of Ecology Evolution and Systematics* 36:467–97.

Kondo, K., Nakamula, T., Piyakarnchana, T., and Mechvichai, W. 1991. Pollination in *Bruguiera gymnorhiza* (Rhizophoraceae) in Miyara River, Ishigaki Island, Japan, and Phangnga, Thailand. *Plant Species Biology* 6:105–10.

Kramer, A. T., Fant, J. B., and Ashley, M. V. 2011. Influences of landscape and pollinators on population genetic structure: Examples from three *Penstemon* (Plantaginaceae) species in the Great Basin. *American Journal of Botany* 98:109–21.

Kunitake, Y. K., Hasegawa, M., Miyashita, T., and Higuchi, H. 2004. Role of a seasonally specialist bird *Zosterops japonica* on pollen transfer and reproductive success of *Camellia japonica* in a temperate area. *Plant Species Biology* 19:197–201.

Ladley, J. J., and Kelly, D. 1995. Explosive New Zealand mistletoe. *Nature* 378:766.

Larson, B. M. H., and Barrett, S. C. H. 2000. A comparative analysis of pollen limitation in flowering plants. *Biological Journal of the Linnean Society* 69: 503–20.

Las-Casas, F. M. G., Azevedo Júnior, S. M., and Dias Filho, M. M. 2012. The community of hummingbirds (Aves: Trochilidae) and the assemblage of flowers in a Caatinga vegetation. *Brazilian Journal of Biology* 72:51–58.

Lindberg, A. B., and Olesen, J. M. 2001. The fragility of extreme specialization: *Passiflora mixta* and its pollinating hummingbird *Ensifera ensifera*. *Journal of Tropical Ecology* 17:323–29.

Linnebjerg, J. F., Hansen, D. M., Bunbury, N., and Olesen, J. M. 2010. Diet composition of the invasive red-whiskered bulbul *Pycnonotus jocosus* in Mauritius. *Journal of Tropical Ecology* 26:347–50.

Liu, Z.-J., Chen, L.-J., Liu, K.-W., Li, L.-Q., Rao, W.-H., Zhang, Y.-T., Tang, G.-D., and Huang, L.-Q. 2013. Adding perches for cross-pollination ensures the reproduction of a self-incompatible orchid. *PLoS ONE* 8:e53695.

Llorens, T. M., Byrne, M., Yates, C. J., Nistelberger, H. M., and Coates, D. J. 2012. Evaluating the influence of different aspects of habitat fragmentation on mating patterns and pollen dispersal in the bird-pollinated *Banksia sphaerocarpa* var. *caesia*. *Molecular Ecology* 21:314–28.

Loehle, C., and Eschenbach, W. 2012. Historical bird and terrestrial mammal extinction rates and causes. *Diversity and Distributions* 18:84–91.

Mackinnon, J., Phillipps, K., and He, F.-q. 2000. *A Field Guide to the Birds of China*. Oxford University Press, Oxford, UK.

Mortensen, H. S., Dupont, Y. L., and Olesen, J. M. 2008. A snake in paradise: Disturbance of plant reproduction following extirpation of bird flower-visitors on Guam. *Biological Conservation* 141:2146–21.

Nabhan, G. P., and Buchmann, S. L. 1997. Services provided by pollinators. In

Nature's Services: Societal dependence on natural ecosystems, ed. G. C. Daily, 133–50. Island Press, Washington.

Newstrom, L., and Robertson, A. W. 2005. Progress in understanding pollination systems in New Zealand. *New Zealand Journal of Botany* 43:1–59.

Norton, D. A. 1991. *Trilepidea adamsii*: An obituary for a species. *Conservation Biology* 5:52–57.

Ollerton, J., Johnson, S. D., and Hingston, A. B. 2006. Geographical variation in diversity and specificity of pollination systems. In *Plant pollinator interactions: From specialization to generalization*, ed. N. M. Waser and J. Ollerton, 283–308. University of Chicago Press, Chicago.

Olsen, K. M. 1996. Pollination effectiveness and pollinator importance in a population of *Heterotheca subaxillaris* (Asteraceae). *Oecologia* 109:114–21.

Ortega-Olivencia, A., Rodríguez-Riaño, T., Pérez-Bote, J. L., López, J., Mayo, C., Valtueña, F. J., and Navarro-Pérez, M. 2012. Insects, birds and lizards as pollinators of the largest-flowered *Scrophularia* of Europe and Macaronesia. *Annals of Botany* 109:153–67.

Ortega-Olivencia, A., Rodríguez-Riaño, T., Valtueña, F. J., López, J., and Devesa, J. A. 2005. First confirmation of a native bird-pollinated plant in Europe. *Oikos* 110:578–90.

Padyšáková, E., Bartoš, M., Tropek, R., and Jane ek, Š. 2013. Generalization versus specialization in pollination systems: Visitors, thieves, and pollinators of *Hypoestes aristata* (Acanthaceae). *PLoS ONE* 8:e59299.

Parra, V., Vargas, C. F., and Eguiarte, L. E. 1993. Reproductive biology, pollen and seed dispersal, and neighborhood size in the hummingbird-pollinated *Echeveria gibbiflora* (Crassulaceae). *American Journal of Botany* 80:153–59.

Paton, D. C. 1993. Honeybees in the Australian environment: Does *Apis mellifera* disrupt or benefit the native biota? *Bioscience* 43:95–103.

———. 2000. Disruption of bird-plant pollination systems in southern Australia. *Conservation Biology* 14:1232–34.

Paton, D. C., and Ford, H. A. 1977. Pollination by birds of native plants in South Australia. *Emu* 77:73–85.

Pattemore, D. E., and Anderson, S. H. 2013. Severe pollen limitation in populations of the New Zealand shrub *Alseuosmia macrophylla* (Alseuosmiaceae) can be attributed to the loss of pollinating bird species. *Austral Ecology* 38:95–102.

Pattemore, D. E., and Wilcove, D. S. 2012. Invasive rats and recent colonist birds partially compensate for the loss of endemic New Zealand pollinators. *Proceedings of the Royal Society B: Biological Sciences* 279:1597–1605.

Patterson, B., Vaillancourt, R. E., and Potts, B. M. 2001. Eucalypt seed collectors: Beware of sampling seedlots from low in the canopy! *Australian Forestry* 64:139–42.

Pauw, A., and Louw, K. 2012. Urbanization drives a reduction in functional diversity in a guild of nectar-feeding birds. *Ecology and Society* 17:27.

Pellmyr, O. 2002. Polllination by animals. In *Plant-animal interactions: An evolu-*

tionary approach, ed. C. M. Herrera and O. Pellmyr, 157–84. Blackwell Science: Malden, MA.

Price, M. V., Campbell, D. R., Waser, N. M., and Brody, A. K. 2008. Bridging the generation gap in plants: Pollination, parental fecundity, and offspring demography. *Ecology* 89:1596–1604.

Price, M. V., and Waser, N. M. 1982. Experimental studies of pollen carryover: Hummingbirds and *Ipomopsis aggregata*. *Oecologia* 54:353–58.

Proctor, M., Yeo, P., and Lack, A. 1996. The natural history of pollination. Timber Press, Portland, OR.

Raju, A. J. S., Rao, S. P., and Rangaiah, K. 2005. Pollination by bats and birds in the obligate outcrosser *Bombax ceiba* L. (Bombacaceae), a tropical dry season flowering tree species in the Eastern Ghats forests of India. *Ornithological Science* 4:81–87.

Ramsey, M. W. 1988. Differences in pollinator effectiveness of birds and insects visiting *Banksia menziesii* (Proteaceae). *Oecologia* 76:119–24.

Renner, S. S., and Ricklefs, R. E. 1995. Dioecy and its correlates in the flowering plants. *American Journal of Botany* 82:596–606.

Richardson, M. B. G., Ayre, D. J., and Whelan, R. J. 2000. Pollinator behaviour, mate choice and the realised mating systems of *Grevillea mucronulata* and *Grevillea sphacelata*. *Australian Journal of Botany* 48:357–66.

Robertson, A. W. 1992. The relationship between floral display size, pollen carryover and geitonogamy in *Myosotis colensoi* (Kirk) Macbride (Boraginaceae). *Biological Journal of the Linnean Society* 46:333–349.

Robertson, A. W., Kelly, D., and Ladley, J. J. 2011. Futile selfing in the tree *Fuchsia excorticata* (Onagraceae) and *Sophora microphylla* (Fabaceae): Inbreeding depression over 11 years. *International Journal of Plant Sciences* 172:191–98.

Robertson, A. W., Ladley, J. J., and Kelly, D. 2005. Effectiveness of short-tongued bees as pollinators of apparently ornithophilous New Zealand mistletoes. *Austral Ecology* 30:298–309.

Robertson, A. W., Ladley, J. J., Kelly, D., McNutt, K. L., Peterson, P. G., Merrett, M. F., and Karl, B. J. 2008. Assessing pollination and fruit dispersal in *Fuchsia excorticata* (Onagraceae). *New Zealand Journal of Botany* 46:299–314.

Rodríguez-Rodríguez, M. C., and Valido, A. 2008. Opportunistic nectar-feeding birds are effective pollinators of bird-flowers from Canary Islands: Experimental evidence from *Isoplexis canariensis* (Scrophulariaceae). *American Journal of Botany* 95:1408–15.

Roubik, D. W. 1995. *Pollination of Cultivated Plants in the Tropics, FAO Agricultural Services Bulletin*. Food and Agriculture Organization of the United Nations, Rome.

Rovere, A. E., Smith-Ramírez, C., Armesto, J. J., and Premoli, A. C. 2006. Breeding system of *Embothrium coccineum* (Proteaceae) in two populations on different slopes of the Andes. *Revista Chilena de Historia Natural* 79:225–32.

Schmid, B., Nottebrock, H., Esler, K. J., Pagel, J., Pauw, A., Bohning-Gaese, K., Schurr, F. M., and Schleuning, M. 2015. Reward quality predicts effects of bird-pollinators on the reproduction of African *Protea* shrubs. *Perspectives in Plant Ecology Evolution and Systematics* 17:209–17.

Schmid, S., Schmid, V. S., Zillikens, A., Harter-Marques, B., and Steiner, J. 2011. Bimodal pollination system of the bromeliad *Aechmea nudicaulis* involving hummingbirds and bees. *Plant Biology* 13:41–50.

Schmidt-Adam, G., Murray, B. G., and Young, A. G. 2009. The relative importance of birds and bees in the pollination of *Metrosideros excelsa* (Myrtaceae). *Austral Ecology* 34:490–98.

Schmidt-Adam, G., Young, A. G., and Murray, B. G. 2000. Low outcrossing rates and shift in pollinators in New Zealand pohutukawa (*Metrosideros excelsa*; Myrtaceae). *American Journal of Botany* 87:1265–71.

Schuchmann, K.-L. 1999. Family Trochilidae (Hummingbirds). In *Handbook of the Birds of the World: Barn-Owls to Hummingbirds*, vol. 5, ed. J. Del Hoyo, A. Elliott, and J. Sargatal, 468–535. Lynx Edicions, Barcelona.

Scofield, D. G., and Schultz, S. T. 2006. Mitosis, stature and evolution of plant mating systems: low-φ and high-φ plants. *Proceedings of the Royal Society B: Biological Sciences* 273:275–82.

Şekercioğlu, Ç. H. 2006. Ecological significance of bird populations. In *Old World Flycatchers to Old World Warblers*, vol. 11, ed. J. del Hoyo, A. Elliot, and D. Christie, 15–51. Lynx Editions, Barcelona.

———. 2012. Bird functional diversity and ecosystem services in tropical forests, agroforests and agricultural areas. *Journal of Ornithology* 153:S153–S161.

Şekercioğlu, Ç. H., Daily, G. C., and Ehrlich, P. R. 2004. Ecosystem consequences of bird declines. *Proceedings of the National Academy of Sciences* 101:18042–47.

Smith, T. B., Freed, L. A., Lepson, J. K., and Carothers, J. H. 1995. Evolutionary consequences of extinctions in populations of a Hawaiian honeycreeper. *Conservation Biology* 9:107–13.

Steadman, D. W. 1995. Prehistoric extinctions of Pacific Island birds: Biodiversity meets zooarcheaology. *Science* 267:1123–31.

Steenhuisen, S.-L., Van der Bank, H., and Johnson, S. D. 2012. The relative contributions of insect and bird pollinators to outcrossing in an african *Protea* (Proteaceae). *American Journal of Botany* 99:1104–11.

Stewart, A. M., and Craig, J. L. 1989. Factors affecting pollinator effectiveness in *Feijoa sellowiana*. *New Zealand Journal of Crop and Horticultural Science* 17: 145–54.

Stiles, F. G. 1978. Ecological and evolutionary implications of bird pollination. *American Zoologist* 18:715–27.

———. 1981. Geographical aspects of bird-flower coevolution, with particular reference to Central America. *Annals of the Missouri Botanical Garden* 68:323–51.

Symes, C. T., Human, H., and Nicolson, S. W. 2009. Appearances can be deceiving: Pollination in two sympatric winter-flowering *Aloe* species. *South African Journal of Botany* 75:668–74.

Szabo, J. K., Khwaja, N., Garnett, S. T., and Butchart, S. H. M. 2012. Global patterns and drivers of avian extinctions at the species and subspecies level. *PLoS ONE* 7:e47080.

Tadmor-Melamed, H., Markman, S., Arieli, A., Distl, M., Wink, M., and Izhaki, I. 2004. Limited ability of palestine sunbirds *Nectarinia osea* to cope with pyridine alkaloids in nectar of tree tobacco *Nicotiana glauca*. *Functional Ecology* 18:844–50.

Thorogood, R., Henry, T., and Fordham, S. 2007. North Island kokako (*Callaeas cinerea wilsoni*) feed on flax (*Phormium tenax*) nectar on Tiritiri Matangi Island, Hauraki Gulf, New Zealand. *Notornis* 54:52–54.

Triplett, S., Luck, G. W., and Spooner, P. 2012. The importance of managing the costs and benefits of bird activity for agricultural sustainability. *International Journal of Agricultural Sustainability* 10:268–88.

Turner, R. C., Midgley, J. J., Barnard, P., Simmons, R. E., and Johnson, S. D. 2012. Experimental evidence for bird pollination and corolla damage by ants in the short-tubed flowers of *Erica halicacaba* (Ericaceae). *South African Journal of Botany* 79:25–31.

Valido, A., Dupont, Y. L., and Olesen, J. M. 2004. Bird-flower interactions in the Macaronesian islands. *Journal of Biogeography* 31:1945–53.

Van Etten, M. L., Tate, J. A., Anderson, S. H., Kelly, D., Ladley, J. J., Merrett, M. F., Peterson, P. G., and Robertson, A. W. 2015. The compounding effects of high pollen limitation, selfing rates and inbreeding depression leave a New Zealand tree with few viable offspring. *Annals of Botany* 116:833–43.

Varassin, I. G., Trigo, J., and Sazima, M. 2001. The role of nectar production, flower pigments and odour in the pollination of four species of *Passiflora* (Passifloraceae) in south-eastern Brazil. *Botanical Journal of the Linnnean Society* 136:139–52.

Vaughton, G. 1992. Effectiveness of nectarivorous birds and honeybees as pollinators of *Banksia spinulosa* (Proteaceae). *Australian Journal of Ecology* 17:43–50.

———. 1996. Pollination disruption by European honeybees in the Australian bird-pollinated shrub *Grevillea barklyana* (Proteaceae). *Plant Systematics and Evolution* 200:89–100.

Wackers, F. L., and Bonifay, C. 2004. How to be sweet? Extrafloral nectar allocation by *Gossypium hirsutum* fits optimal defence theory predictions. *Ecology* 85:1512–18.

Waser, N. M. 1978. Competition for hummingbird pollination and sequential flowering in two Colorado wildflowers. *Ecology* 59:934–44.

———. 1979. Pollinator availability as a determinant of flowering time in Ocotillo (*Fouquieria splendens*). *Oecologia* 39:107–21.

———. 1982. A comparison of distances flown by different visitors to flowers of the same species. *Oecologia* 55:251–57.

Waser, N. M., Campbell, D. R., Price, M. V., and Brody, A. K. 2010. Density-dependent demographic responses of a semelparous plant to natural variation in seed rain. *Oikos* 119:1929–35.

Waser, N. M., Chittka, L., Price, M. V., Williams, N. M., and Ollerton, J. 1996. Generalization in pollination systems, and why it matters. *Ecology* 77:1043–60.

Wenny, D. G., DeVault, T. L., Johnson, M. D., Kelly, D., Şekercioğlu, Ç. H., Tomback, D. F., and Whelan, C. J. 2011. The need to quantify ecosystem services provided by birds. *Auk* 128:1–14.

Weston, K. A., Chapman, H. M., Kelly, D., and Moltchanova, E. V. 2012. Dependence on sunbird pollination for fruit set in three West African montane mistletoe species. *Journal of Tropical Ecology* 28:205–13.

Whelan, C. J., Wenny, D. G., and Marquis, R. J. 2008. Ecosystem services provided by birds. *Annals of the New York Academy of Sciences* 1134:25–60.

Whelan, R. J., Ayre, D. J., and Beynon, F. M. 2009. The birds and the bees: pollinator behaviour and variation in the mating system of the rare shrub *Grevillea macleayana*. *Annals of Botany* 103:1395–1401.

Whelan, R. J., and Burbidge, A. H. 1980. Flowering phenology, seed set and bird pollination of five Western Australian *Banksia* species. *Australian Journal of Ecology* 5:1–7.

Willis, E. O. 2002. Birds at *Eucalyptus* and other flowers in Southern Brazil: a review. *Ararajuba* 10:43–66.

Wood, J. R., Wilmshurst, J. M., Worthy, T. H., Holzapfel, A. S., and Cooper, A. 2012. A lost link between a flightless parrot and a parasitic plant and the potential role of coprolites in conservation paleobiology. *Conservation Biology* 26:1091–99.

Yumoto, T. 1987. Pollination systems in a warm temperate evergreen broad-leaved forest on Yaku Island. *Ecological Restoration* 2:133–45.

CHAPTER FIVE

Seed Dispersal by Fruit-Eating Birds

Daniel G. Wenny, Çağan H. Şekercioğlu, Norbert J. Cordeiro, Haldre S. Rogers, and Dave Kelly

Living birds are highly effective agents in the transportation of seeds. —Charles Darwin (1859)

Seed dispersal, the process by which vectors such as wind, water, or animals move seeds, is an important phase in the life cycle of plants. Dispersal increases the chances that a seed will arrive in a site suitable for establishment rather than landing under the mother plant, where it faces competition with conspecifics, including siblings (Terborgh 2012). The area around a source plant increases with the square of the linear distance away, so the number of possibly suitable sites also increases exponentially (Hamilton and May 1977). The variability of sites also increases with distance from the source, so that some will be unsuitable, but those which are suitable and are not yet occupied by a conspecific offer great opportunities. Because the chance of a given seed's survival to adulthood is so small, movement of some seeds well away from the parent plant is likely beneficial (Howe and Miriti 2004). Therefore, plants have evolved many methods of seed dispersal, including abiotic dispersal by wind and water, and biotic dispersal by birds, mammals, fish, reptiles, and invertebrates. Seed dispersal by animals includes external transport on fur, feathers, or clothing, internal transport through ingestion and excretion (chapter 6), transport by ants, and scatterhoarding by seed-eating vertebrates (chapter 7).

Early work on seed dispersal focused on long-distance dispersal to islands (Darwin 1859; Ridley 1930; Carlquist 1967). Subsequent work identified dispersal "syndromes," suites of plant characteristics associated with particular seed dispersal vectors. Fruits eaten by birds are often red, blue, or black, small- to moderate-sized (< 20 mm diameter), lack a thick outer rind (unless it is dehiscent), and lack odor (van der Pijl 1972). However, field observations show that birds eat a much wider variety of fruits than just the classic "bird fruits" (fig. 5.1; Wenny et al. in prep.) Therefore, the combination of (1) considerable overlap among dispersers in desirable fruit traits, (2) multiple dispersers for individual plant species, (3) the existence of secondary dispersal, and (4) the lack of natural history data for many species makes assigning a plant species to one syndrome based on plant characteristics alone both difficult and ill-advised.

Given that both plants and dispersers benefit from seed dispersal relationships (Wheelwright and Orians 1982; Janzen 1983a; Janzen 1983b; Howe 1984a; Herrera 1985; Herrera 1986), one might predict that coevolution would be common (Snow 1971; McKey 1975). However, most studies conclude that fruit-frugivore relationships are best described as diffuse mutualisms whereby any focal plant species interacts with a set of potential disperser species, and any focal frugivore species may potentially disperse many different plant species, forming complex interaction networks (Levey et al. 1994; Jordano 2000; Herrera 2002; Bascompte and Jordano 2007; Carlo et al. 2007). In addition to the complexity of interactions with dispersers, plants face many other selective pressures that can influence seed and fruit traits, which makes it difficult to attribute plant traits solely to selection from particular dispersers (Herrera 1992; Fischer and Chapman 1993; Jordano 1995). For example, plants must both attract dispersers and repel fruit and seed predators, yet they can exert only limited control over which animal species eat the fruits (Janzen 1983b; Howe 1986) by manipulating size, color, external protection, attachment strength, presentation on the plant, phenology, nutritional composition, and chemistry (Moermond and Denslow 1983; Herrera 1987; Fleming et al. 1993; Lord 2004; Levey et al. 2006; Forget et al. 2007; Lomáscolo et al. 2008; Lomáscolo et al. 2012).

Howe and Smallwood (1982) developed a conceptual framework for testing multiple hypotheses on the benefits of a diffuse mutualism between plants and seed-dispersing animals. Empirical findings supported the ideas that seed dispersal by birds can benefit plants through gene flow (Godoy and Jordano 2001), escape from areas of high mortality (Harms et al. 2000), colonization of new sites (Ne'eman and Izhaki 1996; Shanahan et al. 2001; Richardson et al. 2002), and directed dispersal to favorable sites (Wenny and Levey 1998; Tewksbury et al. 1999; Spiegel and Nathan 2012). Birds remove fruit pulp from seeds during handling, ingestion,

FIGURE 5.1. Representative functional fruit types eaten by birds (not to scale). Berries are fleshy fruits that contain several to many small seeds: (a) *Vaccinium corymbosum* (Ericaceae); (b) *Ilex verticillata* (Aquifoliaceae). Drupes are one-seeded fleshy fruits with a large seed relative to fruit diameter; (c) *Ocotea endresiana* (Lauraceae); (d) *Maesopsis eminii* (Rhamnaceae). Arils are flesh-covered seeds in dehiscent capsules: (e) *Stemmadenia donnell-smithi* (Apocynaceae) with red-legged honeycreeper (*Cyanerpes cyanea*); (f) *Leptonychia usambarensis* (Malvaceae). The white bar at the bottom right of each photo is approximately one centimeter in length. Photos by Norbert J. Cordeiro (d, f), Çağan H. Şekercioğlu (e), and Daniel G. Wenny (a, b, c).

and/or gut processing. Pulp removal can reduce germination inhibitors and speed germination (Samuels and Levey 2005; Robertson et al. 2006; Traveset et al. 2007), and may increase germination success (Traveset et al. 2007; but see Kelly et al. 2010). Gut passage can also remove fungal pathogens and chemical cues used by seed predators (Fricke et al. 2014). All these potential benefits can accrue to plants dispersed by frugivores.

Frugivory by birds is widespread, but the exact number of bird and plant species involved depends on which definitions are adopted. A recent review (Wenny et al., in prep.) tallied all birds that are recorded as eating fruits using a functional (ecological) definition of fruits, not a technical (structural) one, and plant lists at the genus level. They combined information on observations of bird feeding and plant morphology to determine likely dispersal agents. They reported some level of frugivory for more than 4,000 bird species (40% of all bird species; table 5.1), with more than 730 species eating trace amounts, 1,665 being occasional frugivores (up to 30% of the diet), 686 being moderately frugivorous (31–70% of diet), and 1,005 species being highly frugivorous (>70% of diet). The great majority of these species (79%) also ate invertebrates (>10% of diet).

TABLE 5.1 **Number of bird species in each frugivory category, listed by order and family. Frugivory categories are the approximate percentage of fruit in the diet. First category (< 1%) includes species that have been observed eating fruit, but which seldom do so. The other 10 categories should be interpreted as 10 = 1–10%, 20 = 11–20%, etc. See table 5.2 extended online for references (www.press.uchicago.edu/sites/whybirdsmatter/). Taxonomy follows BirdLife International (http://www.birdlife.org/datazone/info/taxonomy).**

Estimated % fruit in diet												
Taxon	**<1%**	**10**	**20**	**30**	**40**	**50**	**60**	**70**	**80**	**90**	**100**	**Total**
Struthioniformes												
Rheidae			1									1
Casuariidae							3					3
Dromaiidae				1			2					3
Apterygidae	5											5
Tinamiformes												
Tinamidae	1	10	2	2	25	2	1					43
Anseriformes												
Anatidae	8	6	1		4							19
Dendrocygnidae		1		1	1							3
Galliformes												
Cracidae			3	1	2	2	5	6	8	1	24	52
Megapodiidae	1	2	3	3	4	1	1					15
Numididae		1	2	5								8
Odontophoridae	5	1	2	4	2	2			1			17
Phasianidae	31	9	37	20	5	5			6			113

TABLE 5.1 *(continued)*

Taxon	<1%	10	20	30	40	50	60	70	80	90	100	Total
Pelecaniformes												
Ardeidae	1											1
Threskiornithidae	1											1
Accipitriformes												
Accipitridae	4	1	1				1					7
Cathartidae	2											2
Otidiformes												
Otididae	1	6	7									14
Mesitornithiformes												
Mesitornithidae	1											1
Cariamiformes												
Cariamidae	1											1
Gruiformes												
Gruidae	1	3										4
Psophiidae								3		3		6
Rallidae	9	9	1	1								20
Charadriiformes												
Charadriidae	3	1	1									5
Laridae	12	6										18
Scolopacidae	7	4	1	1	1							14
Stercorariidae	2											2
Turnicidae				1								1
Pterocliformes												
Pteroclidae	1											1
Columbiformes												
Columbidae	13	6	14	35	10	25		1	36	2	158	300
Opisthocomiformes												
Opisthocomidae		1										1
Musophagiformes												
Musophagidae							1	4	9		10	24
Cuculiformes												
Cuculidae	13	11	20	1			1		1		4	51
Strigiformes												
Strigidae	2											2
Tytonidae		1	1									2
Caprimulgiformes												
Steatornithidae											1	1
Apodiformes												
Trochilidae	2											2
Coliiformes												
Coliidae									6			6
Trogoniformes												
Trogonidae		3	17			5	3		10		2	40
Coraciiformes												
Alcedinidae	4											4
Coraciidae	2											2
Meropidae	1											1
Momotidae	3		3		1							7
Todidae	3											3
Bucerotiformes												
Bucerotidae	8	4	1	4	4	1	15	2	10	5	6	60
Bucorvidae	1	1										2
Phoeniculidae	1	3	2									6

continues

TABLE 5.1 ***(continued)***

Taxon	<1%	10	20	30	40	50	60	70	80	90	100	Total
Piciformes												
Bucconidae	1	1	2									4
Capitonidae	1					1	1	2	13			18
Indicatoridae	4	3	3									10
Lybiidae			1		2	14	5		28		1	51
Megalaimidae					2	2	1	3	12		15	35
Picidae	28	21	37	11	5	4	1		2			109
Ramphastidae							2	21	18		9	50
Semnornithidae					1				1			2
Falconiformes												
Falconidae	5	3	1									9
Psittaciformes												
Cacatuidae	3			9					2			14
Psittacidae	14	18	56	39	23	45	27	2	50		43	317
Strigopidae		1	1	1							1	4
Passeriformes												
Acanthisittidae	2											2
Acanthizidae	9		3									12
Aegithalidae	3	1	1									5
Alaudidae	9	1	1									11
Artamidae	1											1
Bombycillidae									7	1		8
Calcariidae	1											1
Callaeatidae			2		1							3
Campephagidae	9	1	27	10					6			53
Cardinalidae	3	6	13	25	3	3	1	2	1			57
Chloropseidae			5	2	1	2			1			11
Cisticolidae	4		4									8
Cnemophilidae											3	3
Coerebidae			1									1
Colluricinclidae	2	1	2			2			1			8
Conopophagidae	1											1
Corcoracidae	1											1
Corvidae	10	19	25	23	5	1	3		7			93
Cotingidae	5	2	12	3	1	7		3	31	1	34	99
Cracticidae	4	1	3	1								9
Dasyornithidae	1											1
Dicaeidae			1				18	1	24		1	45
Dicruridae	4	1										5
Dulidae								1				1
Emberizidae	31	9	23	43	11	3	1	1				122
Estrilidae	4	1	7	6	3			1				22
Eupetidae	2											2
Eurylaimidae	1		1	1					3		1	7
Falcunculidae	1											1
Formicariidae	2		1	1								4
Fringillidae	17	23	13	10	5	6		1	2	13	17	107
Furnariidae	10			1								11
Hirundinidae	3	1										4
Icteridae	7	17	18	13	7	8	1		4			75
Irenidae									1		1	2
Laniidae	6		2									8

TABLE 5.1 (*continued*)

Taxon	<1%	10	20	30	40	50	60	70	80	90	100	Total
Malaconitidae	4	4	7	1								16
Maluridae	4					1						5
Melanocharitidae	1								7		1	9
Meliphagidae	23	23	23	26	8		3	1		4		111
Mimidae	2	4	12	6	3	5	1				1	34
Mohoidae		2	1									3
Monarchidae	6											6
Motacillidae	6		1	1								8
Muscicapidae	72	4	28	7		1			3		1	116
Nectariniidae	21	7	11	16	1	1						57
Oriolidae				1	1	1	23		1	2	1	30
Orthonychidae	1		1									2
Pachycephalidae	3		7						2			12
Paradisaeidae			1	3	2	14	1		16		3	40
Paridae	9	6	7	9	1							32
Parulidae	18	4	11	3	1							37
Passeridae	6	1	2	1								10
Petroicidae	3											3
Philipettidae							1	1				2
Picathartidae	1											1
Pipridae	2		5						47			54
Pittidae	7											7
Pityriaseidae	1											1
Platysteiridae	2											2
Ploceidae	7	9	11	6	3	5					1	42
Polioptilidae	1											1
Pomatostomidae	1	1										2
Prunellidae	7											7
Ptilonorhynchidae					1		3	2	12		1	19
Pycnonotidae	9	2	14	12	1	2	6	2	66		2	116
Regulidae		1										1
Remizidae	1	2	2									5
Rhabdornithidae				3								3
Rhinocryptidae	2	2	1									5
Sapayoaidae	1											1
Sittidae	6											6
Sturnidae	1	4	14	10	8	17	7	5	29	3	12	110
Sylviidae	29	2	23									54
Thamnophilidae	9	1	3									13
Thraupidae		24	27	9	7	26	9	8	56	21		187
Timaliidae	27	3	65	30	4		2		7		4	142
Troglodytidae	11											11
Turdidae	6	10	49	17	13	12	12	3	23			145
Turnagridae		2										2
Tyrannidae	50	3	84	8	2	3			14	1	2	167
Vangidae	1	1	1									3
Vireonidae	10	5	12	8	1	3					1	40
Zosteropidae	3	4	40	8	20	2	4	4	3			88
Grand Total	**733**	**352**	**849**	**464**	**206**	**234**	**166**	**80**	**587**	**57**	**361**	**4,089**

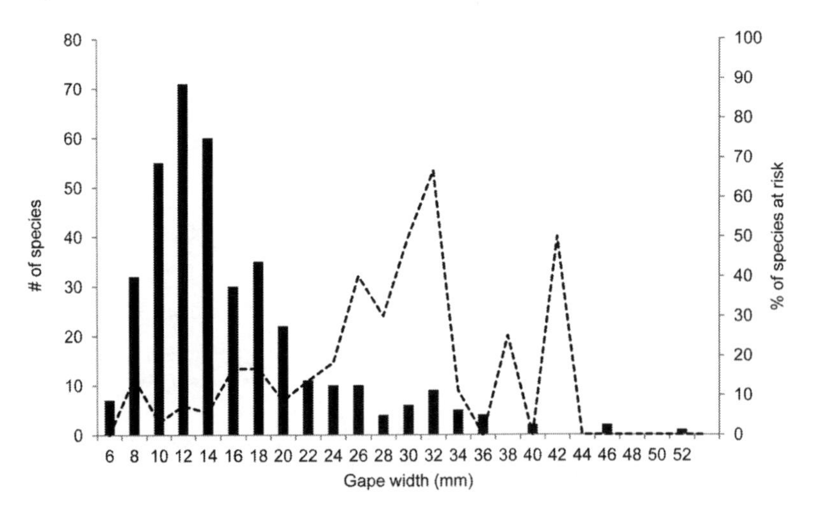

FIGURE 5.2. Frequency distribution of gape widths among 376 species of frugivorous birds. The dashed line indicates the percentage of species in each gape width category that are at risk of extinction (Wenny et al., in prep.). For more details, see supplemental table S5.1, www.press.uchicago.edu/sites/whybirdsmatter/.

The size of a bird's gape limits the maximum size of fruit it can swallow, and from a sample of 376 frugivorous species of known gape, only 2% had gapes of more than 34 mm in width (fig. 5.2). Half the species had gape widths of less than 12 mm, showing that smaller birds are numerically dominant, certainly by species counts and probably also by number of individual birds. The highly frugivorous species tend to have larger gapes and mostly live in the tropics, except for three waxwings (Bombycillidae) in the northern temperate zone and some pigeons (Colombiformes) and parrots (Psittaciformes) in Australasia (Wenny et al., in prep.). Specialist frugivores have received more research attention, yet smaller birds often include some fruit in their diet and may be important dispersers due to their abundance and varied behavior (Cordeiro and Howe 2003).

Fruit-eating birds occur at least seasonally in virtually all terrestrial habitats not covered by ice year-round, including oceanic islands that often lack other dispersers (Sekercioglu 2006a; Whelan et al. 2008), and they disperse the seeds of about 68,900 plant species in more than 1,500 genera from 240 plant families (table 5.2; see also table 5.2 extended online: www.press.uchicago.edu/sites/whybirdsmatter/). This represents about 25% of all seed plant species and more than 50% of all families. In these

TABLE 5.2 **Plant families with fleshy-fruited species likely or possibly dispersed by birds. The first column lists families alphabetically within the four larger groups. The second column lists percentages of genera and species (separated by /) in each family that are documented or possibly dispersed by birds (* > 50%; ** > 90%). The "Documented Genera" column lists the number of genera in each family that contain at least one species consumed (and presumably dispersed) by birds. The "Possible Genera" column, on the other hand, lists genera with seeds or fruits that have characteristics similar to bird dispersed species, but that lack documented observations. The last column lists the number of species within each of the documented and possible genera that could be dispersed by birds. See table 5.2 extended online for references (www.press.uchicago.edu/sites/whybirdsmatter/).**

		Genera		
Family	**% of taxa**	**Documented**	**Possible**	**Species**
Gymnosperms				
Cupressaceae	/	1		68
Cycadaceae	**/**	1		90
Ephedraceae	**/	1		30
Ginkgoaceae	**/**		1	1
Gnetaceae	**/**	1		41
Podocarpaceae	*/*	11	2	162
Taxaceae	**/**	1	5	33
Zamiaceae	**/**	1	8	218
Basal angiosperms				
Amborellaceae	**/**		1	1
Annonaceae	/*	29	9	1400
Austrobaileyaceae	**/**		1	2
Canellaceae	**/**	2	3	21
Chloranthaceae	**/**	4		66
Degeneriaceae	**/**		1	2
Eupomatiaceae	**/**	1		3
Gomortegaceae	**/**		1	1
Hernandiaceae	/*		1	32
Himantandraceae	**/**	1		2
Lauraceae	*/**	37	10	2,589
Magnoliaceae	/**	2		241
Monimiaceae	/*	6	3	104
Myristicaceae	*/**	11	3	368
Piperaceae	**/	2	4	1,015
Schisandraceae	**/**		2	47
Siparunaceae	/**	1		70
Trimeniaceae	**/**		1	5
Winteraceae	*/**	3	2	86
Monocots				
Agavaceae	/	1	1	11
Alstroemeriaceae	/*	2		128
Amaryllidaceae	/	3	2	147
Araceae	/*	14	6	2,154
Arecaceae	/*	70	7	1,995
Asparagaceae	**/**	2	1	354
Asphodelaceae	/		1	18
Asteliaceae	/*	1	1	31
Blandfordiaceae	**/**		1	4
Bromeliaceae	/	1	1	862
Commelinaceae	/	7		288
Costaceae	/*	1		104

continues

TABLE 5.2 ***(continued)***

Family	% of taxa	Genera		Species
		Documented	Possible	
Cyclanthaceae	*/**	6	1	221
Cyperaceae	/	1		92
Dioscoreaceae	/	2		12
Flagellariaceae	**/**	1		4
Haemodoraceae	/	1		2
Hanguanaceae	**/**		1	2
Heliconiaceae	**/**	1		207
Hemerocallidaceae	/	1		38
Hypoxidaceae	/		1	7
Iridaceae	/	2		16
Joinvilleaceae	**/**		1	2
Laxmanniaceae	/	2		16
Liliaceae	/	2	2	13
Marantaceae	/**	3	6	409
Melanthiaceae	/		2	28
Musaceae	**/**	1	2	42
Orchidaceae	/		1	9
Pandanaceae	**/*	2	2	509
Petermanniaceae	**/**		1	1
Philesiaceae	**/**	3		4
Poaceae	/	2		17
Rhipogonaceae	**/**	1		6
Ruscaceae	/**	9	14	453
Smilacaceae	**/**	2		288
Strelitziaceae	**/**	1	2	7
Triuridaceae	/		1	4
Zingiberaceae	/*	5	5	727
Eudicots				
Acanthaceae	/	1		3
Achariaceae	/	3		18
Achatocarpaceae	**/**		2	11
Actinidiaceae	**/**	3		331
Adoxaceae	*/**	2	1	242
Aextoxicaceae	**/**	1		1
Aizoaceae	/	3	1	62
Alseuosmiaceae	**/**	1	2	9
Amaranthaceae	/	9		87
Anacardiaceae	/*	23	3	365
Anisophyllaceae	/**	1		36
Aphloiaceae	**/**	1		2
Apocynaceae	/	12	9	756
Aptandraceae	/		2	6
Aquifoliaceae	**/**	1		414
Araliaceae	/*	20	7	1,281
Argophyllaceae	/	1		6
Asteraceae	/	4		53
Balanopaceae	**/**	1		12
Basellaceae	/*		2	15
Berberidaceae	*/*	5	5	646
Berberidopsidaceae	/*		1	2

TABLE 5.2 ***(continued)***

Family	% of taxa	Genera		Species
		Documented	Possible	
Bignoniaceae	/	1		17
Bixaceae	/	1		5
Boraginaceae	/	6	8	724
Brunelliaceae	**/**	1		60
Burseraceae	*/**	9		550
Buxaceae	*/	1	2	27
Cactaceae	/	16	12	918
Calophyllaceae	/	2	1	198
Campanulaceae	/	6	1	446
Cannabaceae	/*	5		136
Capparaceae	/**	4	1	400
Caprifoliaceae	/	4		230
Cardiopteridaceae	/	1		22
Caricaceae	/	1	1	9
Caryocaraceae	/	1		1
Celastraceae	/**	12	10	841
Centroplacaceae	/*	1		5
Chrysobalanaceae	/*	5	1	377
Cleomaceae	/	2		63
Clusiaceae	*/*	7	2	346
Columelliaceae	/	1		3
Combretaceae	/*	2	1	370
Connaraceae	/**	5	1	176
Convolvulaceae	/	1	1	1
Coriariaceae	**/**	1		16
Cornaceae	**/**	5	3	110
Corynocarpaceae	**/**	1		6
Coulaceae	*/*	1	1	2
Crossosomataceae	/		1	3
Cucurbitaceae	/	13	1	238
Cunoniaceae	/	2	1	23
Curtisiaceae	**/**	1		1
Cyrillaceae	/		1	1
Daphniphyllaceae	**/**	1		35
Dichapetalaceae	/**	1		149
Dilleniaceae	/	7		170
Dipentodontaceae	/**	1		15
Ebenaceae	/*	2		318
Elaeagnaceae	**/**	3		48
Elaeocarpaceae	/*	4		539
Ericaceae	/	29	15	1,834
Erythropalaceae	/**	1	1	35
Erythroxylaceae	**/**	1	3	259
Escalloniaceae	/	1		60
Euphorbiaceae	/	16	7	917
Fabaceae	/	21	9	1,752
Garryaceae	**/**	2		25
Gentianaceae	/	1		72
Gerrardinaceae	**/**		1	2
Gesneriaceae	/	13	4	1,325

continues

TABLE 5.2 ***(continued)***

Family	% of taxa	Genera Documented	Genera Possible	Species
Goodeniaceae	/	1		40
Goupiaceae	**/**	1		3
Griseliniaceae	**/**	1		7
Grossulariaceae	**/**	1		150
Gunneraceae	**/**	1		69
Halophytaceae	**/**		1	1
Helwingiaceae	**/**	1		4
Huaceae	/*		1	2
Humiriaceae	/*	2	2	36
Hydrangeaceae	/	2		14
Hypericaceae	/	2		62
Icacinaceae	/*	7	2	80
Ixonanthaceae	/	1		8
Koeberliniaceae	**/**	1		2
Lacistemataceae	**/**	2		16
Lamiaceae	/	10		1,143
Lardizabalaceae	/*	3	5	39
Lecythidaceae	/	1	2	66
Lepidobotryaceae	**/**	1	1	2
Linaceae	/	2		18
Loganiaceae	/	4	1	81
Loranthaceae	**/**	76		801
Lythraceae	/	1		2
Malpighiaceae	/	3	1	273
Malvaceae	/*	19	2	1,696
Marcgraviaceae	*/*	6		114
Melastomataceae	/*	18	10	1,527
Meliaceae	/*	16	2	414
Melianthaceae	/*	1		7
Memecylaceae	**/**	1	6	450
Menispermaceae	/	17	4	243
Montiniaceae	/		1	1
Moraceae	/	17	3	698
Mutingiaceae	**/**	1	2	3
Myodocarpaceae	/		1	7
Myricaceae	/*	2		43
Myrtaceae	/*	27	9	3,140
Nitrariaceae	*/*	1	1	9
Nyctaginaceae	/	3		8
Ochnaceae	/*	5		369
Octoknomaceae	**/**		1	14
Olacaceae	/	1	2	60
Oleaceae	/*	10	1	543
Onagraceae	/	1		119
Oncothecaceae	**/**		1	2
Opiliaceae	**/**	2	9	36
Paeoniaceae	**/**	1		33
Papaveraceae	/	1		10
Paracryphiaceae	/	1		10

TABLE 5.2 ***(continued)***

Family	% of taxa	Genera		Species
		Documented	Possible	
Passifloraceae	/	2	2	300
Penaeaceae	/	1		10
Pennantiaceae	**/**	1		5
Pentaphragmataceae	**/		1	2
Pentaphylacaceae	*/**	6	1	397
Peraceae	/	2		93
Peridiscaceae	**/**		2	2
Petenaeaceae	**/**		1	1
Phyllanthaceae	/**	12	1	1,957
Phyllonomaceae	**/**	1		4
Phyrmaceae	/	1	1	3
Phytolaccaceae	/	5		35
Picramniaceae	*/*	1	1	44
Picrodendraceae	/	1	2	10
Pittosporaceae	/	3		97
Polemoniaceae	/	1		18
Polygalaceae	/	2		218
Polygonaceae	/	3		204
Primulaceae	/*	9	3	1,109
Proteaceae	/	4	7	236
Putranjivaceae	/**	1	1	207
Ranunculaceae	/	1		8
Resedaceae	/	1		6
Rhabdodendraceae	**/**	1		3
Rhamnaceae	/	13	4	422
Rhizophoraceae	/**	2	1	90
Rosaceae	/*	21	3	1,886
Rousseaceae	**/**	1	3	14
Rubiaceae	/*	67	15	6,852
Rutaceae	/*	22	10	708
Sabiaceae	**/*	1	2	90
Salicaceae	/**	19	10	452
Salvadoraceae	*/*	2		8
Santalaceae	*/*	14	13	609
Sapindaceae	/	29	6	865
Sapotaceae	/*	17	5	760
Sarcolaenaceae	/*	3		37
Schlegeliaceae	**/**	2	2	37
Schoepfiaceae	/		1	25
Scrophulariaceae	/	2	4	254
Simaroubaceae	/	7	2	48
Sladeniaceae	/*	1		2
Solanaceae	/	15	10	1,047
Stachyuraceae	**/**		1	7
Staphyleaceae	/	1		17
Stegnospermaceae	**/**		1	3
Stemonuraceae	/	2		42
Stilbaceae	/	1		3
Stombosiaceae	/*	2		12
Strasburgeriaceae	/	1		1

continues

TABLE 5.2 ***(continued)***

Family	% of taxa	Genera Documented	Genera Possible	Species
Styracaceae	/*	1		120
Surianaceae	/	1		1
Symplocaceae	/*	1		250
Tapisciaceae	**/		2	7
Tetrameristeaceae	**/*		2	4
Thymelaeaceae	/	8	2	368
Torricelliaceae	**/*		3	11
Tovariaceae	**/**	1		2
Tropaeolaceae	**/*	1		74
Ulmaceae	/	2		13
Urticaceae	/	12		542
Verbenaceae	/*	5	1	511
Violaceae	/	2	1	103
Vitaceae	**/**	10	4	1053
Ximeniaceae	/*		1	10
Zygophyllaceae	/	1		5
Total		**1,129**	**448**	**68,976**

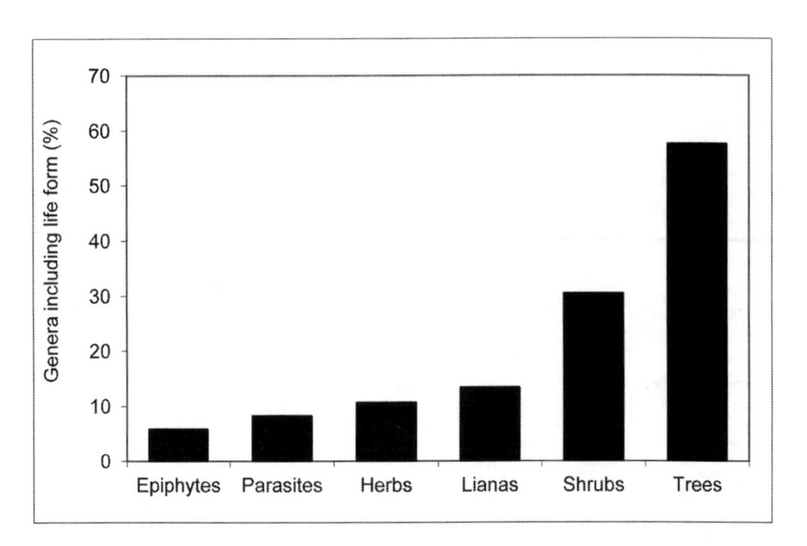

FIGURE 5.3. Proportions of 1,378 fleshy-fruited plant genera that are certainly or probably dispersed by birds, including various plant growth forms (from Wenny et al. in prep). As some genera include more than one growth form, the sum of all categories exceeds 100%.

bird-dispersed genera, 30% include shrub species and 58% include trees (fig. 5.3). Because herbaceous plants are rarely bird-dispersed but usually represent a majority of plant species, bird dispersal generally accounts for less than 25% of the species in regional flora. In tropical forest areas, however, woody species dominate the flora and 65 to 75% of the tree species, and 40 to 60% of the shrubs are bird-dispersed (Wenny et al., in prep).

Birds serve as mobile links within and among habitats (Gilbert 1980; Lundberg and Moberg 2003) and can transport seeds considerable distances. Birds disperse most seeds within 200 m of source trees, and far fewer seeds up to 1.5 km (Mack 1995; Wenny 2000; Clark et al. 2005; Jordano et al. 2007; Weir and Corlett 2007; Kays et al. 2011; Breitbach et al. 2012; Wotton and Kelly 2012; Carlo et al. 2013). Larger bird species are especially important because they disperse large seeds that smaller birds cannot swallow, and because they disperse seeds for longer distances (Jordano et al. 2007; Corlett 2009; Wotton and Kelly 2012). Dispersal distances of 1.5 to 14.5 km by large pigeons, toucans, and hornbills are estimated from models combining gut passage time with the movement patterns of marked birds (Holbrook and Loiselle 2007; Lenz et al. 2011; Wotton and Kelly 2012). The occurrence of numerous bird-dispersed plants on oceanic islands hundreds of kilometers from likely source areas suggests that birds can provide even longer dispersal distances (Carlquist 1967; Vargas et al. 2012). These rare long-distance dispersal events have disproportionate ecological and evolutionary importance (Şekercioğlu 2006a).

At the ecosystem level, dispersal by birds has a large role in shaping plant community composition in many habitats (Howe and Smallwood 1982; Herrera 1985; Jordano 2000). In this chapter we take a global view of birds that eat fruits and disperse the seeds of dependent plant species. Here we present a preliminary analysis of seed dispersal by birds as an ecosystem service, and address two questions: What are the ecological and economic benefits of seed dispersal? How can the value of seed dispersal by birds be estimated? Finally, given the threatened conservation status of many fruit-eating birds, we explore the consequences of bird population declines for the role of such birds as seed dispersers.

Seed Dispersal by Birds as an Ecosystem Service

The Economic Value of Seed Dispersal

Seed dispersal by birds is a globally important process. Based on the number of bird species involved it is the second most important ecological function of birds after insectivory (Sodhi et al. 2011; Wenny et al. 2011). The large number of plant species involved also emphasizes its importance. However, quantifying the benefits from dispersal as an ecosystem service has proven difficult, because most seed dispersal indirectly benefits human society. Thus, while some regional or ecosystem-level estimates include ecosystem services like pollination, seed dispersal is rarely included in these estimates (e.g., Costanza et al. 1997; Kareiva et al. 2011; Ninan and Inoue 2013). Part of the reason dispersal has been overlooked is that, unlike pollination, in which the results of the service (seeds) are easily measured on the parent plant after the service is provided, the benefits of seed dispersal, especially for trees, are spatially distant and often not measurable until decades after the service (Kelly et al. 2004). In addition, for plant species not used directly by humans, the end benefits of dispersal are other ecosystem services that may also be difficult to quantify (Millenium Ecosystem Assessment 2005).

Despite these difficulties, assessing both the direct uses of plants dispersed by birds and the ecological importance of dispersal is vital. Seed dispersal helps regulate plant populations and habitat dynamics, which contributes to our use of valuable timber species, edible fruits, and medicinal plants. It also supports many other ecosystem services, especially all the ecological roles of forests, woodlands, shrublands, and other habitats with plants dispersed by birds. The benefits provided by forests, for example, include carbon sequestration, erosion control, water filtration, and recreation.

Birds disperse seeds from many economically or culturally useful plant species. Across the tropics, 43 to 59% of the genera and 18 to 47% of the volume of timber species are animal-dispersed (Jansen and Zuidema 2001). Among the genera with bird-dispersed species, at least 85 include timber species, at least 182 genera include species that are edible (including spices), 153 have medicinal uses, 146 include ornamental plants, and 84 genera have other uses (Wenny et al., in prep). All rely primarily on natural dispersal and are not grown in plantations. Some important crop species have wild bird-dispersed congeners that could be used to improve crop plant genetics (Smith et al. 1992).

Bird dispersal is responsible for some undetermined portion of forest value, which includes marketable timber and nontimber forest products. In an Amerindian village in Honduras, the value of nontimber forest products was US$18—24 ha^{-1} $year^{-1}$ (Godoy et al. 2000). Fruits and rubber latex harvested by local people from a 1-ha patch of Peruvian rainforest

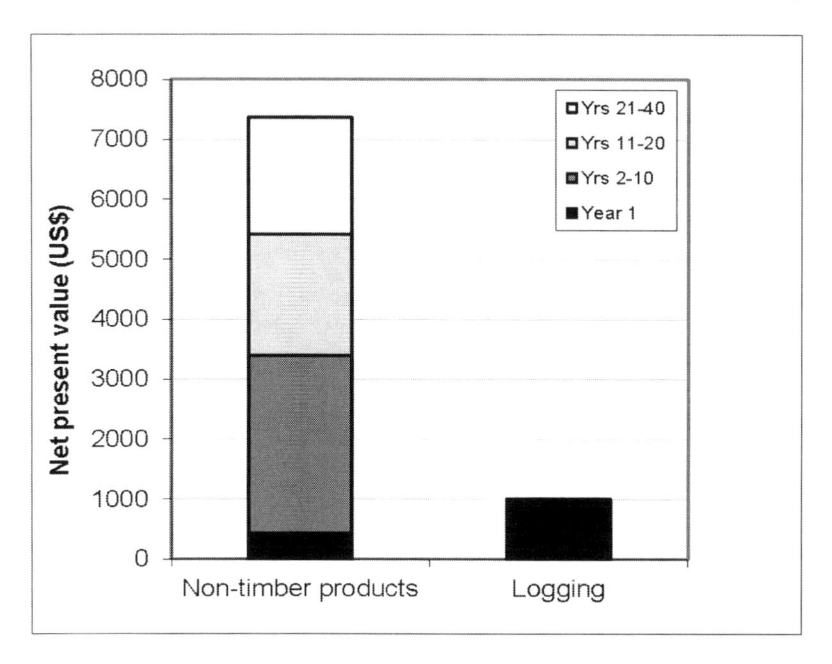

FIGURE 5.4. Economic return from one hectare of Peruvian rain forest over 40 years, as reckoned from natural products (rendered as net present value of income from years 1 through 40, using a 5% discount rate) and from logging (all value obtained in the first year, assuming recovery of timber species over more than 40 years). First-year returns from Peters et al. 1989.

yielded US$422 annually (Peters et al. 1989). Although the first-year return from logging the entire stand (US$1000) was double that of nontimber forest products, the long-term value of nontimber products was seven times as high (in net present value, with a 5% discount rate; fig. 5.4). Over the long term, 90% of income could be derived from fruit and latex, with only 10% from low levels of selective logging (Peters et al. 1989). In both of these examples, the larger immediate payment from timber sales and the very high implicit discount rates due to political and legal uncertainty combine to encourage deforestation even when the net present value of nontimber products is clearly higher than that of timber. While outsiders, especially from developed countries, may value the rain forest for global-scale ecosystem services like carbon sequestration and nonuse values (e.g., see chapter 2), little to none of that value accrues to the local people to support preservation (Godoy et al. 2000). Therefore, empowering local communities is an important way to encourage sustainable resource use and conservation of ecosystem services.

Incense made from gaharu wood in Indonesia provides a concrete illustration of the complexity of ecological processes (Paoli et al. 2001). Gaharu wood is diseased heartwood of *Aquilaria* spp. (Thymeleaceae), which rely on birds for seed dispersal. Fungal pathogens, perhaps transmitted by insects, induce trees to produce aromatic resins in heartwood, which is harvested for incense (Paoli et al. 2001). The gaharu wood in one tree is worth US$129, but harvesting at a sustainable level yields $11 ha^{-1} yr^{-1} (Paoli et al. 2001). Each incense product requires birds to disperse the trees, decades for the tree to grow, insects to spread the pathogen, infection to induce the resin response, and human labor to extract the heartwood. What market price will maintain this entire ecological network to ensure a sustainable supply? Other examples suggest that markets frequently lead to unsustainable use, such as overexploitation of the bird- and primate-dispersed African tree *Prunus africana*, the bark of which is used for prostate drugs (Stewart 2009).

The Ecological Value of Seed Dispersal

The ecological value of avian seed dispersal is better known than the economic value. Birds disperse the seeds of more than 60,000 plant species, are key drivers of primary and secondary succession, and help shape plant community dynamics. Four of their especially important dispersal roles are (1) dispersal within forests to gaps, (2) dispersal to favorable microenvironments in arid habitats, (3) dispersal into abandoned areas assisting forest regeneration, and (4) maintenance of specialized interactions with hemiparasitic mistletoes.

DISPERSAL WITHIN FORESTS. In forests, many birds either feed in forest gaps or move to gaps following feeding. Fruiting plants often produce larger crops in gaps, and some frugivorous birds are also especially active around gaps (Thompson and Willson 1978; Schemske and Brokaw 1981; Levey 1988; Martínez-Ramos and Alvarez-Buylla 1995; Kelly et al. 2000). In deciduous forest in eastern North America, Hoppes (1988) found a bimodal pattern of overall seed rain, with peaks near parent plants and at gap edges. Of bird-dispersed seeds, 50% landed in gaps and gap edges, which comprised only 16.8% of the study area (Hoppes 1988). Many birds sing from perches to maintain territory and attract mates, and gap edges often provide them with desirable perches. For many tropical frugivorous birds with lek breeding systems, males spend most of each day during the breeding season on display perches. Wenny and Levey (1998) showed that dispersal under display perches by three-wattled bellbirds (*Procnias tricarunculata*) enhanced the seedling establishment of a cloud forest tree. Of the seeds dispersed by bellbirds, 52% landed in gaps (under display perches), compared with only 3% of the seeds dispersed by four other bird species. Seedlings in gaps had almost twice the chance of surviving one year. The importance of gaps in forest dynamics can likely be applied to forest edges and corridors (Harvey 2000; Levey et al. 2005; de Melo et al. 2006).

DISPERSAL IN ARID ENVIRONMENTS. In arid and semiarid habitats, the critical recruitment sites for many plant species are not gaps, but the shaded areas beneath shrubs and trees. Under nurse plants, seedlings are protected from heat stress, and soil moisture and nutrient availability are often higher than elsewhere (Fulbright et al. 1995; Nunez et al. 1999). In arid ecosystems, nurse plants are often the only perches available, and thus seed input is locally elevated where seedling establishment is best (Tester et al. 1987; Izhaki et al. 1991; Chavez-Ramirez and Slack 1994). For example, bird-dispersed chilies (*Capsicum annum*) in Arizona typically grow beneath other fleshy-fruited species—especially under *Celtis* trees, which are preferred perches for the main dispersers. The survival of transplanted chilies is higher under *Celtis* than under other nurse plants (Tewksbury et al. 1999). However, dense seed deposits from such nonrandom dispersal increases subsequent density-dependent competition among seedlings (Spiegel and Nathan 2012).

DISPERSAL IN ABANDONED FIELDS AND REGENERATING AREAS. Dispersal into disturbed habitats has become a major focus of research in habitat restoration, because of the extent of anthropogenic disturbance (Vitousek et al. 1997). Birds—and, in the tropics, bats—are particularly important for dispersing early and mid-successional plant seeds to regenerating areas (de la Peña-Domene et al. 2014) because nonflying mammals are much less likely to cross open areas to reach isolated trees or forest fragments (White et al. 2004). Many of these avian dispersers are smaller generalists, because obligate forest species (Lindell et al. 2013) seldom cross open areas to reach isolated fragments (Sieving et al. 1996; Lees and Peres 2009). In particular, large frugivorous birds often disappear from tropical forest fragments (Bregman et al. 2014), thus slowing the spread of larger-seeded and late-successional plant species into regenerating areas (Martinez-Garza and Gonzalez-Montagut 1999). Another common pattern is sharply declining seed rain of bird-dispersed plant species with increasing distance from forest edge (McClanahan and Wolfe 1987; Willson and Crome 1989; Dosch et al. 2007).

Retaining or establishing perches may encourage foraging and increase dispersal into regenerating areas (Deforesta et al. 1984; Debussche and Isenmann 1994; Ne'eman and Izhaki 1996). Perches that serve as recruitment foci include artificial structures (McClanahan and Wolfe 1993; Shiels and Walker 2003; Zanini and Ganade 2005), shrubs (DaSilva et al. 1996; Nepstad et al. 1996; Holl 2002), planted trees (Robinson and Handel 2000; Martinez-Garza and Howe 2003), isolated or remnant trees (Slocum and Horvitz 2000; Carriere et al. 2002; Guevara et al. 2004; Sheldon and Nadkarni 2013), small remnant patches of vegetation (White et al. 2004; Zahawai and Augspurger 2006), or even nonnative plant species (Berens et al. 2008). To speed the recovery of forest after logging, Jansen and Zuidema (2001) suggested that selected fruiting trees be left to attract seed dispersers. Leaving seed trees after logging is common in some north temperate forestry regimes, but apparently is seldom done in tropical forests.

Although perches can increase seed input into regenerating areas, the suitability of such sites for seedling survival is less clear. Woody seedlings may have to deal with grazing animals, fires, compacted soils, and competition from thick grass cover (Duncan and Duncan 2000; Doust et al. 2006; Thaxton et al. 2010), all of which prevent establishment. However, if woody species can establish above the existing matrix, recruitment of additional bird-dispersed species is likely to increase (Holl 2002; de la Peña-Domene et al. 2013). In the Neotropics, bats disperse mainly early-successional plants into open areas (Muscarella and Fleming 2007) while birds disperse both early and late successional species. The important successional role of seed dispersal by birds can be recognized by establishing shelter belts to connect forest fragments and facilitate animal movement (Harvey 2000).

DISPERSAL OF MISTLETOES. While frugivory mutualisms are typically generalized (see above), mistletoes have an obligate need for dispersal because they must find new hosts before the current host dies. The fleshy-fruited mistletoes (i.e., all mistletoes but the Misodendraceae, which use wind, and some Viscaceae, which are ballistic), have a sticky viscin layer to glue the seed to the host branch. The viscin layer is exposed only when a frugivore removes the fruit pulp. Hence, undispersed fruits cannot adhere to host branches and are doomed (Ladley and Kelly 1996; Kelly et al. 2007). Also, the safe site for mistletoes, a small-diameter branch of a suitable host plant, is far more clearly defined than a safe site for a nonparasitic terrestrial plant (Reid 1991; Sargent 1995; Norton and Ladley 1998). Thus, mistletoes may manipulate the delivery site by favoring birds that prefer perching on suitable branches (Reid 1991). Similar specialization has been shown in an epiphytic cactus (Guaraldo et al. 2013), and it probably also occurs in other obligate epiphytes (e.g., *Coussapoa* [Urticaceae]; van Roosmalen 1985).

This specialization reaches its zenith in the Australian *Amyema quandang* (Loranthaceae; Reid 1991). The fruits are dispersed principally by the mistletoebird (*Dicaeum hirundinaceum*) and spiny-cheeked honeyeater (*Acanthagenys rufogularis*). The sticky seeds adhere to the bill when they are regurgitated, or to the cloaca when they are defecated, and must be wiped off onto branches. This means that most seeds are placed onto safe sites. Reid (1991) found that a remarkable 85% of *A. quandang* seeds passed by *D. hirundinaceum* were placed onto branches. However, in other Loranthaceae (e.g. *Peraxilla* spp in New Zealand), this wiping behavior has not been reported; the seeds are voided easily by the birds, and probably only a tiny proportion land by chance on suitable establishment sites (Kelly et al. 2007).

Effective mistletoe dispersal benefits not only the mistletoe but also the whole community. In a remarkable experiment in Australia, Watson and Herring (2012) removed all loranthaceous mistletoes (mainly *Amyema miquellii*) from 17 woodland plots and compared the changes in bird abundance over three years to the corresponding changes that occurred at untreated sites. They found a 21% decrease in bird species richness, and a 35% decrease in woodland-dependent resident birds, including species that do not feed on mistletoes. Watson and Herring (2012) argued that mistletoes increase bird diversity by producing abundant nutrient-rich litter that increases invertebrate biomass. Therefore, the subset of birds that propagate mistletoes indirectly benefits the whole avifauna (and invertebrate fauna). In Arizona, Van Ommeren and Witham (2002) found that mistletoes (*Phoradendron juniperinum*) benefit their juniper (*Juniperus monosperma*) host plants by attracting frugivorous birds that disperse both species. Juniper stands with mistletoe had twice as many juniper seedlings as did stands without mistletoe (van Ommeren and Whitham 2002). These studies suggest that mistletoes act as keystone species (Watson 2001).

Counterbalancing those benefits to various fauna, the mistletoes are hemiparasites and can negatively affect their host plants, largely through excess water demands or deformed host growth. Those best known as North American forestry pests, *Arceuthobium* spp, disperse ballistically but sometimes also adhere to feathers (Punter and Gilbert 1989). Australian loranthaceous mistletoes can harm hosts when large infestations exacerbate water stress, although in wetter New Zealand there is no evidence for negative impacts of the endemic Loranthaceae (Norton and Reid 1997).

Novel Dispersal Associations

Novel bird-plant associations arise when either birds or plants expand their ranges (Wheelwright and Orians 1982; Herrera 1985; Richardson et al. 2000). These novel associations can be beneficial—for example, if exotic birds replace lost natives—or harmful, as when weed spread is facilitated.

Exotic birds can sometimes substitute for native birds and maintain dispersal services (Aslan et al. 2012; Garcia et al. 2014) particularly when native frugivores are rare or absent. In Hawaii, where most native avian frugivores are functionally or globally extinct, the invasive Japanese white-eye (*Zosterops japonicus*) is now the primary disperser for native plants, and likely speeds the restoration of degraded areas by depositing native seeds (Foster and Robinson 2007). In the Bonin Islands of Japan, *Z. japonicus* is also invasive, but, based on fecal samples, it disperses a suite of seed species similar to that dispersed by the native Bonin Islands white-eye (*Apalopteron familiare*; Kawakami et al. 2009). In New Zealand, the silvereye (*Zosterops lateralis*) colonized naturally from Australia in 1856, and has rapidly become the most abundant native bird disperser of native fruits, making 38% of all visits (Kelly et al. 2006).

Exotic plants may provide food to help sustain native birds and thereby facilitate forest regeneration (Rodriguez 2006). In western Kenya, exotic guava (*Psidium guajava*) trees in farmland adjacent to intact forest attracted 40 species of avian frugivores (Berens et al. 2008). Seeds from 12 native (and no exotic) trees and shrubs were found in seed traps, and seedlings from 28 animal-dispersed species were found in plots underneath adult guava trees. Many late-successional species were also found as seedlings, suggesting that guava trees assist native frugivores to promote forest regeneration in abandoned farmland. Unfortunately, however, *P. guajava* is also a highly invasive species that causes widespread negative effects across the tropics (Cronk and Fuller 2001).

While some exotic plants and birds may play beneficial roles in their invaded ranges, most studies have focused on negative impacts (Buckley et al. 2006). Seed dispersal of invasive plants by either native or exotic birds clearly represents an ecosystem disservice. Birds are the most common dispersal agent of invasives, spreading 43% of invasive trees and 61% of invasive shrubs (Richardson and Rejmanek 2011). For example, the invasive tree *Melia azedarach* may rely on bird dispersal to reach beneficial microsites such as high-light and riverine areas (Voigt et al. 2011), while the Asian shrub, *Lonicera maackii*, is dispersed to suitable habitat by American robins (*Tudus migratorius*; Bartuszevige and Gorchov 2006). Birds may also disperse alien plant species that are not zoochorus in their native range (e.g., cockatoos dispersing *Pinus* spp. in Australia (Higgins and Richardson 1998)). Finally, by providing long-distance dispersal, birds allow faster spread of invasive plants (Renne et al. 2002; Buckley et al. 2006).

Exotic birds and exotic plants can create a self-catalysing "invasional meltdown." Exotic Eurasian blackbirds in Australia disperse many weed species (Williams 2006), Japanese white-eyes helped *Myrica faya*, *Bocconia frutescens*, and *Lantana camara* to invade Hawaii (Woodward et al. 1990; Chimera and Drake 2010), and red-whiskered bulbuls (*Pycnonotus jocosus*) in Hawaii disperse seeds of at least seven invasive plant species (Mandon-Dalger et al. 2004). A salient example of how an exotic bird and an exotic weed can facilitate each other's spread was shown for hawthorn (*Crataegus monogyna*, Rosaceae) in New Zealand. Hawthorn spread was slow until the first trees became tall enough to provide nest sites for Eurasian blackbirds (*Turdus merula*). Then the weed-bird system became self-reinforcing, and the invasion accelerated greatly (Williams et al. 2010).

What might give invasive plants an advantage over native plants? Invasive fruits may be more preferred than natives (Sallabanks 1993; Lafleur et al. 2007), particularly when the invasive fruit resembles a native species (Aslan and Rejmanek 2010). This may be due to higher fruit sugar content or more attractive fruiting displays (Knight 1986; Gosper et al. 2005; Aslan and Rejmanek 2010; Jordaan and Downs 2012; Mokotjomela et al. 2013b; Mokotjomela et al. 2013a). Alternatively, the invasive species may fruit during a time of low fruit abundance (Cordeiro et al. 2004; Heleno et al. 2013). Seed size can influence the rate of invasion by exotic plants by affecting which birds can act as seed dispersers (Corlett 1998; Richardson et al. 2000; Gosper et al. 2005). Invasive plants with small seeds that can be consumed by many different frugivores more readily form novel associations in their invaded ranges (Richardson et al. 2000; Renne et al. 2002; Gosper et al. 2005).

Novel bird-plant relationships can produce dilemmas for conservation (Buckley et al. 2006). The widespread invasive *Lantana camara* is dispersed by 28 native bird species. However, these species also disperse fruits or seeds from 202 native plant species that are threatened by the *Lantana* invasion (Turner and Downey 2008). Eradication of *Lantana* may require replacement planting of native plants with similar fruit characteristics and frugivore visitors (Gosper and Vivian-Smith 2009). Similarly, the abundance of native frugivorous birds in a central Pennyslvania landscape was best predicted by the abundance of two invasive shrub species (*Lonicera* spp.), thus leaving a dilemma for land managers tasked with reducing the invasive plants (Whelan and Dilger 1995; Gleditsch and Carlo 2011). Another study even recommended control or eradication of a native bird species, austral thrush (*Turdus falcklandii*), in order to slow the advance of three invasive plant species (Smith-Ramirez et al. 2013).

Conservation and Threats to Frugivory

Frugivores are threatened by human activities, including deforestation, habitat fragmentation, and direct hunting (Pimm et al. 2006). Of 10 avian feeding guilds, frugivores had the fourth highest proportion of threatened species: more than a quarter of primarily frugivorous birds (fruit being >50% of diet) were threatened or near-threatened with extinction (Sekercioglu et al. 2004). Among more than 4,000 bird species recorded by Wenny et al. (in prep.) as eating fruit, threat levels increased significantly for bird species with more highly frugivorous diets, and with larger body size. Larger-bodied birds are at even greater risk from human impact than are smaller birds, because of their greater home range sizes, lower population densities, lower reproductive rates, and greater food value for humans (Sekercioglu 2012). Given the current threat to frugivores, it is important to evaluate how the loss of frugivores might affect plant communities (Garcia and Martinez 2012).

Links between gape size in frugivorous birds and seed size (Wheelwright 1985; Moermond et al. 1986) predispose large-seeded plant species to lose the seed dispersers that are vital to their recruitment and gene flow (Howe 1984b; Terborgh 1986; Bond 1994). A review of 42 studies on disperser loss demonstrated a general pattern: large-seeded plant species have reduced dispersal and higher seed and seedling densities under parent trees (Kurten 2013; Naniwadekar et al. 2015). For six bird-dispersed tree species from India (Sethi and Howe 2009), Tanzania (Cordeiro et al. 2009), and Brazil (Galetti et al. 2013), the seedlings were substantially less abundant in areas devoid of their dispersers than in areas where the dispersers remained. Similar effects may apply to smaller-seeded plants, but those patterns may be partly masked by some shorter-distance dispersal provided by smaller bird species.

Judging the effects of disperser loss on plant recruitment, especially when multiple dispersers are involved, requires assessing both the dispersal quantity (visitation and seed-removal rates), and the dispersal quality (bird effects on seed viability, and whether seeds are delivered to favorable sites (Schupp 1993; Schupp et al. 2010). Large avian frugivores may remove many seeds and transport them in viable condition over longer distances, thus qualifying as effective dispersers (Kinnaird 1998). Smaller birds may also be effective dispersers, especially because they usually occur in greater numbers, but the seed shadows they generate are poorly known.

Case studies show that effects vary with frugivore responses to fragmentation. In Tanzania, a decrease in bird dispersers in forest fragments, in comparison to the number of dispersers in intact forest, reduced the regeneration of *Leptonychia usambarensis* trees (Cordeiro and Howe 2003; Cordeiro et al. 2009). The most frequent visitors were small-bodied generalist foragers, including a thrush and two greenbuls. In the forest fragments, the seedlings were more concentrated under parents and suffered stronger density-dependent mortality, suggesting that disrupted bird-plant mutualisms reduced plant recruitment. In contrast, for *Prunus africana* in western Kenya, seed removal was higher in forest fragments than in continuous forest, likely because of the more heterogeneous matrix habitat and much larger size of forest fragments than in the Tanzanian study (Farwig et al. 2006). By attracting moderately frugivorous species from the surrounding farmland, these forest fragments in Kenya may have more frugivorous birds than are in the fragments in Tanzania. Similarly, near Manaus, Brazil, abundant avian frugivore generalists from surrounding habitats doubled the dispersal rates for *Bocageopsis* trees (Cramer et al. 2007) in forest fragments. Attracting multiple frugivores does not necessarily yield higher rates of seed dispersal. For example, primates and hornbills in Cameroon shared 36 plant species in their diets, but the effective overlap was small because typically either hornbills or primates predominated in eating a given plant species (Poulsen et al. 2002). This limits the extent to which one disperser may be substituted by another. However, if primates were locally eliminated, the dietary overlap with hornbills might allow at least a small amount of seed dispersal to continue (Whitney et al. 1998).

Several studies have measured the impact of changes across the entire disperser fauna, with profound impact from widespread overhunting in tropical forests (Dirzo and Miranda 1991; Wright and Duber 2001). In 101 Amazonian sites at three levels of hunting intensity, 22 of 30 vertebrate species were severely reduced at the most intensively hunted sites, larger animal species declined faster, and frugivores declined faster than either seed predators or herbivores (Peres and Palacios 2007). In Ecuador, frugivore visits to *Virola flexuosa* trees in a hunted site were substantially lower than in a nonhunted site (Holbrook and Loiselle 2007). When large-bodied frugivores are lost for decades, the recruitment of large-seeded tree species decreases, while that of abiotically dispersed trees or lianas increases (Wright et al. 2007; Terborgh et al. 2008; Effiom et al. 2013). Thus, in Amazonia the density and species richness of seedlings and saplings of large-seeded trees were substantially lower in hunted sites than in protected sites (Nuñez-Iturri and Howe 2007; Terborgh et al. 2008). Substitution by other dispersers was unlikely, because all larger fruit-eating animals were equally overhunted. Thus, tropical data often show widespread dispersal limitation of large-seeded tree species.

Outside the tropics, dispersal seems less at risk. Moran et al. (2009) estimated that only 12% of Australian subtropical rain forest plant species lack appropriate dispersers in forest fragments—a smaller proportion than on other continents. They suggest that tolerance of fragmentation by many vertebrates, widespread functional overlap among frugivores, and lack of hunting or disturbance pressures were the main reasons for this difference. Similarly, in the deciduous forests of temperate eastern North America, the lack of hunting and the preponderance of small- to medium-seeded plants (defined as plants having fruits < 10 mm diameter; Wheelwright 1988) dispersed by smaller generalist avian frugivores suggest that dispersal limitation is not widespread. In warm-temperate New Zealand, Kelly et al. (2010) found almost no empirical evidence of dispersal limitation in the native flora, where most native fleshy-fruited plants have small fruits (that is, only 3% of 304 species have fruits >10 mm in diameter) and only three or four widespread native frugivorous birds survive. However, the high disturbance rates in temperate forests and the abundance of generalist frugivores may make temperate forests susceptible to invasive plants, including some dispersed by birds. It remains to be seen whether invasive plants will become a greater problem in the tropics as forest disturbance becomes more widespread.

Research Needs

We know that birds are important seed dispersers, both quantitatively (because many bird species disperse seeds) and qualitatively (because birds often disperse seeds to suitable sites for plant recruitment). However, the role of infrequent frugivores as seed dispersers is poorly known; most dispersal research has focused on highly frugivorous species, but most bird species that eat fruit do so only occasionally or infrequently. Detailed natural history research on avian diets, the consumer assemblages of plant species, and the variation in these interactions is needed to improve our understanding of how avian frugivores process seeds, which is a key aspect of dispersal quality. Increased understanding of the ecology of frugivory and seed dispersal will improve conservation plans. The responses of plant communities to change in frugivores and seed dispersers, on the one hand, and to the introduction of plants and birds, on the other, are key to effective ecosystem management. Finally, a greater understanding of the ecology of fruit-frugivore interactions will allow more refined estimates of seed dispersal as an ecosystem service. Key questions include the following: For a given plant species, what proportion of recruitment can be attributed to bird dispersal in general and to certain species in particular? Within a plant community, how does seed dispersal by birds affect community composition? With the answers to these questions, we can better assess the direct and indirect benefit to humans of avian seed dispersal. We have shown that seed dispersal is one of the most important ecological functions of birds. The challenge is to get that fact more widely appreciated.

Acknowledgments

The avian gape width measurements given here are from specimens at the Museum of Vertebrate Zoology at the University of California, Berkeley. Daniel G. Wenny thanks Carla Cicero for access to the specimens, and Malia Wenny for help with the measurements. The list of specimens used can be found online in table S5.1 at www.press.uchicago.edu/sites/whybirdsmatter/.

References

Aslan, C. E., and Rejmanek, M. 2010. Avian use of introduced plants: Ornithologist records illuminate interspecific associations and research needs. *Ecological Applications* 20:1005–20.

Aslan, C. E., Zavaleta, E. S., Croll, D., and Tershy, B. 2012. Effects of native and non-native vertebrate mutualists on plants. *Conservation Biology* 26:778–89.

Bartuszevige, A. M., and Gorchov, D. L. 2006. Avian seed dispersal of an invasive shrub. *Biological Invasions* 8:1013–22.

Bascompte, J., and Jordano, P. 2007. Plant-animal mutualistic networks: The architecture of biodiversity. *Annual Review of Ecology Evolution and Systematics* 38:567–93.

Berens, D. G., Farwig, N., Schaab, G., and Böhning-Gaese, K. 2008. Exotic guavas are foci of forest regeneration in Kenyan farmland. *Biotropica* 40:104–12.

Bond, W. J. 1994. Do mutualisms matter? Assessing the impact of pollinator and disperser disruption on plant extinction. *Philosophical Transactions of the Royal Society B-Biological Sciences* 344:83–90.

Bregman, T. P., Sekercioglu, C. H., and Tobias, J. A. 2014. Global patterns and predictors of bird species responses to forest fragmentation: Implications for ecosystem function and conservation. *Biological Conservation* 169:372–83.

Breitbach, N., Böhning-Gaese, K., Laube, I., and Schleuning, M. 2012. Short seed-dispersal distances and low seedling recruitment in farmland populations of bird-dispersed cherry trees. *Journal of Ecology* 100:1349–58.

Buckley, Y. M., Anderson, S., Catterall, C. P., Corlett, R. T., Engel, T., Gosper, C. R., Nathan, R., Richardson, D. M., Setter, M., Spiegel, O., Vivian-Smith, G., Voigt, F. A., Weir, J. E. S., and Westcott, D. A. 2006. Management of plant invasions mediated by frugivore interactions. *Journal of Applied Ecology* 43: 848–57.

Carlo, T. A., Aukema, J. E., and Morales, J. 2007. Plant-frugivore interactions as spatially explicit networks: Integrating frugivore foraging with fruiting plant spatial patterns. In *Seed Dispersal. Theory and Its Application in a Changing World*, eds. A. J. Dennis, E. W. Schupp, R. J. Green, and D. A. Westcott, 369–90. Wallingford, UK: CAB International.

Carlo, T. A., Garcia, D., Martinez, D., Gleditsch, J. M., and Morales, J. M. 2013. Where do seeds go when they go far? Distance and directionality of avian seed dispersal in heterogeneous landscapes. *Ecology* 94:301–07.

Carlquist, S. 1967. The biota of long-distance dispersal. V. Plant dispersal to Pacific islands. *Bulletin of the Torrey Botanical Club* 94:129–62.

Carriere, S. M., Andre, M., Letourmy, P., Olivier, I., and McKey, D. B. 2002. Seed rain beneath remnant trees in a slash-and-burn agricultural system in southern Cameroon. *Journal of Tropical Ecology* 18:353–74.

Chavez-Ramirez, F., and Slack, R. D. 1994. Effects of avian foraging and post-foraging behavior on seed dispersal patterns of Ashe juniper. *Oikos* 71:40–46.

Chimera, C. G., and Drake, D. R. 2010. Patterns of seed dispersal and dispersal failure in a Hawaiian dry forest having only introduced birds. *Biotropica* 42: 493–502.

Clark, C. J., Poulsen, J. R., Bolker, B. M., Connor, E. F., and Parker, V. T. 2005. Comparative seed shadows of bird-, monkey-, and wind-dispersed trees. *Ecology* 86:2684–94.

Cordeiro, N. J., and Howe, H. F. 2003. Forest fragmentation severs mutualism between seed dispersers and an endemic African tree. *Proceedings of the National Academy of Sciences of the United States of America* 100:14052–56.

Cordeiro, N. J., Ndangalasi, H. J., McEntee, J. P., and Howe, H. F. 2009. Disperser limitation and recruitment of an endemic African tree in a fragmented landscape. *Ecology* 90:1030–41.

Cordeiro, N. J., Patrick, D. A. G., Munisi, B., and Gupta, V. 2004. Role of dispersal in the invasion of an exotic tree in an East African submontane forest. *Journal of Tropical Ecology* 20:449–57.

Corlett, R. T. 1998. Frugivory and seed dispersal by vertebrates in the Oriental (Indomalayan) Region. *Biological Reviews of the Cambridge Philosophical Society* 73:413–48.

———. 2009. Seed dispersal distances and plant migration potential in tropical east Asia. *Biotropica* 41:592–98.

Costanza, R., d'Arge, R., de Groot, R., Farber, S., Grasso, M., Hannon, B., Limburg, K., Naeem, S., O'Neill, R. V., Paruelo, J., Raskin, R. G., Sutton, P., and van den Belt, M. 1997. The value of the world's ecosystem services and natural capital. *Nature* 387:253–60.

Cramer, J. M., Mesquita, R. C. G., and Williamson, G. B. 2007. Forest fragmentation differentially affects seed dispersal of large and small-seeded tropical trees. *Biological Conservation* 137:415–23.

Cronk, Q. C. B., and Fuller, J. L. 2001. *Plant Invaders: The Threat to Natural Ecosystems*. London: Earthscan Publications.

Darwin, C. 1859. *The Origin of Species by Means of Natural Selection*. New York: Doubleday.

DaSilva, J. M. C., Uhl, C., and Murray, G. 1996. Plant succession, landscape management, and the ecology of frugivorous birds in abandoned Amazonian pastures. *Conservation Biology* 10:491–503.

De la Peña-Domene, M., Martínez-Garza, C., and Howe, H. F. 2013. Early recruitment dynamics in tropical restoration. *Ecological Applications* 23:1124–34.

De la Peña-Domene, M., Martínez-Garza, C., Palmas-Pérez, S., Alonso, E. R., and Howe, H. F. 2014. Birds and bats play different roles in early tropical forest restoration. *PLoS ONE* 9:e104656.

De Melo, F. P. L., Dirzo, R., and Tabarelli, M. 2006. Biased seed rain in forest edges: Evidence from the Brazilian Atlantic forest. *Biological Conservation* 132:50–60.

Debussche, M., and Isenmann, P. 1994. Bird-dispersed seed rain and seedling establishment in patchy Mediterranean vegetation. *Oikos* 69:414–26.

Deforesta, H., Charles-Dominique, P., Erard, C., and Prevost, M. F. 1984. The importance of zoochory during the first stages of forest regeneration after clearcutting in French Guyana. *Revue d'Ecologie-La Terre et la Vie* 39:369–400.

Dirzo, R., and Miranda, A. 1991. Altered patterns of herbivory and diversity in the forest understory: A case study of the possible consequences of contemporary defaunation. In *Plant-Animal Interactions: Evolutionary Ecology in Tropical and Temperate Regions*, ed. P. W. Price, T. M. Lewisohn, G. W. Ferandes, and W. W. Benson, 273–87. New York: John Wiley & Sons.

Dosch, J. J., Peterson, C. J., and Haines, B. L. 2007. Seed rain during initial colonization of abandoned pastures in the premontane wet forest zone of southern Costa Rica. *Journal of Tropical Ecology* 23:151–59.

Doust, S. J., Erskine, P. D., and Lamb, D. 2006. Direct seeding to restore rainforest species: Microsite effects on the early establishment and growth of rainforest tree seedlings on degraded land in the wet tropics of Australia. *Forest Ecology & Management* 234:333–43.

Duncan, R. S., and Duncan, V. E. 2000. Forest succession and distance from forest edge in an Afro-tropical grassland. *Biotropica* 32:33–41.

Effiom, E. O., Nuñez-Iturri, G., Smith, H. G., Ottosson, U., and Olsson, O. 2013. Bushmeat hunting changes regeneration of African rainforests. *Proceedings of the Royal Society B-Biological Sciences* 280 DOI: 20130246 10.1098/rspb.2013.0246.

Farwig, N., Böhning-Gaese, K., and Bleher, B. 2006. Enhanced seed dispersal of *Prunus africana* in fragmented and disturbed forests? *Oecologia* 147:238–52.

Fischer, K. E., and Chapman, C. A. 1993. Frugivores and fruit syndromes: Differences in patterns at the genus and species level. *Oikos* 66:472–82.

Fleming, T. H., Venable, D. L., and Herrera M., L. G. 1993. Opportunism vs. specialization: The evolution of dispersal strategies in fleshy-fruited plants. *Vegetatio* 107/108:107–20.

Forget, P. M., Dennis, A. J., Mazer, S. J., Jansen, P. A., Kitamura, S., Lambert, J. E., and Westcott, D. A. 2007. Seed allometry and disperser assemblages in tropical rainforests: a comparison of four floras on different continents. In *Seed Dispersal. Theory and Its Application in a Changing World*, ed. A. J. Dennis, E. W. Schupp, R. J. Green, and D. A. Westcott, 5–36. Wallingford, UK: CAB International.

Foster, J. T., and Robinson, S. K. 2007. Introduced birds and the fate of Hawaiian rainforests. *Conservation Biology* 21:1248–57.

Fricke, E. C., Simon, M. J., Reagan, K. M., Levey, D. J., Riffell, J. A., Carlo, T. A., and Tewksbury, J. J. 2014. When condition trumps location: Seed consumption by fruit-eating birds removes pathogens and predator attractants. *Ecology Letters* 16:1031–36.

Fulbright, T. E., Kuti, J. O., and Tipton, A. R. 1995. Effects of nurse-plant canopy temperatures on shrub seed germination and seedling growth. *Acta Oecologia* 16:621–32.

Galetti, M. A. et al. 2013. Functional extinction of birds drives rapid evolutionary changes in seed size. *Science* 340:1086–90.

Garcia, D., and Martinez, D. 2012. Species richness matters for the quality of ecosystem services: a test using seed dispersal by frugivorous birds. *Proceedings of the Royal Society B-Biological Sciences* 279:3106–13.

Garcia, D., Martinez, D., Stouffer, D. B., and Tylianakis, J. M. 2014. Exotic birds increase generalization and compensate for native bird decline in plant-frugivore assemblages. *Journal of Animal Ecology* 83:1441–50.

Gilbert, L. E. 1980. Food web organization and conservation of Neotropical diversity. In *Conservation Biology: An Evolutionary-Ecological Perspective*, ed. M. E. Soulé, and B. A. Wilcox, 11–33. Sunderland, MA: Sinauer Associates.

Gleditsch, J. M., and Carlo, T. A. 2011. Fruit quantity of invasive shrubs predicts the abundance of common native avian frugivores in central Pennsylvania. *Diversity and Distributions* 17:244–53.

Godoy, J. A., and Jordano, P. 2001. Seed dispersal by animals: Exact identification of source trees with endocarp DNA microsatellites. *Molecular Ecology* 10: 2275–83.

Godoy, R., Wilkie, D., Overman, H., Cubas, A., Cubas, G., Demmer, J., McSweeney, K., and Brokaw, N. 2000. Valuation of consumption and sale of forest goods from a Central American rain forest. *Nature* 406:62–63.

Gosper, C. R., Stansbury, C. D., and Vivian-Smith, G. 2005. Seed dispersal of fleshy-fruited invasive plants by birds: contributing factors and management options. *Diversity & Distributions* 11:549–58.

Gosper, C. R., and Vivian-Smith, G. 2009. Approaches to selecting native plant replacements for fleshy-fruited invasive species. *Restoration Ecology* 17:196–204.

Guaraldo, A. D., Boeni, B. D., and Pizo, M. A. 2013. Specialized seed dispersal in epiphytic cacti and convergence with mistletoes. *Biotropica* 45:465–73.

Guevara, S., Laborde, J., and Sanchez-Rios, G. 2004. Rain forest regeneration beneath the canopy of fig trees isolated in pastures of Los Tuxtlas, Mexico. *Biotropica* 36:99–108.

Hamilton, W. D., and May, R. M. 1977. Dispersal in stable habitats. *Nature* 269: 578–81.

Harms, K. E., Wright, S. J., Calderon, O., Hernandez, A., and Herre, E. A. 2000. Pervasive density-dependent recruitment enhances seedling diversity in a tropical forest. *Nature* 404:493–95.

Harvey, C. A. 2000. Windbreaks enhance seed dispersal into agricultural landscapes in Monteverde, Costa Rica. *Ecological Applications* 10:155–73.

Heleno, R. H., Ramos, J. A., and Memmott, J. 2013. Integration of exotic seeds into an Azorean seed dispersal network. *Biological Invasions* 15:1143–54.

Herrera, C. M. 1985. Determinants of plant-animal coevolution: The case of mutualistic dispersal of seeds by vertebrates. *Oikos* 44:132–41.

———. 1986. Vertebrate-dispersed plants: Why they don't behave the way they should. In *Frugivores and Seed Dispersal*, ed. A. Estrada, and T. H. Fleming, 5–18. Dordrecht: Dr. W. Junk Publishers.

———. 1987. Vertebrate-dispersed plants of the Iberean peninsula: A study of fruit characteristics. *Ecological Monographs* 57:305–31.

———. 1992. Interspecific variation in fruit shape: Allometry, phylogeny, and adaptation to dispersal agents. *Ecology* 73:1832–41.

———. 2002. Seed dispersal by vertebrates. In *Plant-Animal Interactions*, ed. C. M. Herrera, and O. Pellmyr, 185–208, Oxford, UK: Blackwell Science Ltd.

Higgins, S. I., and Richardson, D. M. 1998. Pine invasions in the southern hemisphere: Modelling interactions between organism, environment and disturbance. *Plant Ecology* 135:79–93.

Holbrook, K. M., and Loiselle, B. A. 2007. Using toucan-generated dispersal models to estimate seed dispersal in Amazonian Ecuador. In *Seed Dispersal. Theory and Its Application in a Changing World*, ed. A. J. Dennis, E. W. Schupp, R. J. Green, and D. A. Westcott, 300–321. Wallingford, UK: CAB International.

Holl, K. D. 2002. Long-term vegetation recovery on reclaimed coal surface mines in the eastern USA. *Journal of Applied Ecology* 39:960–70.

Hoppes, W. G. 1988. Seedfall pattern of several species of bird-dispersed plants in an Illinois woodland. *Ecology* 69:320–29.

Howe, H. F. 1984a. Constraints on the evolution of mutualisms. *American Naturalist* 123:764–77.

———. 1984b. Implications of seed dispersal by animals for tropical reserve management. *Biological Conservation* 30:261–81.

———. 1986. Seed dispersal by fruit-eating birds and mammals. In *Seed Dispersal*, ed. D. R. Murray, 123–89. Sydney: Academic Press.

Howe, H. F., and Miriti, M. N. 2004. When seed dispersal matters. *Bioscience* 54:651–60.

Howe, H. F., and Smallwood, J. 1982. Ecology of seed dispersal. *Annual Review of Ecology and Systematics* 13:201–28.

Izhaki, I., Walton, P. B., and Safriel, U. N. 1991. Seed shadows generated by frugivorous birds in an eastern Mediterranean scrub. *Journal of Ecology* 79:575–90.

Jansen, P. A., and Zuidema, P. 2001. Logging, seed dispersal by vertebrates, and natural regeneration of tropical timber trees. In *The Cutting Edge: Conserving Wildlife in Logged Tropical Forests*, ed. R. Fimbel, J. Robinson, and A. Grajal, 35–59. New York: Columbia University Press.

Janzen, D. H. 1983a. Dispersal of seeds by vertebrate guts. In *Coevolution*, ed. D. J. Futuyma, and M. Slatkin, 232–62. Sunderland, MA: Sinauer Associates.

———. 1983b. Seed and pollen dispersal by animals: Convergence in the ecology of contamination and sloppy harvest. *Biological Journal of the Linnean Society* 20:103–13.

Jordaan, L. A., and Downs, C. T. 2012. Nutritional and morphological traits of invasive and exotic fleshy-fruits in South Africa. *Biotropica* 44:738–43.

Jordano, P. 1995. Angiosperm fleshy fruits and seed dispersers: A comparative analysis of adaptation and constraints in plant-animal interactions. *American Naturalist* 145:163–91.

———. 2000. Fruits and frugivory. In *Seeds: The Ecology of Regeneration in Plant Communities*, ed. M. Fenner, 125–65.

Jordano, P., Garcia, C., Godoy, J. A., and Garcia-Castano, J. L. 2007. Differential contribution of frugivores to complex seed dispersal patterns. *Proceedings of the National Academy of Sciences of the United States of America* 104:3278–82.

Kareiva, P., Tallis, H., Ricketts, T. H., Daily, G. C., and Polasky, S., eds. 2011. *Natural Capital: Theory and practice of mapping ecosystem services*. Oxford, UK: Oxford University Press.

Kawakami, K., Mizusawa, L., and Higuchi, H. 2009. Re-established mutualism in a seed-dispersal system consisting of native and introduced birds and plants on the Bonin Islands, Japan. *Ecological Research* 24:741–48.

Kays, R., Jansen, P. A., Knecht, E. M. H., Vohwinkel, R., and Wikelski, M. 2011. The effect of feeding time on dispersal of *Virola* seeds by toucans determined from GPS tracking and accelerometers. *Acta Oecologica-International Journal of Ecology* 37:625–31.

Kelly, D., Ladley, J. J., and Robertson, A. W. 2004. Is dispersal easier than pollination? Two tests in new Zealand Loranthaceae. *New Zealand Journal of Botany* 42:89–103.

———. 2007. Is the pollen-limited mistletoe *Peraxilla tetrapetala* (Loranthaceae) also seed limited? *Austral Ecology* 32:850–57.

Kelly, D., Ladley, J. J., Robertson, A. W., Anderson, S. H., Wotton, D. M., and Wiser, S. K. 2010. Mutualisms with the wreckage of an avifauna: The status of bird pollination and fruit-dispersal in New Zealand. *New Zealand Journal of Ecology* 34:66–85.

Kelly, D., Ladley, J. J., Robertson, A. W., and Norton, D. A. 2000. Limited forest fragmentation improves reproduction in the declining New Zealand mistletoe *Peraxilla tetrapetala* (Loranthaceae). In *Genetics, Demography and Viability of Fragmented Populations*, ed. A. C. Young and G. Clarke, 241–52. Cambridge: Cambridge University Press.

Kelly, D., Roberston, A., Ladley, J. J., Anderson, S., and McKenzie, R. 2006. The relative (un)importance of introduced animals as pollinators and dispersers of native plants. In *Biological invasions in New Zealand*, ed. R. Allen, and W. Lee, 227–45. Berlin: Springer.

Knight, R. S. 1986. Fruit displays of indigenous and invasive alien plants in the southwestern cape. *South African Journal of Botany* 52:249–55.

Kurten, E. L. 2013. Cascading effects of contemporaneous defaunation on tropical forest communities. *Biological Conservation* 163:22–32.

Ladley, J. J., and Kelly, D. 1996. Dispersal, germination, and survival of New Zealand mistletoes (Loranthaceae): dependence on birds. *New Zealand Journal of Ecology* 20:69–79.

Lafleur, N. E., Rubega, M. A., and Elphick, C. S. 2007. Invasive fruits, novel foods, and choice: An investigation of European starling and American Robin frugivory. *Wilson Journal of Ornithology* 119:429–38.

Lees, A. C., and Peres, C. A. 2009. Gap-crossing movements predict species occupancy in Amazonian forest fragments. *Oikos* 118:280–90.

Lenz, J., Fiedler, W., Caprano, T., Friedrichs, W., Gaese, B. H., Wikelski, M., and Böhning-Gaese, K. 2011. Seed-dispersal distributions by trumpeter hornbills in fragmented landscapes. *Proceedings of the Royal Society B-Biological Sciences* 278:2257–64.

Levey, D. J. 1988. Tropical wet forest treefall gaps and distributions of understory birds and plants. *Ecology* 69:1076–89.

Levey, D. J., Bolker, B. M., Tewksbury, J. J., Sargent, S., and Haddad, N. M. 2005. Effects of landscape corridors on seed dispersal by birds. *Science* 309:146–48.

Levey, D. J., Moermond, T. C., and Denslow, J. S. 1994. Frugivory: An overview. In *La Selva: Ecology and Natural History of a Neotropical Rainforest*, ed. L. A. McDade, K. S. Bawa, H. A. Hespenheide, and G. S. Hartshorn, 282–94. Chicago: University of Chicago Press.

Levey, D. J., Tewksbury, J. J., Cipollini, M. L., and Carlo, T. A. 2006. A field test of the directed deterrence hypothesis in two species of wild chili. *Oecologia* 150:61–68.

Lindell, C., Reid, J., and Cole, R. J. 2013. Planting design effects on avian seed dispersers in a tropical forest restoration experiment. *Restoration Ecology* 21:1–8.

Lomáscolo, S. B., Levey, D. J., Kimball, R. T., Bolker, B. M., and Alborn, H. T. 2012. Dispersers shape fruit diversity in *Ficus* (Moraceae). *Proceedings of the National Academy of Sciences of the United States of America* 107:14668–72.

Lomáscolo, S. B., Speranza, P., and Kimball, R. T. 2008. Correlated evolution of fig size and color supports the dispersal syndromes hypothesis. *Oecologia* 156: 783–96.

Lord, J. M. 2004. Frugivore gape size and the evolution of fruit size and shape in southern hemisphere floras. *Austral Ecology* 29:430–36.

Lundberg, J., and Moberg, F. 2003. Mobile link organisms and ecosystem functioning: Implications for ecosystem resilience and management. *Ecosystems* 6:87–98.

Mack, A. L. 1995. Distance and non-randomness of seed dispersal by the dwarf cassowary (*Casuarius bennetti*). *Ecography* 18:286–95.

Mandon-Dalger, I., Clergeau, P., Tassin, J., Riviere, J. N., and Gatti, S. 2004. Relationships between alien plants and an alien bird species on Reunion Island. *Journal of Tropical Ecology* 20:635–42.

Martinez-Garza, C., and Gonzalez-Montagut, R. 1999. Seed rain from forest fragments into tropical pastures in Los Tuxtlas, Mexico. *Plant Ecology* 145:255–65.

Martinez-Garza, C., and Howe, H. F. 2003. Restoring tropical diversity: Beating the time tax on species loss. *Journal of Applied Ecology* 40:423–29.

Martínez-Ramos, M., and Alvarez-Buylla, E. R. 1995. Seed dispersal and patch dynamics in tropical rain forests: A demographic approach. *Ecoscience* 2:223–29.

McClanahan, T. R., and Wolfe, R. W. 1987. Dispersal of ornithochorous seeds from forest edges in central Florida. *Vegetatio* 71:107–12.

———. 1993. Accelerating forest succession in a fragmented landscape: The role of birds and perches. *Conservation Biology* 7:279–88.

McKey, D. 1975. The ecology of coevolved seed dispersal systems. In *Coevolution of Plants and Animals*, ed. L. E. Gilbert, and P. H. Raven, 159–91. Austin: University of Texas Press.

Millenium Ecosystem Assessment. 2005. *Ecosystems and Human Well-Being: Synthesis*. Washington: Island Press.

Moermond, T. C., and Denslow, J. S. 1983. Fruit choice in Neotropical birds: Effects of fruit type and accessibility on selectivity. *Journal of Animal Ecology* 52:407–20.

Moermond, T. C., Denslow, J. S., Levey, D. J., and Santana C., E. 1986. The influence of morphology on fruit choice in Neotropical birds. In *Frugivores and Seed Dispersal*, ed. A. Estrada, and T. H. Fleming, 137–46. Dordrecht: Dr. W. Junk Publishers.

Mokotjomela, T. M., Musil, C. F., and Esler, K. J. 2013a. Do frugivorous birds concentrate their foraging activities on those alien plants with the most abundant and nutritious fruits in the South African Mediterranean–climate region? *Plant Ecology* 214:49–59.

———. 2013b. Frugivorous birds visit fruits of emerging alien shrub species more frequently than those of native shrub species in the South African Mediterranean climate region. *South African Journal of Botany* 86:73–78.

Moran, C., Catterall, C. P., and Kanowski, J. 2009. Reduced dispersal of native plant species as a consequence of the reduced abundance of frugivore species in fragmented rainforest. *Biological Conservation* 142:541–52.

Muscarella, R., and Fleming, T. H. 2007. The role of frugivorous bats in tropical forest succession. *Biological Reviews* 82:573–90.

Naniwadekar, R., Shukla, U., Isvaran, K., and Datta, A. 2015. Reduced hornbill abundance associated with low seed arrival and altered recruitment in a hunted and logged tropical forest. *PLoS ONE* 10:e0120062.

Ne'eman, G., and Izhaki, I. 1996. Colonization in an abandoned East-Mediterranean vineyard. *Journal of Vegetation Science* 7:465–72.

Nepstad, D. C., Uhl, C., Pereira, C. A., and da Silva, J. M. C. 1996. A comparative study of tree establishment in abandoned pasture and mature forest of eastern Amazonia. *Oikos* 76:25–39.

Ninan, K. N., and Inoue, M. 2013. Valuing forest ecosystem services: What we know and what we don't. *Ecological Economics* 93:137–49.

Norton, D., and Ladley, J. J. 1998. Establishment and early growth of *Alepis flavida* in relation to *Nothofagus solandri* branch size. *New Zealand Journal of Botany* 36:213–17.

Norton, D. A., and Reid, N. 1997. Lessons in ecosystem management from management of threatened and pest loranthraceous mistletoes in New Zealand and Australia. *Conservation Biology* 11:759–69.

Nunez, C. I., Aizen, M. A., and Ezcurra, C. 1999. Species associations and nurse plant effects in patches of high-Andean vegetation. *Journal of Vegetation Science* 10:357–64.

Nuñez-Iturri, G., and Howe, H. F. 2007. Bushmeat and the fate of trees with seeds dispersed by large primates in a lowland rain forest in western Amazonia. *Biotropica* 39:348–54.

Paoli, G. D., Peart, D. R., Leighton, M., and Samsoedin, I. 2001. An ecological and economic assessment of the nontimber forest product gaharu wood in Gunung Palung National Park, West Kalimantan, Indonesia. *Conservation Biology* 15:1721–32.

Peres, C. A., and Palacios, E. 2007. Basin-wide effects of game harvest on vertebrate population densities in Amazonian forests: Implications for animal-mediated seed dispersal. *Biotropica* 39:304–15.

Peters, C., Gentry, A. H., and Mendelsohn, R. 1989. Valuation of an Amazonian rainforest. *Nature* 339:655–56.

Pimm, S., Raven, P., Peterson, A., Sekercioglu, C. H., and Ehrlich, P. R. 2006. Human impacts on the rates of recent, present, and future bird extinctions. *Proceedings of the National Academy of Sciences of the United States of America* 103:10941–46.

Poulsen, J. R., Clark, C. J., Connor, E. F., and Smith, T. B. 2002. Differential resource use by primates and hornbills: Implications for seed dispersal. *Ecology* 83:228–40.

Punter, D., and Gilbert, J. 1989. Animal vectors of *Arceuthobium americanum* seed in Manitoba. *Canadian Journal of Forest Research* 19:865–69.

Reid, N. 1991. Coevolution of mistletoes and frugivorous birds? *Australian Journal of Ecology* 16:457–69.

Renne, I. J., Barrow, W. C., Randall, L. A. J., and Bridges, W. C. 2002. Generalized avian dispersal syndrome contributes to Chinese tallow tree (*Sapium sebiferum*, Euphorbiaceae) invasiveness. *Diversity and Distributions* 8:285–95.

Richardson, B. A., Brunsfeld, J., and Klopfenstein, N. B. 2002. DNA from bird-dispersed seed and wind-disseminated pollen provides insights into postglacial colonization and population genetic structure of Whitebark Pine (*Pinus albicaulis*). *Molecular Ecology* 11:215–27.

Richardson, D. M., Allsopp, N., D'Antonio, C. M., Milton, S. J., and Rejmanek, M. 2000. Plant invasions: The role of mutualisms. *Biological Reviews* 75:65–93.

Richardson, D. M., and Rejmanek, M. 2011. Trees and shrubs as invasive alien species: A global review. *Diversity and Distributions* 17:788–809.

Ridley, H. N. 1930. *The Dispersal of Plants throughout the World*. Ashford UK: Reeve.

Robertson, A. W., Trass, A., Ladley, J. J., and Kelly, D. 2006. Assessing the benefits of frugivory for seed germination: The importance of the deinhibition effect. *Functional Ecology* 20:58–66.

Robinson, G. R., and Handel, S. N. 2000. Directing spatial patterns of recruitment during an experimental urban woodland reclamation. *Ecological Applications* 10:174–88.

Rodriguez, L. F. 2006. Can invasive species facilitate native species? Evidence of how, when, and why these impacts occur. *Biological Invasions* 8:927–39.
Sallabanks, R. 1993. Hierachical mechanisms of fruit selection by an avian frugivore. *Ecology* 74:1326–36.
Samuels, I. A., and Levey, D. J. 2005. Effects of gut passage on seed germination: Do experiments answer the questions they ask? *Functional Ecology* 19:365–68.
Sargent, S. 1995. Seed fate in a tropical mistletoe: The importance of host twig size. *Functional Ecology* 9:197–204.
Schemske, D. W., and Brokaw, N. V. L. 1981. Treefalls and the distribution of understory birds in a tropical forest. *Ecology* 62:938–45.
Schupp, E. W. 1993. Quantity, quality and the effectiveness of seed dispersal by animals. *Vegetatio* 107/108:15–29.
Schupp, E. W., Jordano, P., and Gomez, J. M. 2010. Seed dispersal effectiveness revisited: A conceptual review. *New Phytologist* 188:333–53.
Sekercioglu, C. H. 2006a. Ecological significance of bird populations. In *Handbook of the Birds of the World*, ed. J. del Hoyo, A. Elliott, and D. Christie, 15–51. Barcelona: Lynx Edicions.
———. 2006b. Increasing awareness of avian ecological function. *Trends in Ecology & Evolution* 21:464–71.
———. 2012. Bird functional diversity and ecosystem services in tropical forests, agroforests and agricultural areas. *Journal of Ornithology* 153:S153–S161.
Sekercioglu, C. H., Daily, G. C., and Ehrlich, P. R. 2004. Ecosystem consequences of bird declines. *Proceedings of the National Academy of Sciences U S A* 101: 18042–47.
Sethi, P., and Howe, H. F. 2009. Recruitment of hornbill-dispersed trees in hunted and logged forests of the Indian Eastern Himalaya. *Conservation Biology* 23: 710–18.
Shanahan, M., Harrison, R. D., Yamuna, R., Boen, W., and Thornton, I. W. B. 2001. Colonization of an island volcano, Long Island, Papua New Guinea, and an emergent island, Motmot, in its caldera lake. V. Colonization by figs (*Ficus* spp.), their dispersers and pollinators. *Journal of Biogeography* 28:1365–77.
Sheldon, K., and Nadkarni, N. 2013. The use of pasture trees by birds in a tropical montane landscape in Monteverde, Costa Rica. *Journal of Tropical Ecology* 29:459–62.
Shiels, A. B., and Walker, L. R. 2003. Bird perches increase forest seeds on Puerto Rican landslides. *Restoration Ecology* 11:457–65.
Sieving, K. E., Willson, M. F., and DeSanto, T. L. 1996. Habitat barriers to movement of understory birds in fragmented south-temperate rainforest. *Auk* 113:944–49.
Slocum, M. G., and Horvitz, C. C. 2000. Seed arrival under different genera of trees in a Neotropical pasture. *Plant Ecology* 149:51–62.
Smith-Ramirez, C., Arellano, G., Hagen, E., Vargas, R., Castillo, J., and Miranda, A. 2013. The role of *Turdus falcklandii* (Aves: Passeriformes) as disperser

of invasive plants in the Juan Fernandez Archipelago. *Revista Chilena de Historia Natural* 86:33–48.

Smith, N. J. H., Williams, J. T., Plucknett, D. L., and Talbot, J. P. 1992. *Tropical Forests and their crops*. Ithaca, NY: Cornell University Press.

Snow, D. W. 1971. Evolutionary aspects of fruit-eating by birds. *Ibis* 113:194–202.

Sodhi, N. S., Şekercioğlu, Ç. H., Robinson, S., Barlow, J. 2011. *Conservation of Tropical Birds*. Wiley-Blackwell, Oxford.

Spiegel, O., and Nathan, R. 2012. Empirical evaluation of directed dispersal and density-dependent effects across successive recruitment phases. *Journal of Ecology* 100:392–404.

Stewart, K. 2009. Effects of bark harvest and other human activity on populations of the African cherry (*Prunus africana*) on Mount Oku, Cameroon. *Forest Ecology and Management* 258:1121–28.

Terborgh, J. 1986. Community aspects of frugivory in tropical forests. In *Frugivores and seed dispersal*, ed. A. Estrada and T. H. Fleming, 371–84. Dordrecht: Dr. W. Junk Publishers.

———. 2012. Enemies maintain hyperdiverse tropical forests. *American Naturalist* 179:303–14.

Terborgh, J., Nunez-Iturri, G., Pitman, N. C. A., Valverde, F. H. C., Alvarez, P., Swamy, V., Pringle, E. G., and Paine, C. E. T. 2008. Tree recruitment in an empty forest. *Ecology* 89:1757–68.

Tester, M., Patton, D. C., Reid, N., and Lange, R. T. 1987. Seed dispersal by birds and densities of shrubs under trees in arid south Australia. *Transactions of the Royal Society of South Australia* 111:1–5.

Tewksbury, J. J., Nabhan, G. P., Norman, D., Suzan, H., Tuxhill, J., and Donovan, J. 1999. In situ conservation of wild chilies and their biotic associates. *Conservation Biology* 13:98–107.

Thaxton, J. M., Cole, T. C., Cordell, S., Cabin, R. J., Sandquist, D. R., and Litton, C. M. 2010. Native species regeneration following ungulate exclusion and non-native grass removal in a remnant Hawaiian dry forest. *Pacific Science* 64:533–44.

Thompson, J. N., and Willson, M. F. 1978. Disturbance and the dispersal of fleshy fruits. *Science* 200:1161–63.

Traveset, A., Robertson, A. W., and Rodriguéz-Peréz, J. 2007. A review on the role of endozoochory in seed germination. In *Seed Dispersal: Theory and Its Application in a Changing World*, ed. A. J. Dennis, E. W. Schupp, R. J. Green, and D. A. Westcott, 78–103. Wallingford, UK: CAB International.

Turner, P. J., and Downey, P. O. 2008. The role of native birds in weed invasion, species decline, revetation and reinvasion: Consequences for lantana management. *Proceedings of the Australian Weed Conference* 16:30–32.

Van der Pijl, L. 1972. *Principles of Dispersal in Higher Plants*. Berlin: Springer-Verlag.

Van Ommeren, R., and Whitham, T. G. 2002. Changes in interactions between juniper and mistletoe mediated by shared avian frugivores: Parasitism to potential mutualism. *Oecologia* 130:281–88.

Van Roosmalen, M. G. M. 1985. *Fruits of the Guianan Flora*. Wageningen, Netherlands: Institute of Systematic Botany, Utrecht University.

Vargas, P., Heleno, R., Traveset, A., and Nogales, M. 2012. Colonization of the Galápagos Islands by plants with no specific syndromes for long-distance dispersal: A new perspective. *Ecography* 35:33–43.

Vitousek, P. M., Mooney, H. A., Lubchenco, J., and Melillo, J. M. 1997. Human domination of Earth's ecosystems. *Science* 277:494–99.

Voigt, F. A., Farwig, N., and Johnson, S. D. 2011. Interactions between the invasive tree *Melia azedarach* (Meliaceae) and native frugivores in South Africa. *Journal of Tropical Ecology* 27:355–63.

Watson, D. M. 2001. Mistletoe: A keystone resource in forests and woodlands worldwide. *Annual Review of Ecology and Systematics* 32:219–49.

Watson, D. M., and Herring, M. 2012. Mistletoe as a keystone resource: An experimental test. *Proceedings of the Royal Society B-Biological Sciences* 279: 3853–60.

Weir, J. E. S., and Corlett, R. T. 2007. How far do birds disperse seeds in the degraded tropical landscape of Hong Kong, China? *Landscape Ecology* 22:131–40.

Wenny, D. G. 2000. Seed dispersal of a high quality fruit by specialized frugivores: High quality dispersal? *Biotropica* 32:327–37.

Wenny, D. G., DeVault, T. L., Johnson, M. D., Kelly, D., Sekercioglu, C. H., Tomback, D. F., and Whelan, C. J. 2011. The need to quantify ecosystem services provided by birds. *Auk* 128:1–14.

Wenny, D. G., and Levey, D. J. 1998. Directed seed dispersal by bellbirds in a tropical cloud forest. *Proceedings of the National Academy of Sciences, USA* 95:6204–7.

Wheelwright, N. T. 1985. Fruit size, gape width, and the diets of fruit-eating birds. *Ecology* 66:808–18.

———. 1988. Fruit-eating birds and bird-dispersed plants in the tropics and temperate zone. *Trends in Ecology and Evolution* 3:270–74.

Wheelwright, N. T., and Orians, G. H. 1982. Seed dispersal by animals: Contrasts with pollen dispersal, problems of terminology, and constraints on coevolution. *American Naturalist* 119:402–13.

Whelan, C. J., and Dilger, M. L. 1995. Exotic, invasive shrubs: A paradox revisited. *Natural Areas Journal* 15:296.

Whelan, C. J., Wenny, D. G., and Marquis, R. J. 2008. Ecosystem services provided by birds. *Annals of the New York Academy of Sciences* 1134:25–60.

White, E., Tucker, N., Meyers, N., and Wilson, J. 2004. Seed dispersal to revegetated isolated rainforest patches in North Queensland. *Forest Ecology & Management* 192:409–26.

Whitney, K. D., Fogiel, M. K., Lamperti, A. M., Holbrook, K. M., Stauffer, D. J., Hardesty, B. D., Parker, V. T., and Smith, T. B. 1998. Seed dispersal by *Ceratogymna* hornbills in the Dja reserve, Cameroon. *Journal of Tropical Ecology* 14:351–71.

Williams, P., Kean, J., and Buxton, R. 2010. Multiple factors determine the rate of increase of an invading non-native tree in New Zealand. *Biological Invasions* 12:1377–88.

Williams, P. A. 2006. The role of blackbirds (*Turdus merula*) in weed invasion in New Zealand. *New Zealand Journal of Ecology* 30:285–91.

Willson, M., and Crome, F. H. J. 1989. Patterns of seed rain at the edge of a tropical Queensland rain forest. *Journal of Tropical Ecology* 5:301–8.

Woodward, S., Vitousek, P., Matson, K., F, H., Benvenuto, K., and Matson, P. A. 1990. Use of the exotic tree *Myrica faya* by native and exotic birds in Hawai'i Volcanoes National Park. *Pacific Science* 44:88–93.

Wotton, D. M., and Kelly, D. 2012. Do larger frugivores move seeds further? Body size, seed dispersal distance, and a case study of a large, sedentary pigeon. *Journal of Biogeography* 39:1973–83.

Wright, S. J., and Duber, H. C. 2001. Poachers and forest fragmentation alter seed dispersal, seed survival, and seedling recruitment in the palm *Attalea butyraceae*, with implications for tropical tree diversity. *Biotropica* 33:583–95.

Wright, S. J., Hernandez, A., and Condit, R. 2007. The bushmeat harvest alters seedling banks by favoring lianas, large seeds, and seeds dispersed by bats, birds, and wind. *Biotropica* 39:363–71.

Zahawi, R. A., and Augspurger, C. K. 2006. Tropical forest restoration: Tree islands as recruitment foci in degraded lands of Honduras. *Ecological Applications* 16:464–78.

Zanini, L., and Ganade, G. 2005. Restoration of Araucaria forest: The role of perches, pioneer vegetation, and soil fertility. *Restoration Ecology* 13:507–14.

CHAPTER SIX

Dispersal of Plants by Waterbirds

Andy J. Green, Merel Soons, Anne-Laure Brochet, and Erik Kleyheeg

The widespread distribution of fresh-water plants and of the lower animals, whether retaining the same identical form or in some degree modified, I believe mainly depends on the wide dispersal of their seeds and eggs by animals, more especially by fresh-water birds, which have large powers of flight, and naturally travel from one to another and often distant piece of water. —Charles Darwin (1859)

Humans have had a long and special relationship with waterbirds, particularly as sources of food both in the wild and following domestication (Kear 1990; Green and Elmberg 2014). Today waterbirds remain a great attraction as hunting quarry, and tens of millions of dollars are spent each year by waterfowl hunters in North America alone. Bird watchers and other people visiting nature reservess often search out the major spectacle provided by waterbirds on migration or on their wintering grounds. Management for hunting or conservation interests often focuses on the measures that attract the largest concentrations or diversity of migratory waterbirds. However, managers usually pay little or no attention to the vital role of birds as dispersers of plants and invertebrates that lack their own active means of dispersal, but which can be transported over great distances on the outside or inside of waterbirds.

The dispersal of viable plant units (hereafter "diaspores") may be the most important ecosystem service provided by birds (Şekercioğlu 2006). However, the great majority of diaspore dispersal literature focuses on the dispersal of plants with fleshy fruits by terrestrial birds (chapter 5). Nevertheless, ducks, shorebirds, and other waterbirds play major roles as vectors of passive dispersal for plants, both by internal transport within their guts ("endozoochory"; e.g., see Figuerola and Green 2002a; van Leeuwen

FIGURE 6.1. Eurasian coot (*Fulica atra*) with duckweed (*Lemna gibba* or *L. minor*) attached to its bill. Photo by Nicky Petkov / www.NaturePhotos.eu.

et al. 2012) and by external transport on feathers or skin ("epizoochory" or "ectozoochory"; e.g., see Figuerola and Green 2002b; Brochet et al. 2010a; fig. 6.1). As shown experimentally by Darwin (1859), secondary internal transport can also occur, when diaspores are first ingested by fish or crustaceans that are then predated by piscivorous birds. In addition, plants may be dispersed when waterbirds use them as nest material. Many of the plants dispersed by waterbirds are major components of ecosystems, and provide numerous indirect benefits to humans.

Dispersal is crucial for the regional survival, range expansion, and migration of plant species, especially plants that are confined to spatially discrete habitats in an otherwise unsuitable landscape (Howe and Smallwood 1982). Nonmarine aquatic or wetland habitats are often islandlike in their spatial isolation from one another. Dispersal processes to islands and to inland waterbodies, such as lakes or ponds, are cases in which metapopulation or metacommunity models apply to plant populations (whether island, or aquatic, or both). In both island and island-like aquatic systems, waterbirds are important vectors of passive dispersal for numerous plant species. They are highly mobile on short and long timescales, including frequent daily movements and seasonal long-distance migrations. They are thus particularly good vectors for long-distance dispersal and maintain connectivity between plant populations in different catchments that have no active means of interchange (Amezaga et al. 2002). Their importance as vectors is even greater today, as plants and other organisms need to move to adapt to the rapid changes to natural environments caused by human activity, including climate change. Suitable habitats for plant species change distribution continuously (e.g. as temporary wetlands dry and reflood, as some wetlands are degraded whilst others are created, or as the distribution of the suitable temperature range changes), and waterbirds provide a means by which plant species can track these changes.

Although waterbirds sometimes feed on fleshy fruits and disperse the seeds within, they mainly disperse wetland and terrestrial plants that lack fleshy fruits. Most plant species have what van der Pijl (1972) called "non-adapted diaspores" because they lack a fleshy-fruit and a priori are not obviously adapted for internal transport, while at the same time lacking hooks, barbs, or other apparent adaptations for external transport. Plant families lacking fleshy fruits have often been wrongly assumed to be exclusively dispersed by abiotic means such as wind or water, and for that reason many reviews of "zoochory" or biotic dispersal make no mention of waterbirds as vectors of diaspores (e.g. Tiffney 2004). Like interactions between herbaceous diaspores and large mammals (Janzen 1984), the study of interactions between plants and waterbirds has been rather unfashionable compared with interactions between vertebrates and forest plants with fleshy fruits. This is ironic, since Darwin (1859) paid more attention to the role of waterbirds as dispersal vectors than to the role of frugivores.

Diaspore dispersal by waterbirds is an ancient process that likely dates to the origin of waterbirds in the Early Cretaceous, coinciding with the origin and early radiation of the angiosperms (Soltis et al. 2008; Lockley et al. 2012). Darwin (1859) first drew attention to diaspore dispersal by waterbirds. He realized that their capacity as vectors of passive dispersal, coupled with their frequent and long-distance movements, provided an explanation over evolutionary timescales for the widespread distributions of many aquatic organisms, despite their own limitations for movement. Most of the literature exploring the dispersal of diaspores by waterbirds in detail focuses on aquatic plants. The truly aquatic plants (e.g., pondweeds) have no capacity for wind dispersal (anemochory, van der Pijl 1972), and this increases their dependency on dispersal by animals (zoochory) and water (hydrochory). However, dispersal by water is always limited to areas connected by surface water flows and thus cannot result in dispersal to isolated water bodies or wetlands in different catchments. Waterbirds also disperse diaspores of a variety of terrestrial plants, especially those from moist soils and those whose diaspores end up being washed or blown into wetlands. The boundaries between "aquatic" and "terrestrial" are not rigid or clear-cut, since many important habitats are dynamic and are inundated only for part of the time and most waterbirds are not entirely aquatic. Furthermore, some aquatic plants can often tolerate a short terrestrial phase, and vice versa. Dormant diaspores can remain in seed banks until favorable conditions (whether wetter or drier) arise. Diaspore dispersal by waterbirds may be relatively unimportant in closed forests, but forests cover only 31% of total land area (http://www.fao.org/forestry/28808/en/). Furthermore, forest streams are often frequented by herons, kingfishers or specialized ducks, all of which may have an important role as vectors of dispersal upstream or between catchments. Wetlands cover an estimated 12% of land area (Downing 2009), but their catchment areas are far more extensive.

This chapter focuses on current understanding of plant dispersal by waterbirds. We begin with a review of principal waterbird families and their importance as plant vectors. We then review current understanding of the importance of zoochory for the ecology of plants, effects on seeds of gut passage through the waterbird gut, the plants that are dispersed, and the extent to which seed morphology can predict waterbird dispersal. We next consider the coupling of seed dispersal with seed production, the role of plant-waterbird coevolution, and the seed adaptations that exist for waterbird dispersal. We then focus on establishment success of dispersed diaspores. We conclude by considering how all this dispersal provides benefits to humans.

Waterbirds That Are Diaspore Vectors

No systematic surveys compare the relative importance of all waterbirds in a given region as vectors, and this makes generalizations about the roles of different bird groups difficult. All species can be expected to have some role in external and internal diaspore dispersal. Many waterbirds are dietary generalists. Many species of herons, spoonbills, cormorants, grebes, or terns concentrate on fish, although most of these birds also eat amphibians, reptiles, and large invertebrates such as crayfish. These waterbirds have major potential for secondary dispersal of plant diaspores ingested by their prey. Other waterbirds, such as storks, ibis, ducks, rails, gulls, and shorebirds, are typically omnivorous, eating a range of animals and plant material and varying their diets with age, season, or geographic location. Some birds, such as geese, wigeon, swans, and rallids (e.g. coots or swamphens), are largely herbivorous but also ingest and disperse diaspores. The same is true of shorebirds and flamingos, which are often assumed to be invertebrate predators.

Factors determining the importance of different waterbird species as vectors include their abundance and degree of migratory behavior, both of which are relatively well studied. In terms of numbers of species, populations and individuals, the Anatidae, shorebirds, Rallidae, and Laridae (gulls and terns) are the most important waterbird groups, followed by Ardeidae (herons and egrets; Wetlands International 2012; see fig. 6.2), and each of these groups is considered separately below.

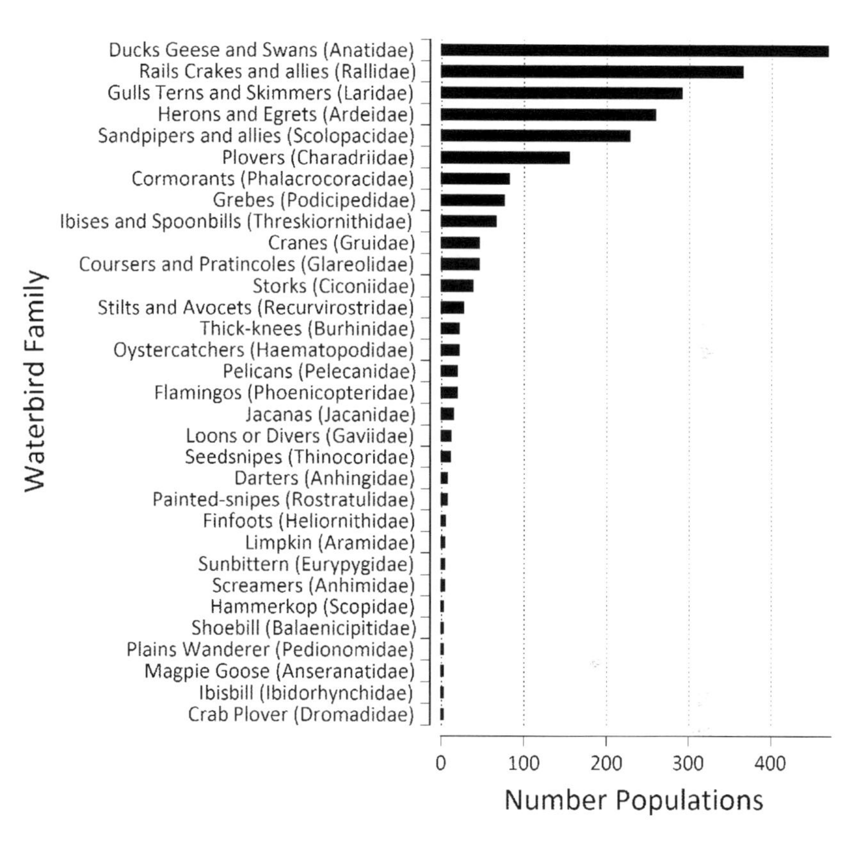

FIGURE 6.2. Overview of bird families included in the term "waterbirds" as applied in this chapter, with the global number of waterbird populations per family (from Wetlands International 2012; details of population size for each population are available online at wpe.wetlands.org).

Anatidae

Ridley (1930) provided extensive lists of plants whose seeds have been found in the guts of many waterfowl, especially Holarctic ducks. He found Cyperaceae particularly dependent on dispersal by ducks, and this is supported by recent literature. Migratory Anatidae are tremendously important diaspore dispersers in the Northern Hemisphere. Continental North America alone supports an estimated 49 million ducks, more than 10 million geese, and around 200,000 swans (USFWS 2012).

Pioneering research by Proctor and coworkers (Proctor 1968; deVlaming and Proctor 1968) has stimulated detailed, systematic, and quantitative studies on seed dispersal by waterbirds, but the great majority have focused on *Anas* ducks, for several reasons. First, *Anas* species are widespread and abundant, rendering them obvious candidates for study. Second, their ecology is well known, owing largely to their importance as hunting quarry, and diet studies have long since established that *Anas* species feed, often to a large extent, on plant seeds. Third, they are easily kept in captivity and legally hunted in large numbers, making them readily available as study objects both pre- and postmortem. For these reasons, most of the examples and analyses further on in this chapter are based on Holarctic *Anas* ducks. Hence, in this section we concentrate on the literature on other Anatidae species.

Investigations of the Pacific black duck (*Anas superciliosa*), grey teal (*Anas gracilis*), and chestnut teal (*Anas castanea*; Green et al. 2008; Raulings et al. 2011) of Australia confirm that both nomadic and seasonal migrant ducks are important endo- and epizoochorous seed vectors. In the Mediterranean region, the nomadic and globally threatened marbled teal (*Marmaronetta angustirostris*; subfamily Aythyinae) is also an important vector (Green et al. 2002; Fuentes et al. 2005). Amongst fish-eating migrants from the Merginae subfamily, red-breasted mergansers (*Mergus serrator*) and buffleheads (*Bucephala albeola*) were shown to carry seeds externally in New Jersey (Vivian-Smith and Stiles 1994).

The true geese (subfamily Anserinae) are more terrestrial than ducks, and in some cases are important for dispersal of plants with fleshy fruits. The endemic and globally threatened Hawaiian goose (*Branta sandvicensis*) disperses berries such as *Coprosma erno-deoides* (Rubiaceae), *Vaccinium reticulatum*, *Styphelia tameiameiae* (Ericaceae), and the Chilean strawberry (*Fragaria chilensis*; Rosaceae), as well as seeds of the alien sow thistle (*Sonchus asper* [Asteraceae]; Guppy 1906). Black et al. (1994) analyzed the feces of this goose and found four kinds of berries (mainly

S. tamaiameiae and *V. reticulatum*) and five types of grass seeds, especially molasses grass (*Melinis minutiflora*). Ridley (1930) reported how snow geese (*Anser hyperboreus*) consumed *Empetrum nigrum* (Ericaceae) berries and *Potamogeton natans* seeds, and greylag geese (*Anser anser*) consumed *Rubus chamaemorus* (Rosaceae) berries.

Geese grazing on grasses and other green plant material often ingest diaspores and disperse them in their feces in a manner analogous to that of herbivorous mammals which ingest seeds along with foliage (Janzen 1984). Barnacle geese (*Branta leucopsis*) are important for endozoochory in Arctic tundra (Bruun et al. 2008) and the Netherlands (Chang et al. 2005), while Canada geese (*B. canadensis*) are vectors of a range of native plants (Morton and Hogg 1989; Neff and Baldwin 2005) and alien grasses (Isaac-Renton et al. 2010). Viable seeds of *Scirpus maritimus*, *S. litoralis* (in Spain: A. J. Green and J. Figuerola, unpublished) have been recorded in the feces of migratory greylags. Seeds of smooth cordgrass (*Spartina alterniflora*), seashore saltgrass (*Distichlis spicata*), and seven other species were found on feet and feathers of brant geese (*Branta bernicla*) in New Jersey (Vivian-Smith and Stiles 1994). Brant geese in Europe feed on *Salicornia europaea* seeds (Summers et al. 1993).

Relatively little is known about diaspore dispersal by true swans (subfamily Anserinae), which may be more important vectors of aquatic plants than geese. Ridley (1930) reported that both mute swans (*Cygnus olor*) and Bewick's swan (*C. columbianus*) consumed *P. natans* seeds. Feces from black swans (*Cygnus atratus*) in Australia contained viable seeds of *Typha* and alien *Medicago* and *Polygonum* (Green et al. 2008).

Other groups of Anatidae are likely important diaspore vectors. Ridley (1930) proposed that whistling ducks (Dendrocygninae) are good vectors, as they ingest many aquatic plant species and range widely in areas like the Caribbean and Indonesia. The high proportion and variety of seeds in the diet of African and Australian *Dendrocygna* (Green et al. 2002) supports this. In South Africa, the spur-winged goose (*Plectropterus gambensis* [Plectropterinae]) ingests a variety of seeds (Halse 1985). The aberrant magpie goose (*Anseranas semipalmata*; often placed in its own family, the Anseranatidae) feeds partly on grass, *Polygonum*, and other seeds (Marchant and Higgins 1990).

Within the Tadorninae (shelduck subfamily), Ridley (1930) reported that the upland goose (*Chloephaga picta*) consumed berries of *Empetrum rubrum* in the Falkland Islands. In Tierra del Fuego, this species and the ashy-headed goose (*C. poliocephala*) are important frugivores (Willson et al. 1997). The shelduck (*Tadorna tadorna*) is a likely vector for *Salicornia* seeds (Viain et al. 2011). In South Africa, the Egyptian goose (*Alopochen aegyptiacus*) is likely to be an important seed vector (Halse 1984). Fossilized feces of the extinct, flightless *Thambetochen chauliodous* from Hawaii, the size of a large swan but more related to shelducks or dabbling ducks, were rich in fern spores (James & Burney 1997).

Rallidae

Eurasian coot (*Fulica atra*; fig. 6.1) are similar to sympatric ducks in their internal dispersal of diaspores, with *Chara* in feces in France (Charalambidou and Santamaria 2005), viable *Ruppia* and *Arthrocnemum* seeds in Spain (Figuerola et al. 2002, 2003), and viable *Typha* seeds in Australia (Green et al. 2008). Diaspores of at least 13 species were present in their upper gut in northeast France (Mouronval et al. 2007), and 13 species in the Camargue, where coot consumed more seeds than gadwall (*Anas strepera*; Allouche and Tamisier 1984). Various saltmarsh seeds have been recorded on the feet and plumage of coot in Spain (Figuerola and Green 2002b), and Cook (1990) proposed that the seeds of the water lily (*Nymphoides peltata*) are dispersed when stuck on the bill or shield of coot.

Common gallinules (*Gallinula galeata*) in Argentina consumed seeds throughout the annual cycle, especially Polygonaceae and Poaceae (Beltzer et al. 1991). Sticky seeds of *Pisonia grandis* were recorded attached to six common moorhens (*Gallinula chloropus*) in the Seychelles, as well as to several seabird species (Burger 2005). Ridley (1930) reported that the water rail (*Rallus aquaticus*) takes *Rosa* and grass seeds, while the buff-banded rail (*Hypotaenidia philippensis*) eats fruits of *Freycinetia banskii* in New Zealand. Rails of the genus *Porphyrio* are major vectors of the genus *Coprosma* (Rubiaceae) within and between New Zealand and Pacific islands, and *Scleria* seeds (Cyperaceae) were in the gut of *Porphyrio* in Fiji (Guppy 1906). A seed of the Japanese chaff flower (*Achyranthes japonica*) was found in the feathers of a Swinhoe's rail (*Coturnicops exquisitus*; Choi et al. 2010). Seeds, especially *Caperonia palustris* (Euphorbiacea), *Thalia geniculata* (Marantaceae), *Eleocharis* (Cyperaceae), *Ludwigia* (Onagraceae) and *Neptunia oleraceae* (Fabaceae), are major food items of the purple gallinule (*Porphyrio martinincus*) in Venezuela (Tárano et al. 1995). Viable seeds of four *Eleocharis* species were recovered from feces of purple swamphen (*P. porphyrio*) in Australia (Bell 2000). Seeds of Typhaceae (*Sparganium ramosum*) and Cyperaceae (*Scirpus* spp. and *Carex divisa*) are a major part of the diet of this species in Spain (Rodríguez and Hiraldo 1975).

Shorebirds

The shorebirds (or "waders" as they are often called in Europe, not to be confused with "wading birds" in the North American sense) are a group of waterbird families with a similar morphology and ecology, within the order Charadriiformes. With migrations often across seas and oceans, their potential importance in diaspore dispersal to oceanic islands has long been recognized. Ridley (1930) reported internal transport of berry seeds by shorebirds, referring to the consumption of *Vaccinium* and *Empetrum* berries by Eurasian curlew (*Numenius arquata*) and *Pluvialis* plovers, and of *Canthium* fruits by bristle-thighed curlew (*N. tahitensis*) in Pacific islands. De Vlaming and Proctor (1968) were the first to demonstrate that Charadriidae shorebirds disperse angiosperms lacking fleshy fruits by internal transport. Proctor (1968) found experimentally that viable seeds that were regurgitated after being retained in the shorebird gizzard for up to 340 hours in killdeer (*Charadrius vociferus*) and 216 hours in least sandpiper (*Calidris minutilla*) remain viable. Proctor (1968) also observed that shorebirds reingest seeds regurgitated by individuals of other species. A given seed may thus be transferred between species and dispersed to and from microhabitats used by different species (e.g., terrestrial habitats). This seed transfer among different species may facilitate effective long-distance dispersal of diaspores. The extraordinary nonstop flights of bar-tailed godwits (*Limosa lapponica*) of over 8,000 km across the Pacific (Gill et al. 2009) underline their potential for long-distance dispersal.

Most shorebirds include seeds in their diet, including 37 of 55 shorebird species in the Western Palaearctic and 26 of 35 species in North America (Green et al. 2002). Seeds of at least 122 genera from 48 families have been recorded in the guts of common snipe (*Gallinago gallinago*; Mueller 1999). Seeds can be the most important food item for various species at certain times of the year, even at stoppage sites during spring and autumn migration (Green et al. 2002). On wintering grounds in Argentina, seeds from at least eight plant families were the only food items recorded for white-rumped sandpiper (*Calidris fuscicollis*; Montalti et al. 2003). External transport by shorebirds is also important but largely unstudied. Darwin (1872) germinated a seed of toad rush (*Juncus bufonius*) removed from mud attached to the leg of a Eurasian woodcock, *Scolopax rusticola*. Bryophyte and algal diaspores were recovered from the plumage of American goldenplover (*Pluvialis dominica*), semipalmated sandipiper (*Calidris pusilla*), and red phalarope (*Phalaropus fulicarius*; Lewis et al. 2014).

Sánchez et al. (2006) examined seed viability following passage through shorebird guts in the field. Viable seeds of *Mesembryanthemum nodiflorum* (Aizoaceae), *Sonchus oleraceus* (Asteraceae) and *Arthrocnemum macrostachyum* (Chenopodiaceae) were frequent in pellets and feces of common redshank, spotted redshank (*Tringa erythropus*), and black-tailed godwit (*Limosa limosa*) during spring and autumn migrations in Spain (Sánchez et al. 2006). Another 11 seed types were recorded at low densities.

Gulls

Calvino-Cancela (2011) reviewed internal transport by gulls. Berries are a major part of the diet of many species, and dispersal of fleshy-fruited plants by gulls greatly influences the development of oceanic island plant communities. However, seeds of plants lacking a fleshy fruit, common in grasslands and cultivated fields, are also dispersed. Seeds have been recorded in the diet of at least 22 gull species, mainly those using inland freshwater habitats. Gulls are not efficient at digesting seeds, and often regurgitate or defecate them intact. Plants dispersed include many genera frequent in duck diets, such as *Polygonum*, *Plantago*, *Chenopodium*, *Rumex*, *Carex* and *Scirpus*. Retention times can exceed 70 hours for defecated seeds and 45 hours for regurgitated seeds, and viability has been demonstrated in several studies. On Surtsey, a volcanic island that appeared in 1963, gulls have brought most of the soil nitrogen, as well as most of the plants (Magnússon et al. 2009).

Gulls help spread alien weeds from agricultural land and garbage dumps to islands. Morton and Hogg (1989) recovered seeds of 23 plant species in pellets and feces of ring-billed gull (*Larus delawarensis*) and herring gull (*L. argentatus*) on an island in Lake Huron. Seeds of nineteen species later germinated. Most (18) of these plant species were exotics (e.g., *Poa annua*, *Chenopodium album*, *Amaranthus retroflexus* and *Taraxacum officinale*) that dominated vegetation around the nesting sites. They also found that herring gull nests contained 15 plant species (only two of which were present in pellets and feces) with viable rootstocks, rhizomes, or seeds.

Earthworms, which themselves ingest and disperse diaspores (Milcu et al. 2006), are common in gull diets (Calvino-Cancela 2011). Gulls may indirectly disperse diaspores within the earthworms they consume.

Herons

Grey heron (*Ardea cinerea*) pellets in Tenerife contained seeds of at least 16 plant species ingested by lizards which are preyed on by the herons, facilitating secondary dispersal of terrestrial plants, although to a lesser extent than by kestrels and shrikes (Rodríguez et al. 2007). Heslop-Harrison (1955) reported viable seeds of *Nuphar lutea* (presumed to have been eaten by a fish) in excreta from a grey heron. Corlett (1998) reported that cattle egrets (*Bubulcus ibis*) eat figs. Seeds of *Achyranthes japonica* were found in the feathers of a Eurasian bittern (*Botaurus stellaris*; Choi et al. 2010).

Other Waterbirds

Arber (1920) reported that Weddell (1849; predating Darwin 1859) observed a tiny, previously unknown floating plant on the feathers of a waterbird in Brazil (a "camichi," a local name for a horned screamer [*Palamedea cornuta*], according to Maximilian 1820). Weddell described this new species as *Wolffia brasiliensis*. Greater flamingos (*Phoenicopterus roseus*) often filter food items from the sediments, and ingest *Ruppia maritima* seeds (Rodríguez-Pérez and Green 2006) and presumably many other diaspores. Holmboe (1900) reported that common cranes (*Grus grus*) consumed *Vaccinium vitis-idea* berries. Australian pelican (*Pelecanus conspicillatus*) feces contained viable diaspores of *Lemna*, *Nitella* and *Typha*, which were presumably first ingested by fish (Green et al. 2008). Ridley (1930) reported occasional seeds in the stomachs of grebes.

Ridley (1930) reported the presence of berries in the diets of skuas, and occasional seeds in petrels. Aoyama et al. (2012) quantified external transport of nine plant species by four species of seabird: the black-footed albatross (*Phoebastria nigripes*), Bulwer's petrel (*Bulweria bulwerii*), wedge-tailed shearwater (*Puffinus pacificus*), and brown booby (*Sula leucogaster*). Carlquist (1967) observed sticky *Boerhavia diffusa* fruits on the feathers of the sooty tern (*Onychoprion fuscatus*).Taylor (1954) and Falla (1960) suggested that various plants reached oceanic islands through external transport on albatrosses and petrels.

The Significance of Passive Dispersal by Waterbirds for Wetland Plant Species

Wetlands often occur spatially scattered throughout otherwise dry (or drier) terrestrial landscapes, so that wetland habitat exists in the form of discrete and isolated patches in an otherwise unsuitable landscape. Human land development has reduced wetland area and increased wetland fragmentation globally, exacerbating the isolation of remaining wetland areas

(Davidson 2014). Loss of connectivity between isolated wetlands and associated changes in land use may result in local and regional species extinctions, loss of regional species dynamics, and increased vulnerability of remnant populations to stresses such as pollution or climate change (Amezaga et al. 2002; Lougheed et al. 2008; Wormworth and Şekercioğlu 2011). Even under natural conditions, plant species must disperse across the landscape to reach new wetland habitats in order to escape from predators, pests, and pathogens and to reach new unoccupied sites to balance the loss of occupied sites. Hence, for the preservation of wetland plant diversity and wetland functions provided by the plant species, dispersal among wetlands is crucial.

Five main dispersal mechanisms have the potential to connect isolated wetlands: wind, water, birds, humans, and other animals. Dispersal by wind is uncommon in submerged or floating plants, as they mostly produce diaspores under or in the water. Many plants (most notably the tall *Epilobium*, *Typha*, and *Phragmites* species and trees such as *Salix*, *Alnus*, and *Betula*) produce seeds in and around wetlands that are well dispersed by wind over distances exceeding tens of kilometers, but the direction of the wind is unpredictable and many seeds are lost (Soons 2006). Hydrochory of diaspores produced under, in, or near the water surface transports diaspores between wet areas likely to offer suitable habitat (Sarneel et al. 2013; Soomers et al. 2013). However, water flows transport diaspores only downstream and between hydrologically connected wetlands (Soons 2006). Animals in search of water may provide a much more targeted (or "directed") means of dispersal among all wetland types, which is essential to maintain the viability of wetland plant populations (Purves and Dushoff 2005; Kleyheeg 2015).

Both mammals and birds are able to transport large numbers of plant diaspores over long distances, but large migratory mammals have been subjected to massive prehistorical extinctions (Janzen 1984), and nowadays are restricted by movement barriers such as fences and inhabited areas, which birds can easily surpass. Given the abundance of waterbirds, their role in the directed dispersal of diaspores among wetlands is critical.

Although aquatic plants disperse via asexual propagules such as rhizomes or stem fragments, these propagules are typically more important for dispersal within a catchment by hydrochory (Santamaria 2002), or for short-distance dispersal by external transport. Dispersal among catchments is predominantly by internal or external transport of seeds. Although small floating plants such as *Lemna* or *Azolla* may be exceptions (fig. 6.1), they have limited resistance to desiccation during external transport (Coughlan et al. 2015).

Despite major advances in taxonomy and the discovery of cryptic species using molecular methods, Darwin's (1859) observation that aquatic plants have particularly broad distributions has stood the test of time. Aquatic plants have lower levels of endemism, and are more likely than terrestrial plants to occur on more than one continent. Within a genus such as *Ranunculus*, which has many European species, aquatic species have a greater latitudinal range and overall area of occupancy than do terrestrial species (Santamaria 2002). The biogeography of aquatic plants thus shows a lasting footprint of long-distance dispersal. Corresponding to their high dispersal ability, natural selection has honed aquatic plants with general purpose genotypes, high stress tolerance and high clonal persistence, all of which allows them to occupy large ranges.

The importance of seed dispersal for aquatic plants varies with the spatial configuration of wetlands in the landscape across fragmentation gradients which include latitudinal and climatic gradients. In Europe, the proportion of temporary aquatic habitats increases from north to south—and with it, the relative importance of sexual reproduction. In northern habitats, plants are more able to spread within a habitat and to persist from year to year in the absence of seed production (often overwintering as rhizomes). The chances of seedling establishment can be extremely low, owing to intense competition with established plants (Santamaria 2002). Hence, seeds of aquatic plants appear adapted for long-distance dispersal to unoccupied habitats rather than local dispersal within an already occupied habitat. In contrast, in the Mediterranean region of southern Europe, aquatic habitats are highly dynamic and often temporary, and suitable microhabitats for a given plant species often change greatly from one season or year to the next. Ponds and lakes often dry out completely in summer or during drought cycles. Hence, diaspores with dormant capacity (seeds or spores) are important at a local scale for survival of drought and for colonization of areas that become suitable in a given year. Seed dispersal *within* a wetland complex becomes more important, and may be more likely to be followed by successful establishment than in northern permanent habitats. Birds are also typically major seed dispersal vectors within a wetland complex, often moving the seeds to sites they could not reach by hydrochory (Figuerola et al. 2003; Brochet et al. 2010a). These latitudinal patterns explain why widespread pondweed species, such as *Potamogeton pectinatus*, invest more in seed production in southern than in northern populations (Santamaria et al. 2005).

De Vlaming and Proctor (1968) emphasized that because most aquatic plants are monoecious, a single viable diaspore dispersed to a new habitat may be sufficient to establish a new population. Similarly Cruden (1966) pointed out that most plants with a distribution suggesting long-distance dispersal by shorebirds are self-compatible. Proctor (1980) showed for the Characeae that only bisexual or parthenogenetic taxa are present on isolated oceanic islands. In contrast, dioecious taxa are restricted to continental land masses and islands within a maximum range of 200 to 300 km, such that repeated dispersal events allow establishment of both sexes.

What Happens to Diaspores in the Waterbird Gut?

For almost all plant species whose seeds are ingested by ducks, some seeds survive gut transit (Brochet et al. 2009). Exceptions appear due to large seed size, which makes gut passage unlikely. Aquatic seeds that seldom survive passage through the waterbird gut include the large, soft seeds of the water lilies *Nymphaea alba*, *Nuphar lutea*, and *Nymphoides peltata* (Smits et al. 1989; Soons et al. 2008), which are favored food items of ducks (Tréca 1981). However, the seeds of these species are adapted for external transport (Smits et al. 1989; Cook 1990). They may also be secondarily dispersed by internal transport when they are in fish ingested by piscivorous birds. Such large seeds may also be regurgitated by waterbirds occasionally without damage (Kleyheeg 2015). During experiments, mallards sometimes regurgitate charophytes before they enter the gizzard (Malone 1966).

In an experimental study in which seeds of 23 wetland plant species were fed to mallards, the proportion of seeds retrieved from feces varied from 0 to 54%, with a negative relationship between seed volume and retrieval (Soons et al. 2008). In the same study, smaller seeds were retained for less time in the gut. In a similar study in which diaspores of eight species were fed to green-winged teal (*Anas crecca*), retrieval varied from 2 to 83% (Brochet et al. 2010b). Wongsriphuek et al. (2008) found that seed retrieval increased with higher fiber content among 10 wetland species fed to mallards. Retrieval was not related to seed size. Van Leeuwen et al. (2012) found in a meta-analysis that larger propagules, including plant seeds, have lower survival during passage through the waterfowl gut; but García-Álvarez et al. (2015) showed that there are exceptions with large, durable seeds, such as the invasive primrose *Ludwigia grandiflora* (Onagraceae).

The fact that many seeds survive gut processing by waterbirds can be explained by optimality modeling (Sibly 1981; van Leeuwen et al. 2012). Even when ducks are consuming a single preferred seed species which provides a high assimilation rate, the diminishing returns from the digestion of the last fractions of seeds renders total digestion suboptimal. Since ducks are highly omnivorous, they may often consume and process a combination of different seed types with other plant or animal food, and retention time may be determined largely by the optimal strategy for digesting foods other than seeds.

Thus, Sibly's (1981) model would predict that seeds mixed with higher-quality food, such as animal pellets, should be retained for shorter periods than those mixed with lower-quality food such as plant leaves, as was observed with digestive markers by Charalambidou et al. (2005). Similarly, when food items are more available in the feeding environment and handling times are reduced, optimality theory predicts that gut passage rate can be increased, thus increasing the rate of diaspore survival. This expectation is supported by field studies, indicating that seed survival increases when ducks ingest seeds at a higher rate (Figuerola et al. 2002; Green et al. 2002).

Waterbirds are likely to select seeds partly on their nutritional quality, which is likely to be positively related to their digestibility and negatively related to their capacity to survive digestion. Thus, the breeding white-faced whistling duck (*Dendrocygna viduata*) and red-billed teal (*Anas erythrorhyncha*) fed largely on *Panicum schinzii* seeds, which had a particularly high fat content (Petrie 1996; Petrie and Rogers 1996). Given their bill morphology, it is difficult for ducks to reject relatively poor quality seeds mixed with other foods (Gurd 2006), and they typically ingest many kinds of seeds simultaneously (Brochet et al. 2012a). The digestive assimilation efficiency of seeds varies among bird species and even between sexes of a given waterbird species (Santiago-Quesada et al. 2009).

Variation in overall retention time between food items such as diaspores is partly related to variation in the time they are retained in the gizzard. Larger items tend to be retained longer, which explains why smaller seeds generally have shorter overall retention times (Soons et al. 2008; Kleyheeg 2015; but see Figuerola et al. 2010). Unlike frugivores, retention time in waterfowl is negatively related to body size, making smaller species better vectors (García-Álvarez et al. 2015).

The chance that a diaspore survives retention in the gizzard is related to the strength of the gizzard and to the amount of grit (small stones) present to crush food (Kleyheeg 2015). The size and quantity of grit varies between individuals and species of waterfowl in a manner related to diet, with herbivorous species having more grit (Figuerola et al. 2005a). Some authors have suggested that shorebirds sometimes ingest hard seeds for the same reason they ingest grit: to help crush other food in the gizzard, rather than for direct nutritional benefit (Green et al. 2002). It is difficult to mimic natural conditions in captivity, and existing studies of seed survival and retention times may be misleading. Wild ducks tend to have larger gizzards and intestines than do captive ones (Charalambidou and Santamaría 2002), such that captive studies may tend to overestimate seed survival while perhaps underestimating retention times. On the other hand, the low activity levels of captive birds compared to those of wild ones, which spend much time swimming or flying, may lead to a major underestimation of the proportion of seeds that survive gut passage, as well as a slight overestimation of retention times (Kleyheeg et al. 2015).

Recent studies addressed the effects of passage through the waterbird gut on germinability (the probability of germination) and germination rate (the time taken to germinate) of seeds. As with terrestrial birds (Traveset 1998), the effects of gut processing by waterfowl vary. In some cases, gut processing increased germinability, but in other cases passage decreased it (Soons et al. 2008; Brochet et al. 2010b; García-Álvarez et al. 2015). The rate of germination is usually, but not always, increased by gut passage (Brochet et al. 2010b; Figuerola et al. 2010; García-Álvarez et al. 2015). The differences between studies are likely related to plant species–specific effects of gut passage on germination capacity.

Which Plants Are Dispersed by Ducks?

Given the shortage of studies that quantify viable diaspores moved by waterbirds in the field, and given that the great majority of diaspore types have some capacity to survive gut passage, diet studies that identify diaspores are of great interest. Many such studies exist for dabbling ducks, and reanalysis of these datasets sheds light on the variation in dispersal processes over space and time, as well as between specific plant and bird species. The frequency of a given seed type in the upper guts of ducks is a strong predictor of its frequency as a viable seed in faeces (Brochet et al. 2009).

We reviewed 70 studies of the diet of dabbling ducks in Europe and found that seeds of at least 445 plant species of 189 genera and 57 families were reported (table 6.1 and supplemental table S6.1 at www.press.uchicago.edu/sites/whybirdsmatter/). The species encompass a wide range of families, from Poaceae (grasses), with no obvious adaptations for any means of dispersal, to Asteraceae (Compositae), with often complex adaptations for wind dispersal (such as the plume of dandelion seeds).

To find general patterns in the plant species dispersed by dabbling ducks, we analyzed 413 of the 444 plant species in Europe, for which quantitative data exist. We analyzed plant traits reflecting species' habitats, seed production, size, and dispersal capacity by wind and water. Species habitat can be estimated using Ellenberg indicator values (Ellenberg et al. 1991), which represent the optimum conditions at which European species occur along an environmental gradient. We looked at species occurrence along the following gradients: nutrient-poor to nutrient-rich (indicated by Ellenberg *N* values), dry to wet (Ellenberg *F*), and shaded to well-lit (Ellenberg *L*). Ellenberg values were taken from the PLANTATT database (Hill et al. 2004; extracted 5 December 2006). Species trait data used were seed production (measured as the number of seeds per individual plant, ramet, or tussock), seed size (measured as seed volume, in mm^3), wind dispersal capacity as approximated by seed terminal velocity (measured as the constant falling rate of a seed in still air, after an initial short phase of acceleration, in ms^{-1}) and water dispersal capacity as approximated by seed buoyancy (measured as the percentage of seeds still floating after one week in water). Trait data were taken from the LEDA databse (Kleyer et al. 2008; extracted 13 July 2010).

Frequency distributions of the Ellenberg values of the species ingested by ducks were compared to those of all plant species for which Ellenberg data are available (fig. 6.3), showing that ducks feed disproportionally on plant species from sites of rich (but not extremely rich) fertility (Ellenberg *N* values 6–8), wet to inundated sites (Ellenberg *F* values 8–12), and habitats on the transition from semishaded to well-lit (Ellenberg *L* value 7). This analysis shows how plant species from wet, relatively nutrient-rich, and relatively open (but not too open) habitats have a greater probability of being dispersed by dabbling ducks.

On the other hand, in terms of numbers of species, fig. 6.3 indicates that most plant species that are present in duck diets, and which therefore are thought to be dispersed by ducks, are not aquatic but rather terrestrial, especially plants of moist soils (Hagy and Kaminski 2012). Small seeds in terrestrial plants are characteristic of early successional, light-rich environments. Whenever these seeds are washed or blown into wetlands, as during storms, they may be ingested and then dispersed by waterbirds. Wetlands are unsuitable habitat for many plant species whose seeds are taken there by rainfall (Gordon and van der Valk 2003). They may then

TABLE 6.1 **Seeds (oogonia for algae) found in digestive tracts of eight dabbling duck species in Europe (gadwall, garganey, mallard, marbled teal, pintail, shoveler, common teal, and wigeon). Taxonomy is after *Flora Europaea* (provided online by the Royal Botanic Garden of Edinburgh, accessed in 2013) for plants eaten. Most of the data for the mallard, pintail, and common teal come from the supporting information table provided online for the paper by Brochet et al. (2012a), supplemented by additional references. See table 6.1 extended online for more details (www.press.uchicago.edu/sites/whybirdsmatter/).**

ALGAE
Characeae
Chara canescens
Chara sp.
VASCULAR PLANTS
Alismataceae
Alisma plantago-aquatica
Baldellia ranunculoides
Sagittaria sagittifolia
Amaranthaceae
Amaranthus albus
Amaranthus deflexus
Amaranthus hybridus
Amaranthus retroflexus
Araceae
Calla palustris
Betulaceae
Alnus glutinosa
Alnus incana
Betula pendula
Betula pubescens
Boraginaceae
Myosotis arvensis
Myosotis scorpioides
Callitrichaceae
Callitriche sp.
Caprifoliaceae
Sambucus nigra
Sambucus racemosa
Viburnum lantana
Caryophyllaceae
Arenaria sp.
Cerastium sp.
Lychnis flos-cuculi
Spergula arvensis
Spergularia marina
Spergularia media
Stellaria holostea
Stellaria media
Ceratophyllaceae
Ceratophyllum demersum
Chenopodiaceae
Arthrocnemum fruticosum
Arthrocnemum macrostachyum
Atriplex hastata

TABLE 6.1 **(continued)**

Atriplex hortensis
Atriplex littoralis
Atriplex patula
Atriplex prostrata
Bassia hirsuta
Beta vulgaris
Chenopodium album
Chenopodium ficifolium
Chenopodium glaucum
Chenopodium murale
Chenopodium polyspermum
Chenopodium rubrum
Chenopodium vulvaria
Halimione pedunculata
Halimione portulacoides
Halocnemum strobilaceum
Salicornia europaea
Salsola soda
Suaeda maritima
Suaeda vera
Suaedea corniculata
Compositae
Artemisia sp.
Aster tripolium
Baccharis halimifolia
Bidens cernua
Bidens frondosa
Bidens tripartita
Chamomilla recutita
Cirsium arvense
Cirsium palustre
Cirsium vulgare
Filaginella uliginosa
Helianthus annuus
Hieracium umbellatum
Inula sp.
Senecio aquaticus
Silybum marianum
Soliva sp.
Convolvulaceae
Calystegia sepium
Convolvulus arvensis
Corylaceae
Carpinus betulus
Cruciferae
Brassica napus
Cochlearia sp.
Coronopus squamatus
Lepidium sp.
Nasturtium microphyllum
Nasturtium officinale

continues

TABLE 6.1 **(*continued*)**

Rapistrum sp.
Rorippa amphibia
Cyperaceae
Carex acuta
Carex acutiformis
Carex aquatilis
Carex arenaria
Carex bohemica
Carex canescens
Carex chordorrhiza
Carex curta
Carex disticha
Carex divulsa
Carex elata
Carex elongata
Carex extensa
Carex flacca
Carex flava
Carex globularis
Carex hirta
Carex hispida
Carex lasiocarpa
Carex limosa
Carex magellanica
Carex nigra
Carex otrubae
Carex ovalis
Carex pallescens
Carex panicea
Carex paniculata
Carex pilulifera
Carex pseudocyperus
Carex riparia
Carex rostrata
Carex tomentosa
Carex trinervis
Carex vesicaria
Carex vulpina
Carex sp.
Cladium mariscus
Cyperus difformis
Cyperus michelianus
Cyperus serotinus
Eleocharis acicularis
Eleocharis multicaulis
Eleocharis ovata
Eleocharis palustris
Eleocharis uniglumis
Eriophorum vaginatum
Fimbristylis sp.
Schoenus nigricans
Scirpus lacustris

TABLE 6.1 **(continued)**

Scirpus litoralis
Scirpus maritimus
Scirpus mucronatus
Scirpus setaceus
Scirpus sylvaticus
Scirpus triqueter
Elaeagnaceae
Elaeagnus angustifolia
Elatinaceae
Elatine hydropiper
Empetraceae
Empetrum nigrum
Equisetaceae
Equisetum fluviatile
Ericaceae
Calluna vulgaris
Vaccinium myrtillus
Vaccinium uliginosum
Vaccinium vitis-idaea
Fagaceae
Quercus faginea
Quercus robur
Geraniaceae
Geranium dissectum
Geranium robertianum
Guttiferae
Hypericum hirsutum
Haloragaceae
Myriophyllum spicatum
Myriophyllum verticillatum
Hippuridaceae
Hippuris vulgaris
Hydrocharitaceae
Hydrocharis morsus-ranae
Vallisneria spiralis
Iridaceae
Iris pseudacorus
Juncaceae
Cyperus serotinus
Juncus acutiflorus
Juncus articulatus
Juncus compressus
Juncus effusus
Juncus filiformis
Juncus gerardi
Juncus inflexus
Juncus littoralis
Luzula spicata
Juncaginaceae
Triglochin maritima

continues

TABLE 6.1 **(*continued*)**

Labiatae
- *Ajuga reptans*
- *Galeopsis speciosa*
- *Galeopsis tetrahit*
- *Lycopus europaeus*
- *Mentha aquatica*
- *Prunella vulgaris*
- *Scutellaria galericulata*
- *Stachys palustris*

Leguminosae
- *Astragalus* sp.
- *Lotus corniculatus*
- *Lotus uliginosus*
- *Medicago arabica*
- *Medicago lupulina*
- *Medicago sativa*
- *Pisum* sp.
- *Trifolium campestre*
- *Trifolium dubium*
- *Trifolium fragiferum*
- *Trifolium pratense*
- *Trifolium repens*
- *Trifolium squamosum*
- *Vicia cracca*

Lemnaceae
- *Lemna gibba*

Lythraceae
- *Lythrum salicaria*

Malvaceae
- *Althaea officinalis*
- *Malva* sp.

Marsileaceae
- *Pilularia* sp.

Menyanthaceae
- *Menyanthes trifoliata*

Najadaceae
- *Najas gracillima*
- *Najas indica*
- *Najas marina*
- *Najas minor*

Nymphaeaceae
- *Nuphar lutea*
- *Nymphaea alba*
- *Nymphoides peltata*

Onagraceae
- *Epilobium hirsutum*
- *Ludwigia peploides*

Oxalidaceae
- *Oxalis* sp.

Papaveraceae
- *Chelidonium majus*
- *Papaver* sp.

TABLE 6.1 **(*continued*)**

Parnassiaceae
Parnassia palustris
Plantaginaceae
Plantago lanceolata
Plantago major
Plantago maritima
Plantago media
Plumbaginaceae
Armeria maritima
Limonium vulgare
Poaceae
Agrostis stolonifera
Alopecurus geniculatus
Alopecurus myosuroides
Alopecurus pratensis
Anthoxanthum odoratum
Apera spica
Arrhenatherum sp.
Avena fatua
Avena sativa
Bromus secalinus
Bromus sterilis
Cynodon dactylon
Digitaria sanguinalis
Echinochloa crus-galli
Eleusine indica
Elymus pungens
Elymus repens
Eragrostis sp.
Festuca arundinacea
Festuca rubra
Glyceria declinata
Glyceria fluitans
Glyceria maxima
Glyceria plicata
Holcus lanatus
Hordeum distichon
Hordeum hystrix
Hordeum marinum
Hordeum secalinum
Hordeum vulgare
Leersia oryzoides
Lolium multiflorum
Lolium perenne
Milium sp.
Oryza sativa
Panicum miliaceum
Parapholis strigosa
Paspalum oaginatum
Paspalum paspalodes
Paspalum vaginatum

continues

TABLE 6.1 **(*continued*)**

Phalaris arundinacea
Phleum pratense
Phragmites australis
Poa annua
Poa bulbosa
Poa pratensis
Poa trivialis
Polypogon sp.
Puccinellia distans
Puccinellia fasciculata
Puccinellia maritima
Secale cereale
Setaria italica
Setaria pumila
Setaria verticillata
Setaria viridis
Sorghum bicolor
Spartina townsendii
Triticum aestivum
Triticum sp.
Zea mays
Polygonaceae
Fagopyrum esculentum
Fallopia convolvulus
Polygonum amphibium
Polygonum aviculare
Polygonum hydropiper
Polygonum lapathifolium
Polygonum minus
Polygonum mite
Polygonum persicaria
Polygonum viviparum
Rumex acetosa
Rumex acetosella
Rumex aquaticus
Rumex conglomeratus
Rumex crispus
Rumex hydrolapathum
Rumex maritimus
Rumex obtusifolius
Rumex palustris
Rumex pulcher
Pontederiaceae
Heteranthera limosa
Heteranthera reniformis
Potamogetonaceae
Potamogeton acutifolius
Potamogeton berchtoldii
Potamogeton gramineus
Potamogeton lucens
Potamogeton natans
Potamogeton nodosus

TABLE 6.1 **(continued)**

Potamogeton obtusifolius
Potamogeton pectinatus
Potamogeton perfoliatus
Potamogeton polygonifolius
Potamogeton pusillus
Potamogeton trichoides
Primulaceae
Glaux maritima
Lysimachia vulgaris
Ranunculaceae
Ranunculus acris
Ranunculus baudotii
Ranunculus bulbosus
Ranunculus flammula
Ranunculus hederaceus
Ranunculus lingua
Ranunculus repens
Ranunculus sardous
Ranunculus sceleratus
Ranunculus trichophyllus
Thalictrum sp.
Resedaceae
Reseda lutea
Reseda luteola
Rosaceae
Cotoneaster sp.
Crataegus laevigata
Crataegus monogyna
Filipendula ulmaria
Fragaria vesca
Potentilla anserina
Potentilla palustris
Prunus cerasus
Prunus spinosa
Pyrus malus
Rosa canina
Rosa multiflora
Rubus arcticus
Rubus chamaemorus
Rubus fruticosus
Rubus sp.
Sorbus aucuparia
Rubiaceae
Galium aparine
Galium palustre
Galium tricornutum
Ruppiaceae
Ruppia cirrhosa
Ruppia maritima
Salicaceae
Salix sp.

continues

TABLE 6.1 **(*continued*)**

Scheuchzeriaceae
  Scheuchzeria palustris
Scrophulariaceae
  Linaria arvensis
  Linaria vulgaris
  Odontites verna
  Rhinanthus minor
  Scrophularia auriculata
  Scrophularia nodosa
  Verbascum sp.
  Veronica anagallis-aquatica
  Veronica beccabunga
  Veronica catenata
  Veronica hederifolia
  Veronica persica
Solanaceae
  Solanum dulcamara
  Solanum lycopersicum
  Solanum nigrum
  Solanum tuberosum
Sparganiaceae
  Sparganium angustifolium
  Sparganium emersum
  Sparganium erectum
  Sparganium minimum
Typhaceae
  Typha latifolia
Umbelliferae
  Anthriscus sylvestris
  Cicuta virosa
  Falcaria vulgaris
  Oenanthe aquatica
  Oenanthe fistulosa
  Torilis japonica
Urticaceae
  Urtica dioica
Valerianaceae
  Valerianella sp.
Vitaceae
  Vitis vinifera
Zannichelliaceae
  Zannichellia palustris
Zosteraceae
  Zostera angustifolia
  Zostera marina
  Zostera noltii

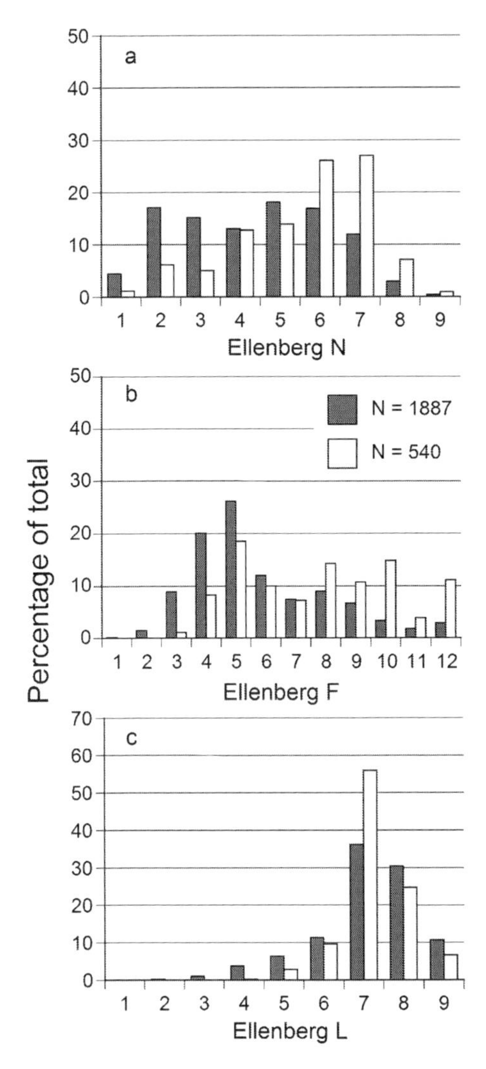

FIGURE 6.3. Analysis of plant species identified from the gut content of *Anas* dabbling ducks in Europe, in comparison to all European plant species for which Ellenberg values are available, showing that ducks feed disproportionately on plant species at (a) sites of rich fertility (Ellenberg *N* values 6–8), at (b) wet to inundated (open water) sites (Ellenberg *F* values 8–12), and at (c) habitats in transition from being semishaded to well lit (Ellenberg L value 7). The gray bars indicate the distribution of Ellenberg values over all plant species; the white bars indicate the distribution for species from duck gut contents. The total number of data points corresponding to dark and light bars for each graph are indicated in the middle panel.

be returned to terrestrial habitats by waterbirds—for example, by defecation on the shoreline or during flight, or when moved to temporary ponds or flooded grasslands that later dry out.

Analysis of the plant species identified from the gut contents of dabbling ducks in comparison to all plant species for which trait data are available in the LEDA database (fig. 6.4), shows that the ducks feed more or less proportionately on species relative to their seed production and seed buoyancy, but disproportionately more on plant species with seeds of relatively

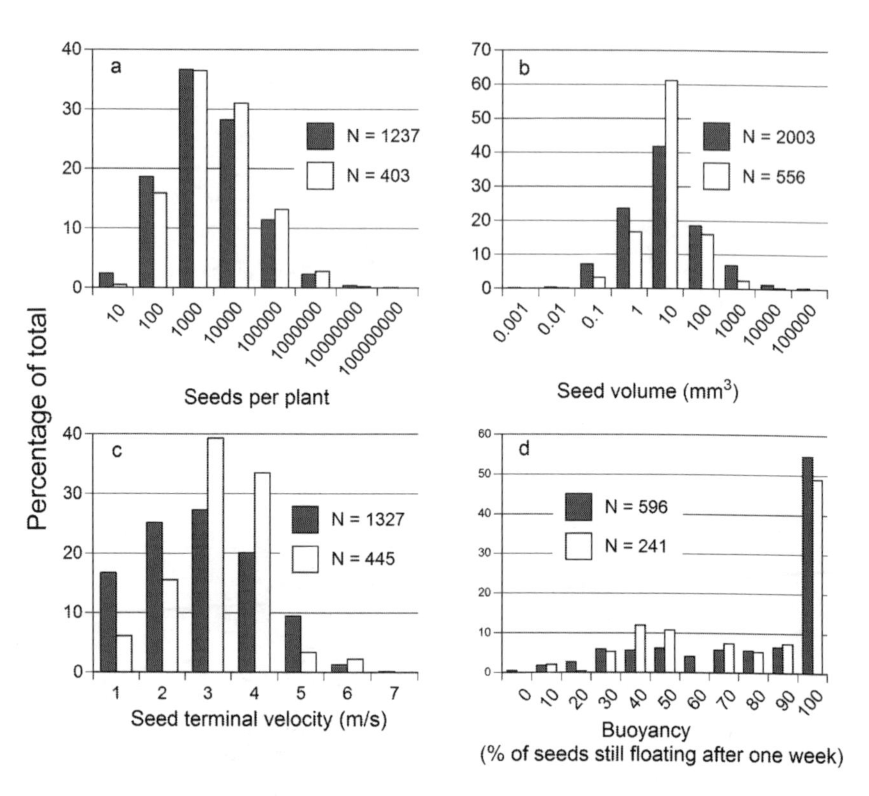

FIGURE 6.4. Analysis of the plant species identified from the gut content of *Anas* dabbling ducks in Europe, in comparison to all European plant species for which dispersal-related trait data are available, showing that the ducks feed more or less proportionately on species relative to their seed production (a) and seed buoyancy (d), but disproportionately on plant species with seeds of relatively small sizes (b: 1–10 mm^3) and high terminal velocities (c: 2–4 m/s). The gray bars indicate the distribution of trait values over all plant species; the white bars indicate the distribution for species from duck gut contents. The total number of data points included in the histograms is indicated in each panel.

small sizes (1–10 mm³) and high terminal velocities (2–4 ms^{-1}). Thus, plant species with relatively small seeds, lacking specific adaptations for wind dispersal, have a greater probability of being dispersed by dabbling ducks.

From this analysis it becomes clear that *Anas* ducks feed on and potentially disperse a very wide range of plant species, and that their role in plant dispersal is very important not only for aquatic species but for moist-soil and terrestrial species within the hydrological catchments of wetlands. This includes many species that on the basis of seed morphology would not be classified by plant ecologists as being primarily dispersed by animals. Such generalizations from measured traits will therefore underestimate a species' potential for dispersal by waterfowl.

Rarefaction analyses (fig. 6.5) show that, in a given study site, ducks are dispersing a high diversity of plant species, and even studies of several hundred duck individuals do not reach an asymptote in taxonomic richness of seeds. How many vascular plant species are being dispersed by ducks in Europe is anybody's guess, and while our review of the literature of dabbling duck diet has identified 445 taxa, the considerable differences between the few localities where detailed studies have been conducted suggests that, on a continental scale, thousands of plant species are being dispersed.

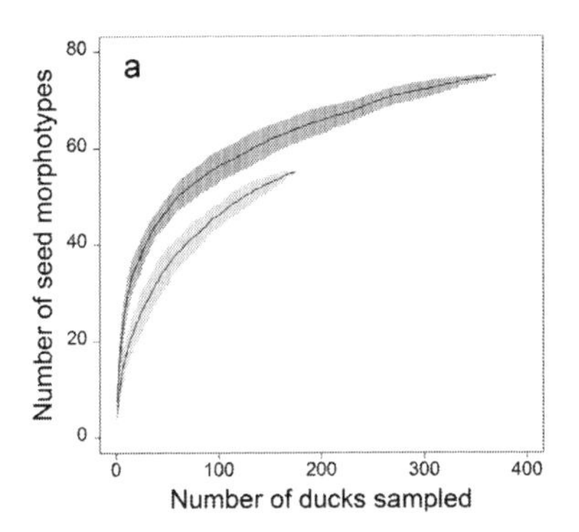

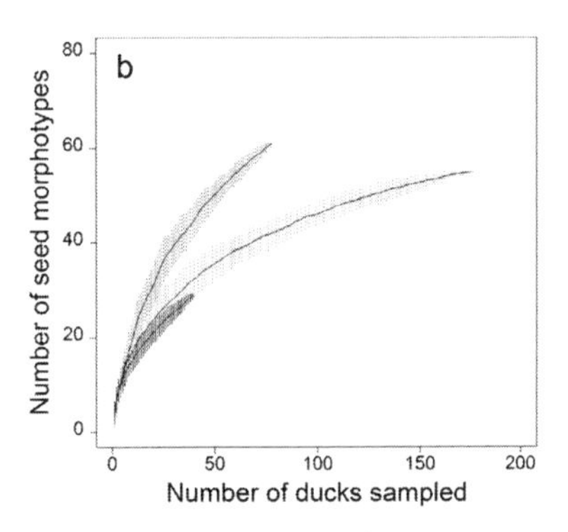

FIGURE 6.5. Rarefaction curves comparing the diversity of seed morphotypes found in the esophagus or gizzard of mallards and green-winged teals, showing means ± s.e. from random permutations of 50 birds: (a) mallard (light) and green-winged teal (dark), from the Camargue in France; (b) mallard from the Camargue (light), a study in the Netherlands (dark), and the Ebro Delta in Spain (black). Note the change in scale on the *X* axis. Data reanalyzed from Brochet et al. 2012a (Camargue), Brochet et al. unpublished (Ebro), and Kleyheeg et al. unpublished (Netherlands). Adjustments were made to the Netherlands data set to ensure comparability (e.g., merging *Carex* species as in the Camargue study).

Can Diaspore Morphology Be Used to Predict Which Diaspores Are Dispersed by Waterbirds?

Much general literature on plant dispersal has been based on the assumption that diaspore morphology provides reliable information as to dispersal means (an idea developed at length by van der Pijl 1972). However, waterbirds throw a spanner in the works of such a concept, as there is no way to reliably predict on the basis of morphology—other than (to some extent) size—which diaspores they will disperse, and the list of plant species potentially dispersed by waterbirds would seem to be enormous (perhaps all non-forest species with diaspores with a volume of less than ca. 10 mm3 which lie in hydrological catchments frequented by waterbirds). On the basis of small size and lack of fleshy fruits, Tiffney (2004) suggested that Cretaceous angiosperms were predominantly dispersed by abiotic means. Yet waterbird communities, especially shorebirds, were well established by the end of the Early Cretaceous (Lockley et al. 1992; Kim et al. 2012), and they were likely to be important diaspore vectors from the very beginning. Tiffney points out that the diaspores of most Cenozoic herbs do not exhibit clear morphological adaptations to vertebrate dispersal (i.e., they lack flesh). However, this cannot be taken as evidence against dispersal by waterbirds. Tiffney argues that biotic dispersal became more widespread and important in the Tertiary than in previous periods, but again by using flesh as the indicator of such dispersal. Likewise, we question Tiffney's assumption that a reduction in diaspore size during climate cooling (during the Tertiary) or with increasing latitude (in modern diaspores) can be taken to indicate a reduced role for biotic dispersal. Janzen (1984) effectively makes the same point by listing many plant genera, classically assumed to disperse by wind or water, which are regularly dispersed by large mammals.

Carlquist (1967) recognized the importance of shorebirds and other waterbirds in the dispersal of plants to oceanic islands, and argued that seed morphology could be used to separate seeds dispersed internally from those dispersed externally. Carlquist and Pauly (1985) later provided experimental support for a link between seed morphology and adhesive capacity for external transport. Nevertheless, many of the plants predicted by Carlquist (1967) to disperse externally are now known to be readily dispersed internally.

On the basis of a diet review and the intensive study of diaspore dispersal by green-winged teals, Brochet et al. (2009, 2010a) found that seed species dispersed by external transport could not be identified by presence of hooks, barbs, or other structures that might readily be interpreted as adaptations for external attachment. Indeed, there was no clear difference in morphology between seeds found on the outside and the inside of green-winged teals, with much overlap between dispersal modes. In summary, it seems that there are no reliable morphological means, other than seed size, to predict which seed types are dispersed by waterbirds, whether internally or externally. Likewise, the extensive literature on plant dispersal modes that makes predictions based on seed morphology (e.g. Tiffney 2004; Thorsen et al. 2009) seems unreliable in that it overlooks dispersal by waterbirds of seeds that lack the classical predictors of zoochory.

When Are Diaspores Dispersed by Waterbirds and in What Direction?

Internal transport of diaspores by waterbirds occurs throughout the annual cycle, even though diaspore production itself is often limited seasonally (Kleyheeg 2015). Ducks and other waterbirds frequently ingest diaspores from the seed bank in wetland sediments, where their availability can remain high even at the end of winter (Green et al. 2002). Although migratory ducks in the Northern Hemisphere typically consume more diaspores and fewer invertebrates in winter, they still ingest a variety of diaspores during the breeding season (Green et al. 2002; Rodríguez-Pérez and Green 2006). Few studies examine diets of ducks living in other climatic regions, and some found that seeds dominate the adult diet during the breeding season (Petrie 1996; Petrie and Rogers 1996). Outside the breeding season, and especially in winter, ducks and many other waterbirds typically undertake regular local movements between sites used for feeding and for resting, and these are often independent waterbodies (Kleyheeg 2015). Since most preening is carried out at the resting sites, diaspores can be carried on feathers or feet to these latter sites before being removed by preening. Thus, there is likely to be directional dispersal from feeding to resting sites both by endo- and epizoochory (Kleyheeg 2015). Studies of duck diet on stopover sites during spring and autumn migration in North America confirm that seeds are ingested in abundance on migration by dabbling and diving ducks, and that, while seeds of some plant species are recorded in the diet in greater abundance in autumn, others are found more in spring (Green et al. 2002).

Figuerola et al. (2002, 2003) conducted a particularly detailed field study of internal transport in Doñana, Spain, Europe's most important wintering site for waterfowl (Rendon et al. 2008). When comparing early winter (November and December, when wintering ducks are still arriving) with late winter (late February, when birds are leaving), they found no consistent difference in rates of seed dispersal, but instead a statistical interaction between season and bird species. For example, numbers of *Ruppia maritima* seeds were higher in late winter in mallard and northern pintail (*Anas acuta*) feces, and were lower in late winter for Eurasian coot, but did not change for northern shoveler (*A. clypeata*). Numbers of *Salicornia* seeds were higher in early winter for pintail, mallard, and coot, but there was no seasonal difference for shoveler.

Brochet et al. (2010a) found no seasonal variation between early and late winter in the overall rates of diaspore dispersal by green-winged teals in the Camargue when considering intact diaspores found at the end of the lower gut. However, the relative composition of different plant taxa in the diet of green-winged teals did vary during the course of the winter (Brochet et al. 2012a). Diaspores of some species (e.g., *Chara* spp.) were more frequently ingested (and hence dispersed) in early winter, while others (e.g., *Echinochloa* sp.) were more frequent in late winter. Likewise, the diet composition of mallards in the Netherlands varied greatly over the course of autumn and winter (Kleyheeg 2015).

In conclusion, seasonality influences both the distance and direction of plant dispersal by migratory waterbirds. Plant dispersal occurs year-round, and rates of dispersal may vary among sites and bird species in a manner specific to each plant species. Spring migration is likely to be particularly important for the dispersal of plants responding to climate change, and there is strong potential for long-distance dispersal by waterbirds during this period, even for plants that produce diaspores in summer or autumn.

Specificity or Redundancy: Potential Coevolution between Waterbirds and Diaspores

Although duck species can differ significantly in the number of diaspores carried for a given plant species, European studies show that, in a given wetland at a given time, different dabbling ducks overlap greatly in the plant species they consume and disperse (fig. 6.6; see also Figuerola et al.

2003). Although some duck species appear consistently more important as vectors (e.g., green-winged teal and mallard being more important than northern pintail in fig. 6.6), this is largely determined by the relative abundance of the duck species at that site. Green-winged teal and northern shoveler seem consistently to disperse a greater variety of seeds than do mallard on an individual basis (Brochet et al. 2009; see also fig. 6.5a), but the sheer abundance of mallard in some wetlands can make them the dominant vector for any plant species (e.g., in the Ebro delta; fig. 6.6). In Australia, Raulings et al. (2011) also found high similarity in the plant species dispersed by three duck species by both endo- and epizoochory.

Diaspore size influences the relative importance of different waterfowl species as vectors. Ducks with finer lamellae in their bills (such as the northern shoveler or the green-winged teal) tend to ingest relatively smaller diaspores than do ducks with coarse lamellae (such as mallard; Brochet et al. 2012b). We reanalyzed data from Brochet et al. 2012a for teal and mallard, taking the 11 diaspore types that were recorded in at least 20% of individuals of at least one of the duck species. The difference in the mean number of diaspores per bird (in the esophagus or gizzard) between mallard and teal was significantly correlated with diaspore mass, both in early and late winter (Spearman's correlation coefficient $rs = 0.57$, 0.56 respectively; $P < 0.001$). This pattern largely explains the differences between mallard and teal for individual plant species. For example, *Potamogeton pectinatus* and *P. nodosus* seeds are much larger than *P. pusillus* seeds, and mallards are particularly important vectors for the first two. Ducks with finer lamellae also tend to disperse a larger number of plant taxa (fig 6.5a and Figuerola et al. 2003).

The high degree of overlap among plant species dispersed by different duck species is also found across different waterbird families, as is shown by comparison of the plants recorded in diets of gulls by Calvino-Cancela (2011) with those recorded in ducks by Brochet et al. (2009). Likewise, the diaspores found in the guts of Eurasian coot and gadwall in the Camargue were very similar (Allouche and Tamisier 1984). Furthermore, seeds dispersed by terrestrial and aquatic bird species overlap. For example, viable *Sonchus oleraceus* seeds are dispersed both by shorebirds (Sánchez et al. 2006) and by Eurasian bullfinch *Pyrrhula vulgaris* (W. E. Collinge in Ridley 1930). Five (*Plantago lanceolata, Ranunculus repens, Rumex crispus, Polygonum aviculare, Galium aparine*) of 17 species recorded to have germinated from songbird droppings by Collinge (in Ridley 1930) were listed in duck diet in Europe by Brochet et al. (2009). Similarly, Heleno et al.

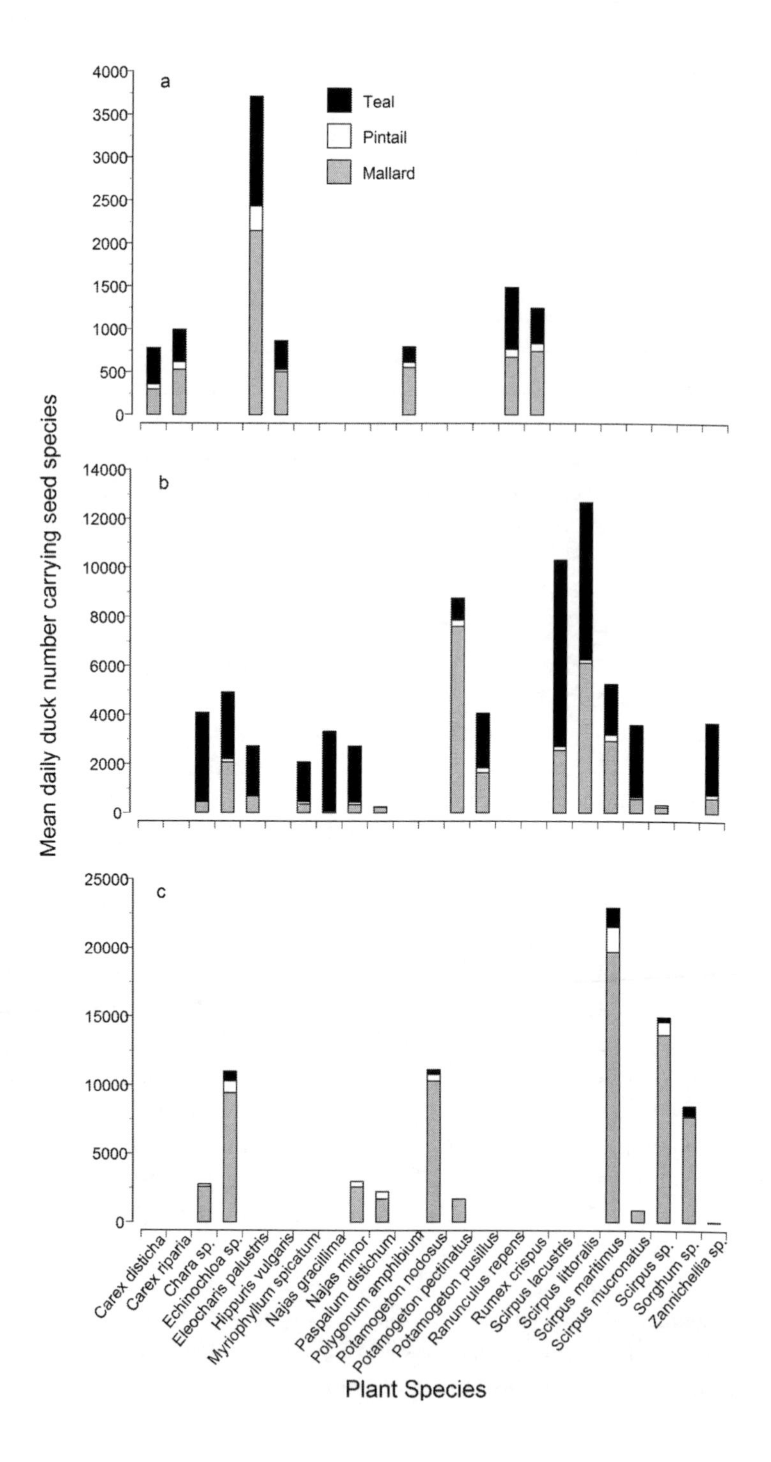

a
b
c
Teal
Pintail
Mallard
Mean daily duck number carrying seed species
Plant Species
Carex disticha
Carex riparia
Chara sp.
Echinochloa sp.
Eleocharis palustris
Hippuris vulgaris
Myriophyllum spicatum
Najas gracillima
Najas minor
Paspalum distichum
Polygonum amphibium
Potamogeton nodosus
Potamogeton pectinatus
Potamogeton pusillus
Ranunculus repens
Rumex crispus
Scirpus lacustris
Scirpus littoralis
Scirpus maritimus
Scirpus mucronatus
Scirpus sp.
Sorghum sp.
Zannichellia sp.

(2011) found that passerines dispersed Juncaceae, Cyperaceae and many other seeds without fleshy fruits.

Janzen (1984) envisaged that, prior to human intervention, large migratory mammals such as bovids, proboscideans, glyptodonts, or antilocaprids would have been the most important vectors of diaspores of many aquatic herbs of shallow seasonal marshes, which are now dependent on waterfowl. It is startling how many plant genera associated with pastures are dispersed by large mammals (Janzen 1984), and are also frequently dispersed by waterfowl (e.g. *Plantago*, *Medicago*, *Chenopodium*, *Carex*, *Juncus*, *Ranunculus*, *Polygonum*, *Atriplex*, *Paspalum*). Janzen speculated that waterfowl "probably offer only a pale shadow of what once could have been massive seed flow by large herbivores." If he is right, that only increases the importance of the role of waterbirds in conservation of modern ecosystems. Sumoski and Orth (2012) experimentally compared the potential of three fish species, a turtle, and the lesser scaup (*Aythya affinis*) to vector seeds of the seagrass *Zostera marina*. The duck was found to have a maximum seed dispersal distance at least 13 times greater than that of any other vector.

Diaspores dispersed by waterbirds are also dispersed by abiotic means, and both may select for diaspore characteristics simultaneously. Diplochory, in which the same individual seed is moved in successive steps, both by birds and by other processes, between the mother plant and germination site (Vander Wall 2004), may be common in waterbird-dispersed plant species. Van der Pijl (1972, 1982) recognized that diplochory as a combination of hydro- and endozoochory by waterbirds occurs frequently among Poaceae and Cyperaceae, and as a combination of zoochory and anemochory among Juncaceae and Cyperaceae. Seeds of dry habitats can be transported by waterbirds after being washed or blown into wetlands

---

FIGURE 6.6. Spatial variation in the dispersal potential of different seeds by different dabbling ducks. The mean number of individuals of green-winged teal (black), mallard (gray), and northern pintail (white) carrying at least one seed of each seed type are presented, on the basis of extrapolating the gut contents by the winter counts for that species, and averaged for the whole winter period. Gut content data taken from (a) Thomas 1982, for the Ouse washes in England (1969–72, esophagus plus gizzard), (b) Pirot 1981, for the Camargue in France (1964–81, esophagus), and (c) Brochet et al. unpublished, for the Ebro delta in Spain (1992–95, oesophagus plus gizzard). Winter counts from the above years were obtained from the British Trust for Ornithology for the Ouse washes, from Tour du Valat for the Camargue, and from Martí and del Moral 2002 for the Ebro Delta. Only seed types present in at least 20% of individuals of one duck species in one study are included.

(Cruden 1966). The presence of seeds of so many nonaquatic plants in the diet of ducks supports this, as does our review of the buoyancy and terminal velocity of those seeds (fig. 6.4).

In summary, the coevolutionary interaction between most plant species and waterbird species is likely to be very "diffuse," with little evidence for potentially tight coadaptation between a given disperser and a given propagule. Waterbirds disperse many diaspore types that are also dispersed by mammals, terrestrial birds or other animals, and/or by wind or water. Moreover, except in landscapes of low species richness such as oceanic islands, a plant regularly dispersed by one waterbird species is likely to be dispersed by several others, with considerable redundancy between the roles of different vectors.

Are Any Diaspores Adapted for Internal Transport by Waterbirds?

Contrary to van der Pijl (1972) and his assumption that waterbirds dispersed "non-adapted diaspores," and despite the diffuse interactions among waterbirds and plants, the relationship between some plants (e.g., some Cyperaceae, pondweeds, and Characeae) and waterbirds seems close enough so that diaspores may be dependent on internal transport by waterbirds, with the possibility of adaptation to these vectors by natural selection. Ridley (1930) proposed that the fleshy red disc in which the stony black achenes of *Scleria sumatrensis* (Cyperaceae) are supported is an adaptation to visually attract birds, possibly rails. Other authors, such as De Vlaming and Proctor (1968) and Morton and Hogg (1989), suggested that small, dry, hard nutlike seeds are adaptations for dispersal by waterbirds. De Vlaming and Proctor (1968) suggested that there is a phylogenetic component to such adaptation, with Cyperaceae, for example, being more adapted than Poaceae and Compositae (Asteraceae).

Small diaspore size, often assumed to be an adaptation for abiotic dispersal, especially by wind (Tiffney 2004), also favors waterbird dispersal (fig. 6.4), and these vectors could potentially also exert selection pressure for small size. For example, Charophyte oospores are particularly small, and were the most abundant diaspore in the intestines of green-winged teals (Brochet et al. 2010a).

Janzen (1984) argued that many herbaceous plants evolved small, hard, numerous seeds with dormancy as adaptations for internal transport by large mammals, even though these are amongst the "non-adapted diaspores" of van der Pijl (1972). Likewise, such diaspore features could favor, or even be selected for, internal transport by waterbirds (Kleyheeg 2015). As with seed dispersal by herbivorous mammals (Janzen 1984), larger seeds tend to have longer gut retention times in waterbirds, and therefore a greater chance of digestion and mortality. Investment in smaller seeds ensures both greater seed production and an increased probability of dispersal to a suitable microhabitat.

As suggested for seed dispersal by herbivorous mammals (Janzen 1984), selection for a seed coat that loses some resistance to germination cues during passage through the waterbird gut depends on whether the potential advantages for rapid germination following dispersal outweigh the potential advantages of delaying germination until some seasonal or successional cue appears in the habitat. In the case of aquatic diaspores, such a cue may include salinity, which itself can vary spatially, seasonally, and annually within a wetland complex. Espinar et al. (2004) found that the influence of gut passage on the germination of *Scirpus litoralis* seeds depends critically on salinity. Passage increased the germination rate at low salinities, but decreased it at high salinities. This response by the plant is potentially adaptive. When subjected to a similar experiment, seeds of *Juncus subulatus* survived gut passage, but exhibited no similar response to salinity variation (Espinar et al. 2006).

The proportion of diaspores destroyed during waterbird gut passage varies greatly (Soons et al. 2008; Brochet et al. 2010b). Nevertheless, even diaspore mortalities of more than 50% during gut passage are not good evidence that those diaspores are not adapted to dispersal by that vector (Janzen 1984). If waterbirds direct diaspores to microhabitats that are suitable for establishment, then high mortalities can be compensated for. Many terrestrial plants have toxic seeds to deter seed predators. The absence of toxic seeds among wetland plants supports the suggestion that these plants are partially adapted to passive dispersal by waterbirds. If, for example, Cyperaceae or pondweed seeds are not adapted for internal transport by waterbirds, then why aren't they toxic so as to avoid seed predation?

However, in the absence of pulp or other tissues that serve to attract animal vectors, it is very hard to establish what selective forces have led to the particular characteristics of diaspores of a given plant species. Not only are diaspores exposed to other modes of biotic and abiotic dispersal, their features are also selected to favor provision of adequate resources for the seedling, and/or survival in the soil or sediment seed bank.

Is Dispersal Effective? Do Seedlings Get Established after Dispersal?

Our knowledge about waterbird-mediated diaspore dispersal has rapidly increased in recent years, yet some aspects remain little studied. An essential question yet to be answered is: How effective is diaspore dispersal by waterbirds? Seed dispersal effectiveness can be expressed as "the number of new adult plants produced by the dispersal activities of a disperser" (Schupp 1993). This is the result of a series of events (or components of dispersal): ingestion of a certain number of seeds, digestion of a proportion thereof in the digestive tract, excretion in a certain habitat type, seed survival and germination, and plant survival until reproduction. Each of these events is hard to quantify under field conditions (see Herrera et al. 1994 for an example from a terrestrial system). To follow the whole cascade for individual seeds is essentially impossible. Although some separate components are now reasonably well studied for waterbird-plant systems, others are not. In particular, research on establishment success following seed dispersal is in its infancy, and has to date focused only on aquatic plants.

Aquatic habitats, compared to terrestrial habitats, provide ecological conditions that are relatively unsuitable for seed germination and seedling establishment (Santamaria 2002). In permanent habitats, clonal populations may persist for many years, even centuries, in the complete absence of seedling establishment. Asexual propagules, such as plant fragments, may disperse more successfully by hydrochory than seeds, owing to their higher establishment capacities. Their passage through the waterbird gut and subsequent dispersal must affect a seed's chances of success, but so far most research has focused simply on documenting the influence of gut passage on germination under laboratory conditions, and specifically on germination rate and germinability.

Competition in aquatic plant populations peaks during the growing season. Seeds might compensate for their poor competitive ability in comparison to established plants or vegetative propagules (which are produced only during the growing season) by their early arrival into a habitat via waterbirds and by early germination, which allows them to reach greater size before competition for light and other resources intensifies. Fennel pondweed (*P. pectinatus*) seeds grown in mesocosms (in the absence of competition or herbivory) were more likely to germinate early in winter when they had passed through a duck gut, although this early start had no effect on the plants' size at the end of the growing season (Figuerola et al. 2005b). Wigeongrass (*Ruppia maritima*) seeds planted in a marsh frequented by a high density of wintering waterfowl were less likely to produce mature plants when they had passed through the guts of ducks, because earlier germination left them more exposed to herbivory (Figuerola and Green 2004). Gut passage has been shown to accelerate germination in various other plant species, although this effect can be masked by exposure of seeds to cold temperatures prior to experiments (Brochet et al. 2010b), and may generally be dependent on water salinity (Espinar et al. 2004). In short, much more research is needed on the consequences of seed dispersal by waterbirds and gut passage for the establishment success of plants.

Conclusions and Benefits to People

Waterbirds are major vectors for a wide variety of plants outside of closed forest habitats, both in wetland and in terrestrial habitats. Tens of thousands of plant species worldwide are likely to benefit from waterbird dispersal for colonization of new habitats, directed dispersal to suitable but hydrologically unconnected sites, gene flow, enhanced germination, and escape from areas of high mortality. Nevertheless, a lack of basic research currently makes it impossible to estimate how many plant species are dispersed by waterbirds and how many waterbird species are effective dispersers for each taxonomic group of plants. Indeed, the limits to which plants can be effectively dispersed by waterbirds are still unclear. Other animals are often alternative vectors for the same plant species, but migratory waterbirds are often highly abundant and are uniquely able to disperse over long distances. Given the recent extinctions and losses of large mammals connecting habitat patches, the role of waterbirds in the dispersal of plants across and between landscapes is likely to have increased—and to increase further in the future. Under the ongoing fragmentation of natural habitats, the ability of waterbirds to fly across barriers is critical. The distances that diaspores are moved by waterbirds remain unclear and little studied, but maxima of hundreds or thousands of kilometers are likely for many migratory species (Viana et al. 2013; Kleyheeg 2015).

Diaspore dispersal by waterbirds is an ecosystem service—specifically, a "supporting service," according to a common classification (MEA

2005)—that is vital for the maintenance of plant biodiversity and connectivity between populations. Anatidae and other waterbirds play an essential role in the colonization and regeneration of new and restored wetlands by aquatic flora and fauna. Waterbirds play a vital role in maintaining connectivity between aquatic communities in isolated aquatic systems, and thus in maintaining species and genetic diversity (Amezaga et al. 2002). In many cases, plants that are dependent on waterbirds for their dispersal are keystone species—notably pondweeds and other submerged plants. The benefits these plants provide to humans are largely indirect, through control of soil erosion and sedimentation, flood prevention, water purification, carbon sequestration, providing essential habitat for fish, and so on. However, all of these are essential to humans. Direct benefits can also be provided—for example, from the Juncaceae and Cyperaceae that are traditionally used by humans for thatching. There may also be costs, because waterbirds can often be effective at spreading alien plants, including weeds.

In the past, passive dispersal of wetland plants by birds probably enabled the quick recolonization of extensive areas following glacial retreat (Santamaria 2002). Now, and increasingly in the future, plants require waterbirds as vectors if they are to colonize areas that become suitable under climate change. Waterbirds are already shifting their distributions in response to climate change (Visser et al. 2009; Godet et al. 2011), though they may track the changes with time lags that potentially have negative consequences for their own population viability. There is already evidence that waterbirds are enabling the colonization of polar regions by new plant species (Klein et al. 2008). It remains to be seen which plants will be able to shift their distributions fast enough via birds to avoid a crash in population range and size. What seems certain is that many species would have much less chance of shifting ranges if it were not for waterbirds. The ecosystem services that waterbirds provide by dispersing plants have an economic value, although no case studies have yet estimated it (Green and Elmberg 2014). One potential way of valuing part of the dispersal service by waterbirds would be to calculate the replacement costs of manually planting the wetland plant species that become established in and around created or restored wetlands after arriving via birds; those costs alone would be extremely high. In addition, the costs of replacing dikes that are vital in preventing floods, or in containing water in fish ponds, and which are protected from wave erosion by vegetation brought by birds, should be estimated. Recent progress in accounting for ecosystem services provided by migratory species (Semmens et al. 2011) is relevant, since many migratory waterbirds that provide dispersal services cross international borders. For example, ducks dispersing plant diaspores to new habitats in the United States breed largely in Canada. The valuation of such dispersal services by waterbirds is an important avenue for future research (Green and Elmberg 2014). However, to facilitate such valuation, much basic research is still needed to improve our understanding of which plants are dispersed by waterbirds, and where.

References

Allouche, L., and Tamisier, A. 1984. Feeding convergence of gadwall, coot and other herbivorous waterfowl species wintering in the Camargue: a preliminary approach. *Wildfowl* 35:135–42.

Amezaga, J. M., Santamaria, L., and Green, A. J. 2002. Biotic wetland connectivity: Supporting a new approach for wetland policy. *Acta Oecologica–International Journal of Ecology* 23:213–22.

Aoyama, Y., Kawakami, K., and Chiba, S. 2012. Seabirds as adhesive seed dispersers of alien and native plants in the oceanic Ogasawara Islands, Japan. *Biodiversity and Conservation* 21:2787–2801.

Arber, A. 1920. Water plants: A study of aquatic angiosperms. Cambridge: Cambridge University Press.

Bell, D. M. 2000. The ecology of coexisting *Eleocharis* species. PhD thesis, University of New England, Armidale.

Beltzer, A. H., Sabattini, R. A., and Marta, M. C. 1991. Ecología alimentaria de la polla de agua negra *Gallinula chloropus galeata* (Aves: Rallidae) en un ambiente lenítico del río Paraná medio, Argentina. *Ornitología Neotropical* 2:29–36.

Black, J. M., Prop, J., Hunter, J. M., Woog, F., Marshall, A. P., and Bowler, J. M. 1994. Foraging behaviour and energetics of the Hawaiian Goose *Branta sandvicensis*. *Wildfowl* 45:65–109.

Brochet, A. L., Guillemain, M., Fritz, H., Gauthier-Clerc, M., and Green, A. J. 2009. The role of migratory ducks in the long-distance dispersal of native plants and the spread of exotic plants in Europe. *Ecography* 32:919–28.

———. 2010a. Plant dispersal by teal (*Anas crecca*) in the Camargue: Duck guts are more important than their feet. *Freshwater Biology* 55:1262–73.

Brochet, A. L., Guillemain, M., Gauthier-Clerc, M., Fritz, H., and Green, A. J. 2010b. Endozoochory of Mediterranean aquatic plant seeds by teal after a period of desiccation: Determinants of seed survival and influence of retention time on germinability and viability. *Aquatic Botany* 93:99–106.

Brochet, A. L., Dessborn, L., Legagneux, P., Elmberg, J., Gauthier-Clerc, M., Fritz, H., and Guillemain, M. . 2012b. Is diet segregation between dabbling ducks due to food partitioning? A review of seasonal patterns in the Western Palearctic. *Journal of Zoology* 286:171–78.

Brochet, A. L., Mouronval, J. B., Aubry, P., Gauthier-Clerc, M., Green, A. J., Fritz, H. and Guillemain, M. 2012a. Diet and feeding habitats of Camargue dabbling ducks: What has changed since the 1960s? *Waterbirds* 35:555–76.

Bruun, H. H., Lundgren, R., and Philipp, M. 2008. Enhancement of local species richness in tundra by seed dispersal through guts of muskox and barnacle goose. *Oecologia* 155:101–10.

Burger, A. E. 2005. Dispersal and germination of seeds of Pisonia grandis, an Indo-Pacific tropical tree associated with insular seabird colonies. *Journal of Tropical Ecology* 21:263–71.

Calvino-Cancela, M. 2011. Gulls (Laridae) as frugivores and seed dispersers. *Plant Ecology* 212:1149–57.

Carlquist, S. 1967. Biota of Long-Distance Dispersal: V. Plant Dispersal to Pacific Islands. *Bulletin of the Torrey Botanical Club* 94:129–62.

Carlquist, S., and Pauly, Q. 1985. Experimental studies on epizoochorous dispersal in Californian plants. *Aliso* 11:167–77.

Chang, E. R., Zozaya, E. L., Kuijper, D. P. J., and Bakker, J. P. 2005. Seed dispersal by small herbivores and tidal water: Are they important filters in the assembly of salt-marsh communities? *Functional Ecology* 19:665–73.

Charalambidou, I., and Santamaría, L. 2002. Waterbirds as endozoochorous dispersers of aquatic organisms: A review of experimental evidence. *Acta Oecologica* 23:165–76.

———. 2005. Field evidence for the potential of waterbirds as dispersers of aquatic organisms. *Wetlands* 25:252–58.

Charalambidou, I., Santamaria, L., Jansen, C., and Nolet, B. A. 2005. Digestive plasticity in Mallard ducks modulates dispersal probabilities of aquatic plants and crustaceans. *Functional Ecology* 19:513–19.

Choi, C.-Y., Nam, H.-Y., and Chae H.-Y. 2010. Exotic seeds on the feathers of migratory birds on a stopover island in Korea. *Journal of Ecology and Field Biology* 33:19–22.

Cook, C. D. K. 1990. Seed dispersal of *Nymphoides peltata* (S.G. Gmelin) O. Kuntze (Menyanthaceae). *Aquatic Botany* 37:325–40.

Corlett, R. T. 1998. Frugivory and seed dispersal by vertebrates in the Oriental (Indomalayan) Region. *Biological Reviews of the Cambridge Philosophical Society* 73:413–48.

Coughlan, N. E., Kelly, T. C., and Jansen, M. A. K. . 2015. Mallard duck (*Anas platyrhynchos*)-mediated dispersal of Lemnaceae: A contributing factor in the spread of invasive *Lemna minuta*? *Plant Biology* 17:108–4.

Cruden, R. W. 1966. Birds as agents of long-distance dispersal for disjunct plant groups of temperate Western Hemisphere. *Evolution* 20:517–32.

Darwin, C. 1859. *The Origin of Species by Means of Natural Selection*. 1st ed. London: John Murray.

———. 1872. *The Origin of Species by Means of Natural Selection*. 6th ed. London: John Murray.

Davidson, N. C. 2014. How much wetland has the world lost? Long-term and recent trends in global wetland area. *Marine and Freshwater Research* 65: 934–41.

De Vlaming, V., and Proctor, V. W. 1968. Dispersal of aquatic organisms: Viability of seeds recovered from the droppings of captive killdeer and mallard ducks. *American Journal of Botany* 55:20–26.

Dehorter, O., and Guillemain, M. 2008. Global diversity of freshwater birds (Aves). *Hydrobiologia* 595:619–26.

Downing, J. A. 2009. Plenary lecture global limnology: Up-scaling aquatic services and processes to planet Earth. *Verh Internat Verein Limnol* 30:1149–66.

Ellenberg, H., Weber, H. E., Düll, R., Wirth, V., Werner, W., and Paulissen, D. 1991. Zeigerwerte von Pflanzen in Mitteleuropa. *Scripta Geobotanica* 18:1–248.

Espinar, J. L., Garcia, L. V., Figuerola, J., Green A. J., and Clemente, L. 2004. Helophyte germination in a Mediterranean salt marsh: Gut-passage by ducks changes seed response to salinity. *Journal of Vegetation Science* 15:315–22.

———. 2006. Effects of salinity and ingestion by ducks on germination patterns of Juncus subulatus seeds. *Journal of Arid Environments* 66:376–83.

Falla, R. A. 1960. Oceanic birds as dispersal agents. *Proceedings of the Royal Society London Series B* 152:655–59.

Figuerola, J., Charalambidou, I., Santamaria, L., and Green, A. J. 2010. Internal dispersal of seeds by waterfowl: Effect of seed size on gut passage time and germination patterns. *Naturwissenschaften* 97:555–65.

Figuerola, J., and Green, A. J. 2002a. Dispersal of aquatic organisms by waterbirds: A review of past research and priorities for future studies. *Freshwater Biology* 47:483–94.

———. 2002b. How frequent is external transport of seeds and invertebrate eggs by waterbirds? A study in Donana, SW Spain. *Archiv Fur Hydrobiologie* 155: 557–65.

———. 2004. Effects of seed ingestion and herbivory by waterfowl on seedling establishment: A field experiment with wigeongrass *Ruppia maritima* in Donana, south-west Spain. *Plant Ecology* 173:33–38.

Figuerola, J., Green, A. J., and Santamaria, L. 2002. Comparative dispersal effectiveness of wigeongrass seeds by waterfowl wintering in south-west Spain: Quantitative and qualitative aspects. *Journal of Ecology* 90:989–1001.

———. 2003. Passive internal transport of aquatic organisms by waterfowl in Doñana, south-west Spain. *Global Ecology and Biogeography* 12:427–36.

Figuerola, J., Mateo, R., Green, A. J., Mondain-Monval, J. Y., Lefranc, H., and Mentaberre, G. 2005a. Grit selection in waterfowl and how it determines exposure to ingested lead shot in Mediterranean wetlands. *Environmental Conservation* 32:226–34.

Figuerola, J., Santamaria, L., Green, A. J., Luque, I., Alvarez, R., and Charalambidou, I. 2005b. Endozoochorous dispersal of aquatic plants: Does seed gut passage affect plant performance? *American Journal of Botany* 92:696–99.

Fuentes, C., Green, A. J., Orr, J., and Olafsson, J. S. 2005. Seasonal variation in species composition and larval size of the benthic chironomid communities in brackish wetlands in southern Alicante, Spain. *Wetlands* 25:289–96.

García-Álvarez, A., van Leeuwen, C. H. A., Luque, C. J., Hussner, A., Vélez-Martín, A., Pérez-Vázquez, A., Green, A. J., and Castellanos, E. M. 2015. Internal transport of alien and native plants by geese and ducks: An experimental study. *Freshwater Biology* 60:1316–29.

Gill, R. E., Tibbitts, T. L., Douglas, D. C., Handel, C. M., Mulcahy, D. M., Gottschalck, J. C., Warnock, N., et al. 2009. Extreme endurance flights by landbirds crossing the Pacific Ocean: Ecological corridor rather than barrier? *Proceedings of the Royal Society B–Biological Sciences* 276:447–58.

Godet, L., Jaffre, M., and Devictor, V. 2011. Waders in winter: Long-term changes of migratory bird assemblages facing climate change. *Biology Letters* 7:714–17.

Gordon, E., and van der Valk, A. G. 2003. Secondary seed dispersal in *Montrichardia arborescens* (L.) schott-dominated wetlands in Laguna Grande, Venezuela. *Plant Ecology* 168:177–90.

Green, A. J., and Elmberg, J. 2014. Ecosystem services provided by waterbirds. *Biological Reviews* 89:105–22.

Green, A. J., Figuerola, J., and Sanchez, M. I. 2002. Implications of waterbird ecology for the dispersal of aquatic organisms. *Acta Oecologica–International Journal of Ecology* 23:177–89.

Green, A. J., Jenkins, K. M., Bell, D., Morris, P. J., and Kingsford, R. T. 2008. The potential role of waterbirds in dispersing invertebrates and plants in arid Australia. *Freshwater Biology* 53:380–92.

Guppy, H. B. 1906. *Observations of a Naturalist in the Pacific between 1896 and 1899. Vol. 2. Plant-dispersal.* London: Macmillan.

Gurd, D. B. 2006. Filter-feeding dabbling ducks (*Anas* spp.) can actively select particles by size. *Zoology* 109:120–26.

Hagy, H. M., and Kaminski, R. M. . 2012. Apparent seed use by ducks in moist-soil wetlands of the Mississippi alluvial valley. *Journal of Wildlife Management* 76:1053–1061.

Halse, S. A. 1984. Diet, body condition, and gut size of Egyptian geese. *Journal of Wildlife Management* 48:569–73.

———. 1985. Diet and size of the digestive organs of spur-winged geese. *Wildfowl* 36:129–34.

Heleno, R. H., Ross, G., Everard A., Memmott J., and Ramos, J. A. 2011. The role of avian "seed predators" as seed dispersers. *Ibis* 153:199–203.

Herrera, C. M., Jordano, P., Lopez Soria, L., Amat, J. A. 1994 Recruitment of a mast-fruiting, bird-dispersed tree: Bridging frugivore activity and seedling establishment. *Ecological Monographs* 64:315–44.

Heslop-Harrison, Y. 1955. *Nuphar* Sm. *Journal of Ecology* 43:342–64.

Hill, M. O., Preston, C. D., and Roy, D. B. 2004. *PLANTATT: Attributes of British and Irish plants: Status, size, life history, geography and habitats*. Huntington, U.K.: Centre for Ecology and Hydrology

Holmboe, J. 1900. Notizen über die endozische Samenverbreitung der Vögel. *Nyt magazin for naturuidenskaberne* 38.

Howe, H. F., and Smallwood, J. 1982. Ecology of seed dispersal. *Annual Review of Ecology and Systematics* 13:201–28.

Isaac-Renton, M., Bennett, J. R., Best, R. J., and Arcese, P. 2010. Effects of introduced Canada geese (Branta canadensis) on native plant communities of the Southern Gulf Islands, British Columbia. *Ecoscience* 17:394–99.

James, H. F., and Burney, D. A. . 1997. The diet and ecology of Hawaii's extinct flightless waterfowl: Evidence from coprolites. *Biological Journal of the Linnean Society* 62:279–97.

Janzen, D. H. 1984. Dispersal of small seeds by big herbivores: Foliage is the fruit. *American Naturalist* 123:338–53.

Kear, J. 1990. *Man and Wildfowl*. London: T & A.D. Poyser.

Kim, J. Y., Lockley, M. G., Seo, S. J., Kim, K. S., Kim, S. H., and Baek, K. S. 2012. A paradise of mesozoic birds: The world's richest and most diverse cretaceous bird track assemblage from the Early Cretaceous Haman formation of the Gajin Tracksite, Jinju, Korea. *Ichnos-an International Journal for Plant and Animal Traces* 19:28–42.

Klein, D. R., Bruun, H. H., Lundgren, R., and Philipp, M. 2008. Climate change influences on species interrelationships and distributions in high-Arctic Greenland. *Advances in Ecological Research* 40:81–100.

Kleyer, M., Bekker R. M., Knevel, I. C., Bakker, J. P., Thompson, K., Sonnenschein, M., Poschlod, P., et al. 2008. The LEDA Traitbase: A database of life-history traits of the Northwest European flora. *Journal of Ecology* 96:1266–74.

Kleyheeg, E. 2015. Seed dispersal by a generalist duck: Ingestion, digestion and transportation by mallards (*Anas platyrhynchos*). PhD thesis, Utrecht University, the Netherlands.

Kleyheeg, E., van Leeuwen, C. H., Morison, M. A., Nolet, B. A., and Soons, M. B. 2015. Bird-mediated seed dispersal: Reduced digestive efficiency in active birds modulates the dispersal capacity of plant seeds. *Oikos*. 124:899–907.

Lewis, L. R., Behling, E., Gousse, H., Qian, E., Elphick, C. S., Lamarre, J. F., Bety, J., Liebezeit, J., Rozzi, R., and Goffinet, B. 2014. First evidence of bryophyte diaspores in the plumage of transequatorial migrant birds. *PeerJ* 2:e424.

Lockley, M. G., Lim, J. D., Kim, J. Y., Kim, K. S., Huh, M., and Hwang, K. G. 2012. Tracking Korea's early birds: A review of cretaceous avian ichnology and its implications for evolution and behavior. *Ichnos* 19:17–27.

Lougheed, V. L., McIntosh, M. D., Parker, C. A., and Stevenson, R. J. 2008. Wetland degradation leads to homogenization of the biota at local and landscape scales. *Freshwater Biology* 53:2402–13.

Magnússon, B., Magnússon, S. H., and Fridriksson, S. 2009. Developments in plant colonization and succession on Surtsey during 1999–2008. *Surtsey Research* 12: 57–76.

Malone, C. R. 1966. Regurgitation of food by mallard ducks. *Wilson Bulletin* 78:227–28.

Marchant, S., and Higgins, P. J. 1990. *Handbook of Australian, New Zealand and Antarctic Birds*, vol. 1. Melbourne: OUP.

Martí, R., and del Moral, J. C. 2002. *La invernada de aves acuáticas en España*. Madrid: Organismo Autónomo Parques Nacionales, Ministerio de Medio Ambiente.

Maximilian, *Prince of Wied-Neuwied. 1820: Travels in Brazil in the Years 1815, 1816, 1817*. London: Henry Colburn & Co.

Milcu, A., Schumacher, J., and Scheu, S. 2006. Earthworms (*Lumbricus terrestris*) affect plant seedling recruitment and microhabitat heterogeneity. *Functional Ecology* 20:261–68.

Millennium Ecosystem Assessment (MEA). 2005. *Ecosystems and Human Well-Being: Wetlands and Water*. Washington: World Resources Institute.

Montalti, D., Arambarri, A. M., Soave, G. E., Darrieu, C. A., and Camperi, A. R. 2003. Seeds in the diet of the white-rumped sandpiper in Argentina. *Waterbirds* 26:166–68.

Morton, J. K., and Hogg, E. H. 1989. Biogeography of island floras in the Great Lakes. 2. Plant dispersal. *Canadian Journal of Botany / Revue Canadienne de Botanique* 67:1803–20.

Mouronval, J. B., Guillemain, M., Canny, A., and Poirier, F. 2007. Diet of non-breeding wildfowl Anatidae and Coot Fulica atra on the Perthois gravel pits, northeast France. *Wildfowl* 57:68–97.

Mueller, H. 1999. Common snipe, *Gallinago gallinago*. In *The Birds of North America*, A. Poole, and F. Gill, ed., No. 417. Philadelphia: The Academy of Natural Sciences and Washington, D.C.: The American Ornithologists' Union.

Nathan, R., Perry, G., Cronin, J. T., Strand, A. E., and Cain, M. L. 2003. Methods for estimating long-distance dispersal. *Oikos* 103:261–73.

Neff, K. P., and A. H. Baldwin. 2005. Seed dispersal into wetlands: Techniques and results for a restored tidal freshwater marsh. *Wetlands* 25:392–404.

Petrie, S. A. 1996. Red-billed teal foods in semiarid South Africa: A North-Temperate contrast. *Journal of Wildlife Management* 60:874–81.

Petrie, S. A., and Rogers, K. H. 1996. Foods consumed by breeding white-faced whistling ducks (*Dendrocygna viduata*) on the Nyl river floodplain, South Africa. *Gibier Faune Sauvage* 13:755–71.

Pirot, J. Y. 1981. Partage alimentaire et spatial des zones humides camarguaises par cinq espèces de canards en hivernage et en transit. PhD thesis, Pierre et Marie Curie University, Paris.

Proctor, V. W. 1968. Long-distance dispersal of seeds by retention in digestive tract of birds. *Science* 160:321–22.

Purves, D. W., and Dushoff, J. 2005. Directed seed dispersal and metapopulation response to habitat loss and disturbance: Application to *Eichhornia paniculata*. *Journal of Ecology* 93:658–69.

Raulings, E., Morris, K., Thompson, R., and Mac Nally, R. 2011. Do birds of a feather disperse plants together? *Freshwater Biology* 56:1390–1402.

Rendon, M. A., Green, A. J., Aquilera, E., and Almaraz, P. 2008. Status, distribution and long-term changes in the waterbird community wintering in Donana, south-west Spain. *Biological Conservation* 141:1371–88.

Ridley, H. N. 1930. *The Dispersal of Plants throughout the World*. Ashford, Kent, UK: L. Reeve & Co.

Rodríguez, R., and Hiraldo, F. 1975. Régimen alimenticio del Calamón (*Porphyrio porphyrio*) en las marismas del Guadalquivir. *Doñana Acta Vertebrata* 2:201–13.

Rodriguez, A., Rodriguez, B., Rumeu, B., and Nogales, M. 2007. Seasonal diet of the grey heron *Ardea cinerea* on an oceanic island (Tenerife, Canary Islands): Indirect interaction with wild seed plants. *Acta Ornithologica* 42:77–87.

Rodríguez-Pérez, H., and Green, A. J. 2006. Waterbird impacts on widgeongrass *Ruppia maritima* in a Mediterranean wetland: Comparing bird groups and seasonal effects. *Oikos* 112:525–34.

Sánchez, M. I., Green, A. J., and Castellanos, E. M. 2006. Internal transport of seeds by migratory waders in the Odiel marshes, south-west Spain: Consequences for long-distance dispersal. *Journal of Avian Biology* 37:201–6.

Santiago-Quesada, F., J. A. Masero, N. Albano, A. Villegas, and J. M. Sanchez-Guzman. 2009. Sex differences in digestive traits in sexually size-dimorphic birds: Insights from an assimilation efficiency experiment on black-tailed godwit. *Comparative Biochemistry and Physiology a. Molecular & Integrative Physiology* 152:565–68.

Sarneel, J. M., Beltman, B., Buijze, A., Groen, R., and Soons, M. B. 2013. The role of wind in the dispersal of floating seeds in slow flowing or stagnant water bodies. *Journal of Vegetation Science* 25:262–74.

Schupp, E. W. 1993. Quantity, quality and the effectiveness of seed dispersal by animals. *Vegetation* 107/108:15–29.

Şekercioğlu, Ç. H. 2006. Increasing awareness of avian ecological function. *Trends in Ecology & Evolution* 21:464–71.

Semmens, D. J., Diffendorfer, J. E., Lopez-Hoffman, L., and Shapiro, C. D. 2011. Accounting for the ecosystem services of migratory species: Quantifying migration support and spatial subsidies. *Ecological Economics* 70:2236–42.

Sibly, R. M. 1981. Strategies of digestion and defecation. In *Physiological Ecology: An Evolutionary Approach to Resource Use*, ed. C. R. Townsend and P. Calow, 109–39. Sunderland, MA: Sinauer.

Smits, A. J. M., Vanruremonde, R., and Vandervelde, G. 1989. Seed dispersal of 3 nymphaeid macrophytes. *Aquatic Botany* 35:167–80.

Soltis, D. E., Bell, C. D., Kim, S., and Soltis, P. S. 2008. Origin and early evolution of angiosperms. *Year in Evolutionary Biology 2008* 1133:3–25.

Soomers, H., Karssenberg, D., Soons, M. B., Verweij, P., Verhoeven, J. T. A., and Wassen, M. J. 2013. Wind and water dispersal of wetland plants across fragmented landscapes. *Ecosystems* 16:434–51.

Soons, M.B. 2006. Wind dispersal in freshwater wetlands: Knowledge for conservation and restoration. *Applied Vegetation Science* 9:271–78.

Soons, M. B., van der Vlugt, C., van Lith, B., Heil, G. W., and Klaassen, M. 2008. Small seed size increases the potential for dispersal of wetland plants by ducks. *Journal of Ecology* 96:619–27.

Sumoski, S. E., and Orth, R. J. . 2012. Biotic dispersal in eelgrass *Zostera marina*. *Marine Ecology Progress Series* 471:1–10.

Summers, R. W., Stansfield, J., Perry, S., Atkins, C., and Bishop, J. 1993. Utilization, diet and diet selection by brent geese *Branta bernicla bernicla* on saltmarshes in Norfolk. *Journal of Zoology* 231:249–73.

Tárano, Z., Stahl, S., and Ojasti, J. 1995. Feeding ecology of the purple gallinule (*Porphyrula martinica*) in the central Llanos of Venezuela. *Ecotropicos* 8:53–61.

Taylor, B. W. 1954. An example of long-distance dispersal. *Ecology* 35:569–72.

Thomas, G. J. 1982. Autumn and winter feeding ecology of waterfowl at the Ouse Washes, England. *Journal of Zoology* 197:131–72.

Thorsen, M. J., Dickinson, K. J. M., and Seddon, P. J. 2009. Seed dispersal systems in the New Zealand flora. *Perspectives in Plant Ecology Evolution and Systematics* 11:285–309.

Tiffney, B. H. 2004. Vertebrate dispersal of seed plants through time. *Annual Review of Ecology Evolution and Systematics* 35:1–29.

Traveset, A. 1998. Effect of seed passage through vertebrate frugivores' guts on germination: A review. *Perspectives in Plant Ecology Evolution and Systematics* 1/2:151–90.

Tréca, B. 1981. Régime alimentaire du Dendrocygne veuf (*Dendrocygna viduata*) dans le delta du Sénégal. *L'Oiseau et la Revue Française d'Ornithologie* 51:219–38.

United States Fish and Wildlife Service (USFWS). 2012. *Waterfowl Population Status, 2012*. Washington: US Department of the Interior.

Van der Pijl, L. 1972. *Principles of Dispersal in Higher Plants*. 2nd edition. Berlin: Springer-Verlag.

———. 1982. *Principles of Dispersal in Higher Plants*. 3rd edition. Berlin: Springer-Verlag.

Van Leeuwen, C. H. A., Van der Velde, G., Van Groenendael, J. M., and Klaassen, M. 2012. Gut travellers: Internal dispersal of aquatic organisms by waterfowl. *Journal of Biogeography* 39:2031–40.

Vander Wall, S. B., and Longland, W. S. 2004. Diplochory: Are two seed dispersers better than one? *Trends in Ecology & Evolution* 19:155–61.

Viain, A., F. Corre, P. Delaporte, E.Joyeux, and P. Bocher. 2011. Numbers, diet and feeding methods of common shelduck *Tadorna tadorna* wintering in the estuarine bays of Aiguillon and Marennes-Oléron, western France. *Wildfowl* 61:121–41.

Viana, D. S., Santamaria, L., Michot, T. C., and Figuerola, J. 2013. Migratory strategies of waterbirds shape the continental-scale dispersal of aquatic organisms. *Ecography* 36:430–38.

Visser, M. E., Perdeck, A. C., van Balen, J. H., and Both, C. 2009. Climate change leads to decreasing bird migration distances. *Global Change Biology* 15:1859–65.

Vivian-Smith, G., and Stiles, E. W. 1994. Dispersal of salt marsh seeds on the feet and feathers of waterfowl. *Wetlands* 14:316–19.

Weddell, H. A. 1849. Observations sur une espece nouvelle du genre *Wolffia* (Lemnacees). *Annales des Sciences Naturelles (Botanique)* 12:155–73.

Wetlands International. 2012. *Waterbird Population Estimates, Fifth Edition: Summary Report*. Wageningen, the Netherlands: Wetlands International.

Willson, M. F., Traveset, A., and Sabag, C. 1997. Geese as frugivores and probable seed-dispersal mutualists. *Journal of Field Ornithology* 68:144–46.

Wongsriphuek, C., Dugger, B. D., and Bartuszevige, A. M. 2008. Dispersal of wetland plant seeds by mallards: Influence of gut passage on recovery, retention, and germination. *Wetlands* 28:290–99.

Wormworth, J., Şekercioğlu, Ç. H. 2011. *Winged Sentinels: Birds and Climate Change.* Cambridge University Press, New York.

CHAPTER SEVEN

Seed Dispersal by Corvids

Birds That Build Forests

Diana F. Tomback

Birds of the family Corvidae, which comprises the familiar jays, magpies, crows, and ravens, occur nearly worldwide (Goodwin 1976; Clements 2007; Dittman and Cardiff 2009). The corvids include species that are common human commensals in urban and rural landscapes, as well as species that occupy remote wildlands (Goodwin 1976; Beletsky 2006). Many corvids store excess food for later use, concealing it in places such as crevices and under objects (Turcek and Kelso 1968; Goodwin 1976; Vander Wall 1990). This chapter focuses on the seed dispersal services provided by granivorous birds of the family Corvidae. Seed storage by these corvids leads to plant regeneration. Granivorous corvids scatter-hoard—that is, they cache seeds by burying one or more seeds in many different locations (Morris 1962; Vander Wall 1990).

Seed dispersal by corvids is "directed dispersal," whereby seeds are placed in selected microsites often favoring seedling establishment (Howe and Smallwood 1982; Wenny 2001). In contrast, seed dispersal by frugivores and waterfowl is far more opportunistic and random (chapters 5 and 7). Whereas avian frugivores (chapter 5) and waterfowl (chapter 6) disperse the seeds of numerous plants but especially understory woody, herbaceous, aquatic, weedy and exotic species, the granivorous corvids disperse the seeds of native forest trees and shrubs. Consequently, corvid seed dispersal shapes the composition and distribution of major broad-leaved and needle-leaved forest communities of the north temperate zone, including the high elevation cloud forests of Mexico and Central America.

Relatively few of the known corvid-tree genera are present in the southern hemisphere (Richardson and Rundel 1998; Vander Wall 2001).

Several corvid genera are dependable seed dispersers for widely distributed woody plant taxa, especially the genus *Pinus* (pine trees) of the Gymnosperm family Pinaceae and several genera of the Angiosperm family Fagaceae, resulting in coevolution and coadaptation (e.g., Bossema 1979; Tomback and Linhart 1990; Vander Wall 2001). These processes have shaped both bird and plant traits through reciprocal selection pressures, resulting in coevolved, mutualistic interactions. In contrast to most, often diffuse, seed dispersal interactions involving fleshy fruits and birds (chapter 5), interactions between corvid species and particular forest tree and shrub species are more specialized. In the case of nutcrackers (*Nucifraga* spp.) and the stone pines with large, wingless seeds, the birds are obligate seed dispersers (Tomback and Linhart 1990).

Seed-storing corvids help shape forest composition and distribution across landscapes, facilitate response to changing climate, and regenerate communities after disturbance (e.g., Tomback and Linhart 1990; Johnson et al. 1997; Gómez 2003). Individuals of the same corvid species may travel across community types and disperse the seeds of different forest species, thus serving as important mobile links connecting forest ecosystems (Johnson and Adkisson 1985; Tomback and Linhart 1990; Tomback and Kendall 2001). Globally, temperate forest ecosystems deliver important goods and ecosystem services to humans, including climate regulation, carbon sequestration, clean water, nutrient cycling, waste treatment, food production, building materials, and recreational opportunities. Costanza et al. (1997) estimated the global value of ecosystem services and products from the combined area (2.95 billion ha) of temperate and boreal zone forests as $894 billion annually; 210 million hectares of temperate forests alone are in the United States, delivering about $64 billion in goods and services (not adjusted for current dollar values; Pimentel et al. 1997; Krieger 2001). Examples below illustrate the economic valuation of corvid seed dispersal services, and demonstrate the consequences of declining seed dispersal services of an obligate corvid mutualist for a widespread forest tree.

General Corvid Traits

The Corvidae (Order Passeriformes; ca. 25 genera, ca. 130 species) inhabit every continent except Antarctica and nearly every environment, from tropical and desert, and coastal and island, to high mountains (Goodwin 1976; Beletsky 2006; Dittman and Cardiff 2009; Gill and Donsker 2014). Larger than most songbirds, corvids range in length from 20 to 71 cm; the largest songbird is the common raven (*Corvus corax*; Goodwin 1976; Beletsky 2006). Shared morphological traits important to corvids' life histories include a long, sturdy bill, an expandable esophagus (gular pouch) or throat (buccal cavity), and strong legs and feet. Key behaviors include transportation of food items in the throat or gular pouch, food storage, exploratory behavior with rapid learning, and omnivorous and sometimes predatory foraging habits (Amadon 1944; Goodwin 1976). All food-storing corvid species may use spatial memory with varying degrees of acuity to relocate their food stores (e.g., Waite 1985; Bunch and Tomback 1986; Verbeek 1997). Crows (*Corvus* spp.) and their allies, for example, demonstrate the variety and plasticity of corvid foraging behavior. Rooks (*Corvus frugilegus*), carrion crows (*Corvus corone*), and magpies (*Pica pica*) in Great Britain harvest acorns from forest remnants, carrying one in the throat and one in the bill, and cache them in agricultural fields. In urban areas, American crows (*Corvus brachyrhynchos*) drop nuts and fruits onto hard pavement to crack the shells or husks, and also drop these objects on paved roads, so as to use motor vehicles as "nutcrackers" (Maple 1974; Grobecker and Pietsch 1978). Similarly, northwestern crows (*Corvus caurinus*) and other species drop clams on rocks to break their shells, and also store a variety of intertidal foods (James and Verbeek 1983; Richardson and Verbeek 1986). New Caledonian crows (*Corvus moneduloides*) make specialized tools from the leaves of *Pandanus* spp. to probe for arthropods and other prey (Hunt 1996; 2000).

Specialized Corvid Traits for Seed Dispersal

A few corvid species possess specialized life histories based on a seed or nut diet and greater dependence on stored food for survival during food scarcity and for feeding their young, ultimately leading to effective seed dispersal (Vander Wall and Balda 1981; Johnson and Webb 1989; Tomback and Linhart 1990). These species have evolved a specialized bill morphology which enables them to grip, pull, or pry out nuts and seeds, or to tear open cones to access seeds. For example, the sturdy, long, sharp, and slightly decurved bill of the spotted or Eurasian nutcracker (*Nucifraga caryocatactes*) varies in depth and width geographically, potentially in relation to the hardness of the "shells" of its major seed sources, which

FIGURE 7.1. Clark's nutcracker (*Nucifraga columbiana*) illustrates how basic corvid morphological traits have become specialized for seed harvest, transport, and caching. (a) Note the long, sharp bill and the sublingual pouch (black arrow) which holds seeds. (b) The nutcracker's sturdy feet are used to grip cones and stabilize the bird as it uses the long, slender bill to pull seeds from opening limber pine (*Pinus flexilis*) cones. Photos by Diana F. Tomback.

include conifer seeds and hazelnuts (*Corylus avellana*) (Turcek and Kelso 1968; Roselaar 1994). A small protrusion (rhamphothecal bulge) on the ventral lower mandible may provide structural support for hammering on seeds or cones. Several New World jays, including scrub jays (*Aphelocoma* spp.) and blue jays (*Cyanocitta cristata*), have modifications of their cranium and lower jaw articulation adapted for pounding on seeds with their bills (Zusi 1987).

For transporting seeds and nuts, these corvids have either an expandable esophagus (e.g., the Eurasian jay, *Garrulus glandarius*, and pinyon jay, *Gymnorhinus cyanocephalus*); an enlarged buccal cavity (e.g., crows and ravens, *Corvus* spp., and magpies, *Pica* spp.); or a sublingual pouch (nutcrackers, *Nucifraga* spp.; Turcek and Kelso 1968; Bock et al. 1973). Eurasian jays may transport as many as 9 acorns or 15 beech nuts at a time. Pinyon jays may carry more than 50 pinyon seeds (Turcek and Kelso 1968; Vander Wall and Balda 1981). The sublingual pouch, which holds 100 or more pine seeds, enables nutcrackers to harvest seeds from remote stands of trees or forest communities and to scatter-hoard (cache) them within their home ranges (Vander Wall and Balda 1977; Mattes 1978; Tomback 1978; Lorenz et al. 2011). In addition, strong legs and feet enable nutcrackers and pinyon jays to grip branches or cones while foraging for seeds (Tomback 1978) (fig. 7.1).

The tree nut dispersers scatter-hoard single nut caches both inside and outside their territories (Bossema 1979; Darley-Hill and Johnson 1981; Cristol 2005). Nutcrackers store 1 to 15 or more pine seeds per cache, with means of three or four seeds (Tomback 1978; Mattes 1978). Pinyon jays, western scrub-jays (*Aphelocoma californica*), and Steller's jays (*Cyanocitta stelleri*) primarily make single seed caches, but pinyon jays will sometimes cache up to seven seeds (Vander Wall and Balda 1981). Nutcrackers and pinyon jays store seeds in many areas within an expansive home range, whereas scrub jays and Steller's jays store seeds within territories.

The breeding biology of the specialized pine seed dispersers reflects the importance of seed stores for survival and reproduction. The pinyon jays and the nutcrackers feed their nestlings and juveniles pine seeds retrieved from scatter hoards made the previous fall, which allows early nesting (Mewaldt 1956; Ligon 1978; Roselaar 1994). This early phenology may enable sufficient development of juveniles by late summer so they can store seeds themselves (Vander Wall and Balda 1977; Tomback 1978). Both nutcracker sexes develop an incubation patch, allowing males to incubate eggs and brood young while females retrieve seeds from their own caches (Mewaldt 1952). In some populations, pinyon jays breed in late summer in response to a new and abundant pinyon pine cone crop (Ligon 1974).

Scatter-hoarding many caches each year is associated with a well-developed spatial memory, particularly among the pine seed dispersers. Nutcrackers and pinyon jays cache seeds within large home ranges and varied terrain, with individual birds potentially making thousands of caches in good seed years (e.g., see Vander Wall and Balda 1977; Ligon 1978; Tomback 1982; Hutchins and Lanner 1982). Steller's jays and western scrub-jays are less dependent on seeds, and cache within more restrictive home ranges (Vander Wall and Balda 1981). Under controlled experimental conditions, both nutcrackers and pinyon jays recovered their own caches with higher accuracy than scrub jays, although all three species used spatial memory (Balda and Kamil 1989).

Lastly, although the specialized granivores consume mast seeds and/or conifer seeds year-round, they also feed opportunistically on arthropods and other invertebrates, plant material, carrion, small vertebrates, and eggs (Tomback 1978; Giuntoli and Mewaldt 1978; Vander Wall and Balda 1981). For food specialists dependent on an episodic food source (seed crops), omnivory may also be the strategy that permits specialization, enabling birds to survive seasons or even years when preferred foods are scarce (Tomback and Linhart 1990).

Adaptive Values, Effective Dispersal, and Landscape-Scale Impacts

Adaptive Values for Disperser and Plant

The adaptive value of food storage to an individual has been discussed at length elsewhere (e.g., Vander Wall 1990 and references therein). For selection to favor food hoarding, an individual must experience an increase in fitness from this activity, despite the energetic cost of food storage and retrieval and potential cache losses (e.g., Andersson and Krebs 1978; Brodin and Ekman 1994; Gendron and Reichman 1995). Hitchcock and Houston (1994) suggest that even a small quantity of reserved food can be highly critical to individual or group survival during a short but severe period of food scarcity.

One important characteristic of corvid seed dispersal is the transportation of seeds over distances of several kilometers or more from parent trees. The benefits of long-distance seed dispersal to plant conservation, biodiversity, and response to climate change are increasingly recognized (Soons and Ozinga 2005; Trakhtenbrot et al. 2005). For example, in naturally patchy landscapes or landscapes fragmented by development, long-distance seed dispersal by corvids maintains forest patches and regional populations over time, thus creating metapopulation structures (e.g., Johnson and Adkisson 1985; Webster and Johnson 2000).

As with seed dispersal by fruit-eating birds (chapter 5), corvid seed dispersal provides escape from seed predators and density-dependent mortality, colonization of open sites, and directed dispersal to especially suitable sites. In fact, seed dispersal by nutcrackers is one of the clearest examples of directed dispersal, whereby seeds are placed in sites supporting germination and establishment (e.g., Wenny 2001). The benefits of seed dispersal to plants are often complex to assess, since different life stages of a plant may have different requirements for optimal success, and dispersal by birds may thus influence population demographics (e.g., Howe and Miriti 2004).

Effective Seed Dispersal and Constraints

Whether a corvid species becomes an effective seed disperser for a given plant species may depend on placement and construction of hoards or caches (e.g., Tomback and Linhart 1990; Johnson et al. 1997; Kunstler

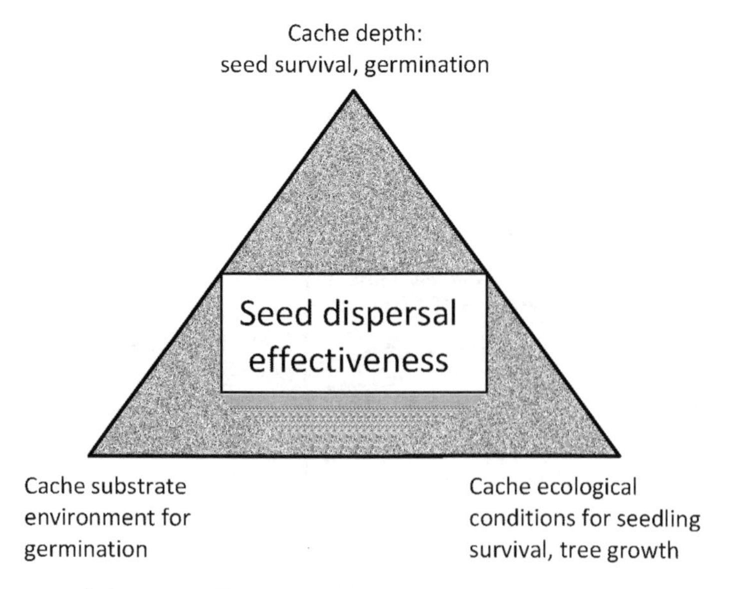

FIGURE 7.2. Cache compatibility triangle. See text for explanation.

et al. 2007). For the bird, cache site selection and cache preparation may be constrained by the energetic costs of transporting seeds, selecting sites that will be accessible in the future, and making each cache. The latter behavior may involve trade-offs between cache depth and the number of seeds cached in order to minimize olfactory stimuli and pilfering (e.g., Tomback 1978; Vander Wall 1993). Plants may also adapt to disperser behavior (e.g., Linhart and Tomback 1990; Galetti et al. 2013). Three determinants of corvid dispersal effectiveness are represented here by a cache compatibility triangle (fig. 7.2). (1) Cache depth is sufficient to protect seeds from predation, desiccation, and solar radiation, but not so deep that the radicle cannot emerge above ground (e.g., Vander Wall 1993). (2) The cache substrate and immediate environment potentially support germination and seedling survival. (3) General ecological conditions or community successional status in which caches are made determine the potential for seed and seedling survival, establishment, and tree growth, such as under a canopy gap, at the edge of a community or in an early seral stage of a community (e.g., Howe and Miriti 2004; Kunstler et al. 2007).

The effectiveness of seed dispersal also depends on the proportion of caches placed in suitable sites, the size and frequency of the seed crop, and the size of the disperser population in relation to seed availability, which impacts predispersal predation as well as the proportion of caches retrieved. Corvid seed dispersers are also seed predators, and large populations of dispersers and poor seed production may result in little to no plant recruitment. A similar effect comes from annual variation in seed production. Large seed and mast crops usually occur every few years (Sork 1993; Kelly 1994; Koenig and Knops 2005), and intervening small seed crops risk nearly complete seed predation (e.g., Tomback 1988; Siepielski and Benkman 2007).

Tree Regeneration and Afforestation at Landscape Scales

Corvids transport seeds from parent trees to caching sites over distances as great as 4 km for jays and crows and approximately 32 km for nutcrackers (e.g., Vander Wall and Balda 1977; Darley-Hill and Johnson 1981; Gómez 2003; Pons and Pausas 2007a; Lorenz et al. 2011). They may disperse seeds across different elevations and communities, including fragmented or disconnected landscapes (e.g., Tomback 1978, 2005; Johnson et al. 1997; Lorenz et al. 2011). Corvids often cache seeds in open habitats, such as canopy gaps or abandoned fields and other disturbed areas, and especially recently burned terrain (e.g., Darley-Hill and Johnson 1981; Tomback 1986; Johnson et al. 1997; Tomback et al. 2001a). For early successional and shade-intolerant tree species, scatter-hoarding by corvids leads to forest development, and caches under canopies maintain late successional forests (Tomback 1978; Johnson et al. 1997; Tomback et al. 2001a). Seed dispersal by corvids facilitates rapid distributional changes in response to climate change. Cached seeds along the edges of forest communities, above upper treeline, and below lower treeline may enable forests to rapidly track climate as it warms or cools (Johnson and Webb 1989; Johnson et al. 1997; Tomback 2005).

Finally, seed dispersal by corvids creates distinct fine-scale and landscape-scale population genetic structure, which differs from that produced by wind-dispersed tree species (Tomback and Linhart 1990; Rogers et al. 1999; Bruederle et al. 2001). Seed dispersal by corvids across long distances also homogenizes genetic diversity over a regional scale.

Dispersal of Tree Nuts versus Pine Seeds

Corvid granivores evolved along two different trajectories: dependence on tree nuts or mast from broad-leaved trees (especially those in family Fagaceae), and dependence on large seeds from needle-leaved trees (family Pinaceae, genus *Pinus*). These specializations involve different corvid species, different seed caching strategies, and different forest communities. Several corvids bridge the two groups, foraging on and storing either seed type when available. Some widely distributed corvid generalists harvest and store the seeds of both tree groups opportunistically, but are less dependent on their scatter hoards for survival and reproduction.

Tree Nuts and their Seed Dispersers

The seed of a nut tree is technically a fruit comprising a large, single seed or "nut" encased in a protective leathery sheath or wood-like shell, ranging from about 10 to 50 mm in length (fig. 7.3). The nut is enclosed within a husk, which dries out and opens at seed maturity. Nuts accumulate under trees, available to seed dispersers and seed predators, or may be removed directly from trees (Bossema 1979; DeGange et al. 1989; Vander Wall 2001). The tree nuts usually dispersed by corvids mature from late August through November (table 7.1). Several genera, such as *Quercus*, *Fagus*, and *Castanea*, are common in temperate broad-leaved communities. Tree nuts are protected by tough shells that can be opened by corvids, but which deter other seed predators (Vander Wall 2001). Both nut size and shell hardness influence foraging choices of corvids, which are selective if more than one nut tree species produces a nut crop locally (Bossema 1979; Pons and Pausas 2007b). Furthermore, some of the tree nuts contain secondary metabolites, such as tannins. Tannins are present in acorns but vary in concentration with oak species (e.g., Moore and Swihart 2006). They potentially limit the digestive efficiency of proteins by corvids and may limit the rate of acorn consumption, which in turn benefits plants by reducing seed predation (e.g., Johnson et al. 1993; Fleck and Tomback 1996).

The most specialized dispersers of tree nuts, and the most thoroughly studied, are Eurasian jays and blue jays. These species harvest and scatter-hoard tree nuts in late summer and fall, and retrieve these nuts later in the year (e.g., Bossema 1979; Darley-Hill and Johnson 1981). Other New and Old World jays, magpies, and *Corvus* spp. harvest and scatter-hoard tree nuts when available, but opportunistically take other kinds of foods as well (fig. 7.4). Magpies, crows, and ravens appear to be the most omnivorous and opportunistic corvids, They survive where conifer seeds and tree nuts are not readily available by feeding on carrion, small vertebrates, eggs, arthropods, grains, fruits, and other plant materials (Kilham 1984; Waite 1985; Birkhead 1991).

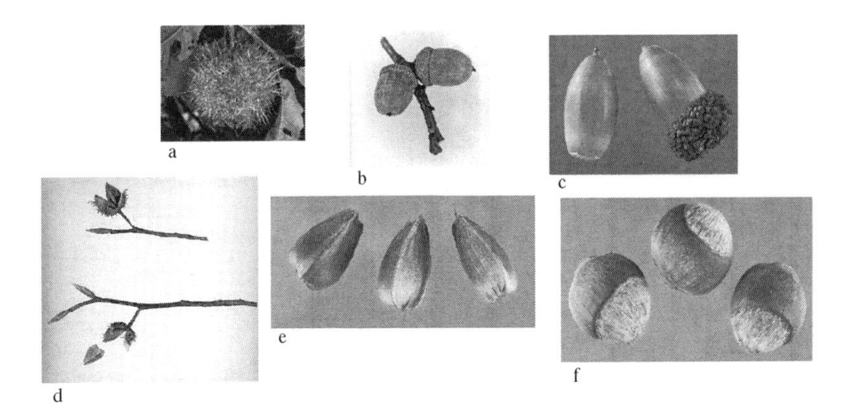

FIGURE 7.3. (a) Chestnut of American chestnut (*Castanea dentata*); photo by Doug Goldman. (b) Acorns of northern red oak (*Quercus rubra*); photo by W. D. Brush. (c) Acorns of white oak (*Quercus alba*); photo by Steve Hurst. (d) Beechnuts of American beech (*Fagus grandifolia*); photo by W. D. Brush. (e) Beechnuts of American beech (*Fagus grandifolia*); photo by Steve Hurst. (f) Hazelnuts of American hazelnut (*Corylus americana*); photo by Steve Hurst. All photos hosted by the USDA-NRCS PLANTS database.

TABLE 7.1 **Tree nuts dispersed by corvids. Information compiled primarily from Turcek and Kelso 1966, Johnson and Webb 1989, Vander Wall 2001,* and references therein.**

Family	**Genus**	**Nut type**	**Size* (mm)**	**~ No. spp.***	**Distribution***
Juglandaceae	*Juglans*	Walnut, butternut	40–80	21	North and South America, Asia
	Carya	Hickory nut, pecan	20–45	18	North America, Asia
Fagaceae	*Quercus*	Acorn	10–50	350–450	North and South America, Europe, Asia, Malaysia
	Fagus	Beechnut	10–15	10	North America, Europe, Asia
	Lithocarpus	Acorn	10–50	300	Asia, Indonesia
	Notholithocarpus	Acorn	10–50	1	North America
	Castanea	Chestnut	20–35	11	North America, Europe, north Africa, Asia
	Castanopsis *Chrysolepis*	Chinquapin nut	10–15	30	Western North America, Asia
Betulaceae	*Corylus*	Hazelnut, filbert	10–15	15	North America, Europe, Asia

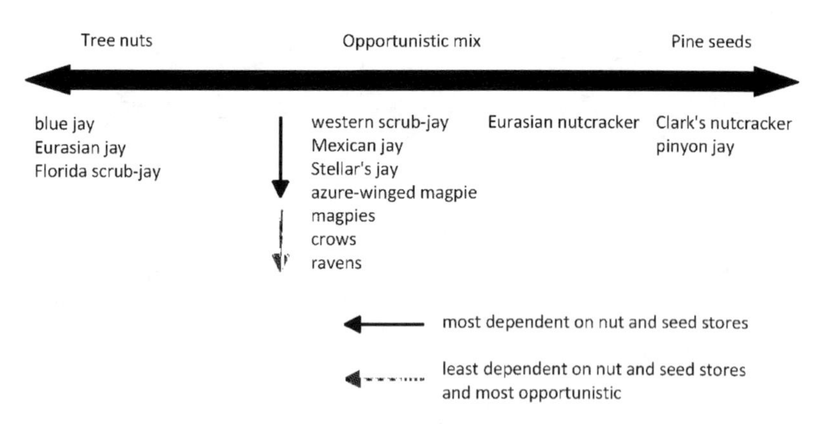

FIGURE 7.4. Seed dispersal by corvids, mapped on a continuum of use of tree nuts versus pine nuts, and by dependency on nuts or seed caches during the birds' annual cycle.

Pine Seeds and Their Seed Dispersers

Pine seeds (genus *Pinus*, family Pinaceae) usually develop over two growing seasons in female strobili (pine cones), which comprise a woody center axis supporting whorls of scales (Mirov 1967). Typically, two seeds develop on the dorsal surface of each cone scale. The seeds generally mature during the second growing season, encased in a hard seed coat. Cones open in most pines, with scales pulling away from the axis as strands of fibrous tissue dry (Krugman and Jenkinson 1974). Whereas the majority of pines have seeds with a woody, membranous wing that facilitates wind dispersal, some species have relatively large seeds with the wing reduced in size or absent. These pines include pinyon pines, Italian stone pine, or umbrella pine (*P. pinea*), and the high elevation five-needle white pines known as "stone pines" (table 7.2; fig. 7.5). This condition is associated with primary or secondary seed dispersal by birds and mammals, where animals take seeds directly from cones (primary) or from the ground after seed fall (secondary) (Tomback and Linhart 1990; Lanner 1998; Tomback et al. 2011). Pine seeds typically are dispersed from late August through December (Krugman and Jenkinson 1974).

The most specialized corvid dispersers of pine seeds are nutcrackers and pinyon jays, which depend year-round on fresh and stored conifer seeds and travel long distances to collect and cache seeds (fig. 7.4; Balda and Bateman 1971; Ligon 1978; Mattes 1978; Tomback 1978; Lorenz and Sullivan 2009). Several North American corvids, including Mexican jays

(*Aphelocoma wollweberi*), western scrub-jays, and Steller's jays, harvest and cache pine seeds, acorns, and other tree nuts, depending on habitat and availability (Brown 1994; Greene et al. 1998; Peterson 1993). For example, all three jays inhabit pine-oak forests on sky islands (forested mountaintops separated by arid, lower elevation habitat) as well as lower-elevation associations of pinyon pine and oak in the southwest and throughout the mountains of Mexico, and they may use both resources (e.g., Martínez-Delgado et al. 1996; Greene et al. 1998), as will azure-winged magpies (*Cyanopica cyanus*) of Spain and Portugal. Different populations of western scrub-jays specialize in eating and caching acorns or pinyon seeds, and are characterized by different bill morphologies (Peterson 1993). Even within the small range (Santa Cruz Island) of the island scrub-jay (*Aphelocoma insularis*), bill morphology differs between jays inhabiting pine and oak

FIGURE 7.5. (a) Whorl of whitebark pine (*Pinus albicaulis*) cones. Photo by Diana F. Tomback. (b) Whitebark pine cone with scales removed to expose large, wingless seeds. The cones do not open, retaining seeds for dispersal by Clark's nutcrackers. Photo by Don Pigott. (c) Open cone of ponderosa pine (*Pinus ponderosa*), which has winged seeds. Photo by Susan M. McDougall, hosted by the USDA-NRCS PLANTS database. (d) Winged seeds of sand pine (*Pinus clausa*). Photo by Steve Hurst, hosted by the USDA-NRCS PLANTS database. (e) Open cone of Colorado pinyon (*Pinus edulis*), showing seed retention for animal dispersal. Photo by Al Schneider, hosted by the USDA-NRCS PLANTS database.

TABLE 7.2 **Pine species observed to have seeds dispersed by Clark's nutcrackers ([CN]), Eurasian nutcrackers ([EN]), azure-winged magpies ([AM]), pinyon jays ([PJ]), western scrub-jays ([WSJ]), Mexican jays ([MJ]), Steller's jays ([SJ]), and common ravens ([CR]). Steller's jays and western scrub-jays, which are widely distributed, may disperse seeds of any pine within or near their territories. Clark's nutcrackers also cache the seeds of Douglas fir (*Pseudotuga menziesii*); some populations of Eurasian nutcrackers cache spruce cones (*Picea* spp.), but also hazlenuts (*Corylus avellana*) and walnuts (*Juglans* spp.). Corvids may disperse the seeds of pines in both tables A and B encountered within their ranges, as well as some smaller-seeded species not listed here. In addition, scrub-jays and Steller's jays may opportunistically cache the seeds of three related pines with heavy, armored cones and large seeds: Coulter pine (*P. coulteri*), gray pine (*P. sabiniana*), and Torrey pine (*P. torreyana*; Johnson et al. 2003). The Eurasian winged seeds that are dispersed by corvids may be underrepresented here. Seed weights are from Tomback and Linhart (1990).**

A. North America: wingless-seed pines

Scientific name	**Pine name**	**Seed mass (g)**[1]	**Distribution**
P. albicaulis	Whitebark[CN, SJ]	0.175	Western United States, western Canada
P. flexilis	Limber[CN, SJ]	0.093	Western United States, southwestern Canada
P. strobiformis	Southwestern white[CN, SJ]	0.168	Southwestern United States, northern Mexico
P. monophylla	Single-leaf pinyon[CN, PJ]	0.409	Eastern California, Great Basin
P. edulis	Colorado pinyon[CN, PJ, MJ, WSJ, SJ]	0.239	Southwestern United States
P. cembroides	Mexican pinyon[WSJ, MJ, CR]	0.412	Southwestern United States, northern Mexico
P. discolor	Border pinyon[WSJ, MJ, CR]	——	Southwestern United States, northern Mexico
P. quadrifolia	Parry pinyon[PJ]	0.472	Southern California, Baja California
B. North America: winged-seed pines			
P. lambertiana	Sugar pine[CN, SJ]	0.216	California, Oregon
P. ayacahuite	Mexican white	——	Southern Mexico, Central America
P. aristata	Rocky Mountain bristlecone[CN]	0.016	Colorado, New Mexico, Arizona
P. longaeva	Great Basin bristlecone[CN, WSJ, PJ]	0.010	Eastern California, Nevada, Utah
P. balfouriana	Foxtail	0.027	California
P. jeffreyi	Jeffrey[CN, PJ]	0.123	Southern Oregon, California, Baja California

continues

TABLE 7.2 *(continued)*

Scientific name	Pine name	Seed mass (g)[1]	Distribution
P. ponderosa	Ponderosa[CN, SJ, PJ]	0.037	Southern British Columbia, western United States, northern Mexico
C. Eurasia: wingless-seed pines			
P. cembra	Swiss stone[EN]	0.227	Alps, Carpathian Mts.
P. sibirica	Siberian stone[EN]	0.252	Ural Mts., western Siberia, northern Mongolia
P. koraiensis	Korean[EN]	0.553	Korea, eastern China, southeastern Siberia, Japan
P. pumila	Japanese stone[EN]	0.092	Northern Mongolia, Siberia, Korea, Japan
P. armandii	Armand[EN]	0.284	China, Burma
P. bungeana	Lacebark[EN?]	———	Northern China
P. gerardiana	Chilgoza[EN]	0.412	Eastern Afghanistan, northern India, Pakistan
P. parviflora	Japanese white[EN]	0.116	Japan
P. pinea	Italian stone[AM*]	0.756	Spain, Portugal, southern France, western Italy, southern Greece, western Turkey, Lebanon
D. Eurasia: winged-seed pines			
P. peuce	Balkan or Macedonian[EN]	0.041	Southwestern Serbia, Montenegro, Kosovo, Albania, Macedonia, western Bulgaria, northern Greece
P. wallichiana	Himalayan[EN]	0.050	Eastern Afghanistan, western Pakistan, Himalayas, southwestern China

*dispersed also by human cultivation

habitat (Langin et al. 2015). Magpies (*Pica* spp.), crows, and ravens are highly opportunistic in feeding habits and able to survive without tree nut or pine seed caches (fig. 7.4; e.g., Richardson and Verbeek 1986; Birkhead 1991), although they cache both when available, potentially contributing to regeneration (Martínez-Delgado et al. 1996; Cristol 2005).

Tree Nut Dispersal

Case History: Blue Jay Dispersal of Acorns and Beech Nuts

Blue jays range across southern Canada to east central British Columbia, and across the eastern United States to the plains and farmlands east of the Rocky Mountains (AOU 1998; Tarvin and Woolfenden 1999). They occupy deciduous forest and open woodland where oaks and beeches are abundant, including urbanized areas. The following information comes from studies of acorn (*Quercus* spp.) dispersal by blue jays in Blacksburg, Virginia (Darley-Hill and Johnson 1981); acorn dispersal in mixed habitat of abandoned fields, forest, and woodland in Iowa (Johnson et al. 1997); and beech nut (*Fagus grandifolia*) and acorn dispersal from woodland surrounded by agricultural land in southeastern Wisconsin (Johnson and Adkisson 1985). Jays harvested and cached tree nuts for nearly 30 days or more in all three study areas, beginning at the end of August or in early September.

While harvesting acorns, the jays selected only healthy nuts, and preferred trees with small or moderately-sized acorns with soft shells—preferences confirmed experimentally (Moore and Swihart 2006). Jays in Blacksburg each transported one to five acorns (mean of 2.2), including one acorn in the bill, for distances of 100 m to 1.9 km. In total, the jays transported and cached about 54% of a monitored *Q. palustris* acorn crop. At caching areas, they deposited all transported acorns in a pile, and then scatter-hoarded one acorn at a time. Caching behavior varied with local substrate: In Iowa, the jays either pushed each acorn into soft soil or moss, or placed an acorn on hard soil and covered it with litter or other plant material. In Blacksburg, acorns were placed on the soil surface and covered with litter, but some were hammered into the substrate and then covered. The areas in Blacksburg selected by jays for scatter-hoarding acorns were analogous to edge and early successional habitat. Numerous oak seedlings were found in jay caching areas, all of which were distant from oak trees. In Iowa, nuts were primarily cached in woodlands and in woodland-grassland ecotones. After a prescribed burn, jays placed caches in open grassland habitat. They selected habitat and cache sites that were suitable for nut germination and seedling establishment. Johnson et al. (1997) noted that scatter-hoarding in recent natural burns would generate patches of even-aged oak forests in prairie habitat, and that oak woodland could develop slowly on former agricultural land as a result of blue jay acorn dispersal.

In Wisconsin, jays preferentially dispersed green, unripe beech nuts, selecting only sound nuts (Johnson and Adkisson 1985). The onset of dispersal coincided with easy removal of beech nuts from burs and caps from acorns. The jays carried beech nuts in their gular pouches and held the stem of a beech bur, which contains two beech nuts, in their bill. Each bird moved nuts from source trees to breeding territories, transporting 3 to 14 beech nuts (mean of 7 nuts) per trip from tens of meters to as far as 4 km. Johnson and Adkisson (1985) estimated that 13,000 trips were made by jays in their study area, scatter-hoarding around 100,000 beech nuts during the dispersal period. They found that the long distances traveled by jays could overcome the patchy distribution of suitable habitat for breeding and caching, as well as the fragmenting of their habitat caused by landscape features such as waterways, highways, and small settlements.

In summary, blue jays were shown to be effective dispersers of tree nuts on a landscape scale, dispersing seeds to forest edge, recent burns, and across fragmented landscapes. They traveled far greater distances between nut trees and cache sites than squirrels typically traverse, crossing different habitats and over barriers that would limit the movements of all rodents (Stapanian and Smith 1986; Johnson and Webb 1989 and references therein). Consequently, blue jays significantly impact the distribution and population structure of nut trees.

The Role of Blue Jays in Post-Pleistocene Tree Migration

Pollen records have provided information on the rate of migration of nut trees out of Pleistocene refuges to present-day distributions as the Laurentian ice sheet retreated over the past 18,000 years (e.g., Davis 1981; Woods and Davis 1989). Long-distance nut dispersal by jays may help resolve Reid's paradox, the discrepancy between known and theoretical rates of plant migration (Johnson and Webb 1989; Clark 1998). Given the typical and maximal distances of nut dispersal by jays, the reconciliation of migration rates for nut tree species, including *Quercus*, *Fagus*, *Carya*, *Castanea*, and *Corylus*, of eastern North American forests appears feasible. Powell and Zimmermann (2004) applied Clark's (1998) modeling approach to blue jay dispersal of oaks, and to nutcracker dispersal of stone pines. Calculated tree migration rates were consistent with observed rates of migration in the pollen record (Clark 1998; Powell and Zimmermann 2004).

Quantifying Ecosystem Services: What Are Eurasian Jays Worth?

Most valuation studies of ecosystem services have examined the replacement costs of general life support functions such as flood control, crop pollination, and drinking water supply (e.g., Costanza et al. 1997; Farley and Costanza 2010). Few studies have focused on the value of the services provided by single species. However, quantifying the economic value of a bird species provides an important argument for its conservation (chapter 2). Acorn dispersal by resident Eurasian jays has maintained oaks over time in Stockholm National Urban Park, which supports one of the largest urban forests of old-growth oaks (*Quercus robur* and *Quercus petrea*) in Europe (Hougner et al. 2006). Created in the 18th century, the 2,700-ha recreation park adjacent to Stockholm holds significant biodiversity. Oaks, some at least 500 years old, comprise about 18% of the forest trees in the park, and about 85% of the oaks result from natural regeneration processes (Hougner et al. 2006). Some of the oldest oaks in the park were originally planted by caretakers, and these ancient trees are often solitary and "charismatic" in appearance (Hougner et al. 2006); thus, visitors appear to value the oaks regardless of origin.

Hougner et al. (2006) calculated the replacement costs for the seed dispersal services of a pair of jays. The replacement costs are here defined as the costs of human-made substitutes, based on a series of assumptions (see supplemental table S7.1 at www.press.uchicago.edu/sites/whybirds matter/). According to previous work (see references in Hougner et al. 2006), 84 jays in Stockholm Urban National Park collectively may hide as many as one million acorns per year in the park. An estimated 464,069 natural saplings in the park result from jay seed dispersal (supplementary table S7.1). The average number of first-year seedlings from jay dispersal is 54/ha. To replicate the total number of natural saplings in the entire park, 33,148 seedlings must be planted per year over 14 years. Human labor may achieve this in two ways: sowing acorns (seeding), or planting oak seedlings. The number of acorns sown must compensate for only 60% germination among commercial acorns, but must also be sown at a natural density (supplementary table S7.1). The replacement cost for a pair of jays, based on sowing acorns over 14 years, is about $4,900. Alternatively, planting oak seedlings is more labor-intensive, and the replacement cost for a pair of jays, based on this method, is $22,500. Given that the aggregated oak forest occupies only about 100 ha total within the park, the replacement costs for a pair of jays amounts to $2,100/ha to $9,500/ha, whether sowing or planting, respectively. This value scales up to $210,000 to $950,000 for the entire park.

Pine Seed Dispersal

Case History: Clark's Nutcracker Dispersal of Pine Seeds

Clark's nutcrackers (*Nucifraga columbiana*) are resident in the higher mountains of western North America. Their dependable range extends from the north central coastal mountains and intermountain ranges of British Columbia, and the Rocky Mountains of British Columbia and Alberta, south to New Mexico and east central Arizona, with an isolated population in Nuevo León, Mexico. In poor seed years, however, the birds wander widely (Tomback 1998). The annual cycle of Clark's nutcracker centers on the use of fresh and stored pine seeds in conjunction with a suite of adaptations as discussed above (Tomback 1978, 1998). Nutcrackers begin feeding on unripe pine seeds in mid-July (Tomback 1978; Hutchins and Lanner 1982). Beginning as early as mid-August, they harvest and store the seeds of pines with large, wingless seeds (table 7.2), which mature earlier than the seeds of other pines (e.g., Tomback 1978; Tomback 1994; table 2 in Krugman and Jenkinson 1974). Wingless seeds are harvested more efficiently, because the nutcrackers must remove the wings from winged seeds before placing them in their sublingual pouch (Tomback 1978; Torick 1995). By late September, nutcrackers harvest and scatter-hoard seeds from winged-seed pines (table 7.2; Giuntoli and Mewaldt 1978; Tomback 1978; Torick 1995; Lorenz et al. 2011). Scatter-hoarding in general may continue through late fall until seed crops are depleted.

The maximum load of seeds carried by nutcrackers in their sublingual pouch represents about 20% of their body mass (fig. 7.1; Vander Wall and Balda 1981). With good cone production, estimates of maximum numbers of seeds cached per individual bird are as follows: Colorado pinyon (*Pinus edulis*), 33,000 (Vander Wall and Balda 1977); single-leaf pinyon (*Pinus monophylla*), 17,900 (Vander Wall 1988); limber pine (*Pinus flexilis*), 16,300 (Vander Wall 1988); and whitebark pine (*Pinus albicaulis*), 35,000 (Tomback 1982) and 98,000 (Hutchins and Lanner 1982). A portion of the stored seeds may feed nestlings and juveniles, and a portion may be taken by cache-raiding rodents (e.g., Vander Wall and Balda 1977; Tomback 1978; 1982). Seeds not retrieved by the next summer or subsequent summers may germinate in response to snowmelt and summer rains (McCaughey and Tomback 2001; Tomback et al. 2001a). The proportion of seeds remaining for germination will vary, depending on the size of the local nutcracker population in relation to cone production the previous fall, and the population size of seed-eating rodents, which will pilfer seeds from caches (e.g., Hutchins and Lanner 1982; Pansing 2014).

Nutcrackers cache seeds by pushing them directly into loose substrate, such as pumice or gravel, or by making a shallow trench with sideswiping motions of the bill, placing seeds in the trench, and then covering the trench with soil and sometimes an object such as a pebble or empty cone (Tomback 1978; 2001). Nutcrackers scatter-hoard seeds in most forest terrain near source trees (Tomback 1978), but also store seeds several kilometers from source trees in communal storage slopes, which tend to be south-facing and steep, and in recently burned terrain and clearcuts (Vander Wall and Balda 1977; Tomback 1978; Hutchins and Lanner 1982; Tomback et al. 2001a). Nutcrackers will cache seeds above treeline and in lower-elevation forest, such as pinyon-juniper communities. Their caches are placed at the base of trees or among tree roots, next to rocks and fallen trees, or at the base of plants, both in open and dense forest (Tomback 1978; Hutchins and Lanner 1982). Nutcrackers store some caches in trees, under bark and in cracks and holes (Tomback 1978; Lorenz et al. 2011). The importance of these caching sites may vary geographically, thus providing access to caches in areas of heavy snow or reducing seed loss to ground squirrels and mice.

Many of the cache sites selected by nutcrackers support seed germination across a number of pine species, but particularly for the wingless-seed pines (Tomback 1982; Tomback et al. 2001a; Pansing 2014). Individual nutcrackers typically harvest and store seeds from more than one pine species, moving from higher to lower elevations for foraging (Tomback 1978; Lorenz and Sullivan 2009). If a regional cone crop failure occurs across conifer species, nutcrackers may emigrate and travel to other regions (e.g., Fisher and Myres 1980; Vander Wall et al. 1981).

Recent genetic analyses of nutcracker populations indicate little or no genetic structure, suggesting high levels of gene flow and mobility among individuals (Dohms and Burg 2013; Fike et al. unpublished data). Consequently, nutcrackers serve as important mobile links for seed dispersal, connecting pine populations. Declines in regional nutcracker populations affect potential tree regeneration, particularly in the primarily nutcracker-dependent pines (Tomback and Kendall 2001). Clark's nutcrackers thus constitute a keystone species throughout the western United States and

Canada, and are essential to the regeneration of many forest community types.

Coevolved Relationship between Clark's Nutcracker and Whitebark Pine

Both whitebark pine and Clark's nutcracker are believed to be descended from ancestral forms that crossed the Bering Strait land bridge from Eurasia 0.6 to 1.3 million years ago or earlier (Lanner 1990; Krutovskii et al. 1994). After encountering other North American pines with large seeds, the nutcracker expanded its range widely throughout the west (Tomback 1983).

Whitebark pine, a widely distributed subalpine and treeline species, is the only North American representative of the "stone pines," which are adapted for seed dispersal by nutcrackers (Tomback 1983; Lanner 1990; Tomback and Linhart 1990). This group includes the Swiss (*P. cembra*), Siberian (*P. sibirica*), and Japanese stone (*P. pumila*) pines, and Korean pine (*P. koraiensis*; table 7.2). All have large, wingless seeds, indehiscent cones (cones that do not open), and vertically oriented branches with branch tips bearing cone whorls—an arrangement that facilitates access to cones by nutcrackers (Lanner 1982, 1990). The seeds of whitebark pine and Siberian stone pine have seed coat, embryo, and storage reserve characteristics adapted to completing maturation and maintaining viability in a soil seed bank—that is, in below-ground caches (Tillman-Sutela et al. 2008). Seeds of the stone pines typically require more than one winter of dormancy before germinating (McCaughey 1993; Tomback et al. 2001a).

Impact of Seed Dispersal by Nutcrackers on Whitebark Pine

Addressed above are the tree, cone, and seed morphologies that either arose early in the evolution of the nutcracker-stone pine interaction, or were preadaptations facilitating the interaction (Tomback and Linhart 1990). Dependency on Clark's nutcrackers for seed dispersal has further influenced the biology and ecology of whitebark pine (see overviews in Tomback 2001, 2005).

Nutcrackers often place more than one seed within a cache, resulting in a seedling cluster, which can grow into a multistemmed tree—each stem a different individual, as confirmed by genetic analysis (Linhart and Tomback 1985; Bruederle et al. 2001). This growth form is also common in other nutcracker-dispersed pines (Tomback et al. 1993; Carsey and Tomback 1994). At the landscape scale, the distances traveled by nutcrackers for seed caching exceed the distances traveled by wind-dispersed seeds (Tomback et al. 1990), thus resulting in high gene flow and little genetic differentiation among adjacent populations (e.g., Bruederle et al. 1998; Rogers et al. 1999). At this scale, nutcrackers determine by their caching choices where whitebark pine occurs on the landscape, including steep slopes, open forests, open terrain, clearcuts, and recent burns (e.g., Vander Wall and Balda 1977; Tomback 1978; Tomback 1986). With scatter hoards below the whitebark pine zone and above the treeline, whitebark pine elevational distribution can respond rapidly to climate change, whether cooling or warming (Tomback 2001). Finally, because nutcrackers cache seeds in recently burned terrain, whitebark pine recruits rapidly after fire (Tomback et al. 1990; 2001a).

Post-Pleistocene Tree Migration

After glacial retreat, nutcrackers dispersed whitebark pine from glacial refugia. Richardson et al. (2002a) identified three potential refugia in the northern and western regions of whitebark pine's range, on the basis of different mitochondrial DNA haplotypes. They speculated that boundaries among haplotypes might indicate geographical limits to seed dispersal. For example, the abrupt transition between two haplotypes at Snoqualmie Pass separated by 30 km may represent a dispersal boundary between populations (Richardson et al. 2002b). Similarly, Mitton et al. (2000) identified eight mitochondrial DNA haplotypes for limber pine, and several likely refugia in the southwestern United States. Tomback (2005, p. 194) notes that both studies illustrate the historical importance and limitations of nutcracker seed dispersal in expanding the ranges of these pines when conditions become more favorable.

Loss of Corvid Seed Dispersal Services: Whitebark Pine as a Case History

Given the importance of the specialized tree nut and pine seed dispersers for forest development, regeneration, and expansion, any perturbation that restricts these interactions will be reflected in altered community composition and dynamics over time. The rapid decline of whitebark pine is altering the dynamics of seed dispersal by Clark's nutcrackers. The decline lowers regional carrying capacity for nutcrackers, and it threatens the future composition and function of high-elevation forests across a broad region.

Whitebark Pine as a High-Elevation Keystone and Foundation Species

Whitebark pine ranges from the southern Sierra Nevada north through the Cascades and Coastal Ranges through central British Columbia, and from the central Wyoming north to about 55° latitude in British Columbia and Alberta (Arno and Hoff 1990). Throughout its range, whitebark pine functions as both a foundation and a keystone species, promoting community stability through ecosystem processes and multiple interactions and through promoting biodiversity (Ellison et al. 2005; Tomback et al. 2001b; Tomback and Achuff 2010).

Whitebark pine communities vary in composition across latitudinal and longitudinal gradients and across successional stages, representing diverse forest habitats (e.g., Arno 2001; Tomback and Kendall 2001). These diverse forest types provide nesting sites and high elevation shelter for many wildlife species, including birds of prey, carnivores, deer (*Odocoileus* spp.), and elk (*Cervus elaphus*) (Tomback et al. 2001b; Tomback and Kendall 2001). The large, nutritious whitebark pine seeds are an important food for granivorous birds, squirrels, and mice, as well as an important prehibernation food for grizzly (*Ursus arctos*) and black bears (*U. americanus*) in the Greater Yellowstone Area (Kendall 1983; Mattson et al. 1992).

A key pioneer after disturbance, whitebark pine promotes community development and stability. With seed dispersal by nutcrackers and hardy seedlings tolerant of poor soils, whitebark pine rapidly regenerates after fire (e.g., Tomback 1986; Tomback et al. 1990; Tomback et al. 2001a). As whitebark pine establishes, it mitigates the conditions for other conifers and plants, thus leading to community development (e.g., Tomback et al. 2001a). On harsh upper subalpine sites, whitebark pine facilitates the establishment and survival of less hardy conifers (Callaway 1998). At treeline on the Rocky Mountain eastern front and in other Rocky Mountain regions, whitebark pine is the most common solitary conifer and the most common windward tree island initiator (Resler and Tomback 2008; Tomback et al. 2014). Through its foundational role in development and maintenance of high-elevation forest, it provides crucial supporting ecosystem functions and services.

Whitebark pine also provides important regulating, provisioning, and cultural services of direct benefit to humans. In the upper subalpine zone and at treeline, whitebark pine communities stabilize loose, rocky substrates and protect against avalanches. They shade snowpack, protracting snowmelt and enabling downstream flows throughout summer, which benefits ranchers and farmers (Arno and Hammerly 1984; Farnes 1990). Whitebark pine seeds and inner bark were historically an important food for Native Americans (Moerman 1998). The tree's rugged, wind-sculpted forms on harsh sites inspire and define the high-mountain recreational experience for skiers, hikers, and backpackers (Tomback and Achuff 2010).

Decline in Whitebark Pine and Altered Seed Dispersal Dynamics

In 2011, the US Fish and Wildlife Service concluded that whitebark pine merited listing as threatened or endangered under the Endangered Species Act, but the listing was precluded by higher-priority actions (US Fish and Wildlife Service 2011). Canada listed whitebark pine federally as endangered under the Species at Risk Act in 2012 (Government of Canada 2012). The two major threats to whitebark pine identified in both status reports were white pine blister rust, a disease caused by the nonnative fungal pathogen *Cronartium ribicola*, and widespread outbreaks of native mountain pine beetles (*Dendroctonus ponderosae*). Other threats noted were fire suppression, which delays the renewal of successional communities, and climate change.

The blister rust pathogen was inadvertently introduced a century ago to the Pacific Northwest, and its subsequent spread throughout much of western North America is decimating whitebark pine and related species, which have low levels of natural resistance (Schwandt et al. 2010; Tomback and Achuff 2010). Infections in tree canopies kill the cone-bearing branches, stem infections eventually kill the trees, and infected seedlings and saplings die within a few years. The trees often lose their reproductive capacity years before they succumb to the rust. The highest blister rust infection levels and mortality are in the Northern Divide Ecosystem of the Rocky Mountains, which includes Glacier National Park and adjacent regions in Montana, and Waterton Lakes National Park in Alberta and surrounding regions (Kendall and Keane 2001; Smith et al. 2008).

Mountain pine beetle outbreaks are a natural disturbance factor in mature western pine forests, creating openings and renewing successional communities (Romme et al. 1986). In the late 1990s, however, mountain pine beetle populations entered an unusual outbreak phase, moving from their primary lodgepole pine (*Pinus contorta*) hosts into high-elevation five-needle white pines, especially whitebark pine (Gibson et al. 2008). The outbreaks attained unprecedented scale and mortality, a result of warming temperatures and drought (Logan and Powell 2001; Logan et al. 2010), and have reduced or nearly eliminated whitebark pine cone production in many high-elevation forests, particularly in the Greater Yellowstone Area and across the Northwest (Gibson et al. 2008; Schwandt et al. 2010). Furthermore, many trees potentially resistant to the blister rust pathogen were killed by mountain pine beetles during the outbreaks, thus reducing the likelihood of selection for resistance (Tomback and Achuff 2010).

Decline in cone production dramatically decreases the likelihood of nutcracker seed dispersal. Working in western Montana and Idaho in stands with high and low blister rust damage, McKinney and Tomback (2007) found that the relative rate of predispersal seed predation, especially by red squirrels (*Tamiasciurus hudsonicus*), which cut down whitebark pine cones for winter food, was far greater in the higher blister rust–infected stands. Given the low cone production in these stands, predispersal seed predation eliminated most cones, essentially precluding seed dispersal. Expanding this study to the Greater Yellowstone Area, which has lower blister rust infection rates, and the Northern Divide Ecosystem, with the highest blister rust infection rates, McKinney et al. (2009) determined that cone production of less than 130 cones/ha essentially eliminates the likelihood of nutcracker seed dispersal. This situation is analogous to the functional extinction of bird pollinators and the impact on bird-dependent plants in New Zealand (chapter 4). Cone production of more than 1,000 cones/ha increases the probability of nutcracker seed dispersal to more than 0.8. Barringer et al. (2012) tested these findings by comparing whitebark pine health and the likelihood of nutcracker visitation to whitebark pine stands between the Greater Yellowstone Area and the Northern Divide Ecosystem. They found that nutcrackers visited stands at all levels of cone production, even below 130 cones/ha, but that the likelihood of nutcracker visitation increased as cone production increased. They also found that the likelihood of nutcracker visitation was more than 0.75 with more than 1,000 cones/ha. Reflecting regional differences in whitebark pine health, the density of whitebark pine regeneration was 26 times greater and cone production was 43 times greater in plots in the southern study region than in the northern study region (Barringer et al. 2012).

Overall, these studies indicate that whitebark pine damage and mortality result in the loss of nutcracker seed dispersal services. Nutcrackers are sensitive to rates of energy gain and forage in forest communities with more rewards (e.g., Tomback 1978; Vander Wall 1988). The low nutcracker visitation in the Northern Divide Ecosystem suggests that, in the absence of management intervention, whitebark pine will not rebound naturally (McKinney et al. 2009; Barringer et al. 2012).

Trophic Cascades and Whitebark Pine Restoration

The loss of whitebark pine has many implications. First of all, the decline of cone production locally alters whitebark pine interactions with red squirrels and nutcrackers: red squirrels take increasingly higher proportions of the remaining cone crop relative to nutcrackers (McKinney and Tomback 2007; 2011). The likelihood of regeneration declines over time, and ultimately whitebark pine experiences local extirpation and replacement by other trees, such as fir and spruce, thus leading to less biodiverse forest communities. As communities lose whitebark pine, the keystone and foundational ecosystem functions and services mentioned above decline (Tomback and Achuff 2010; Tomback et al. 2011).

The consequences of whitebark pine losses reverberate up and down trophic levels, altering community structure and composition and ultimately reducing associated plant, mycorrhizal, and animal biodiversity (chapter 3; Tomback and Kendall 2001; Mohatt et al. 2008). Reduced nutcracker visitation from declining whitebark pine results in a negative feedback loop, limiting future whitebark pine regeneration. A failure of whitebark pine recovery from disease and pine beetle predation reduces structural diversity and biodiversity, as well as regional nutcracker carrying capacity (Tomback and Achuff 2010; McKinney and Tomback 2011).

In regions such as the Northern Divide Ecosystem, with high blister rust infection levels, damage, and mortality (e.g., Smith et al. 2008), the small number of surviving trees with blister rust resistance are unlikely to receive nutcracker seed dispersal services. This is an unusual situation in which genes for blister rust resistance may not increase in frequency over time, because whitebark pine depends on nutcrackers to disperse seeds, and the birds only go where there is sufficient energy reward. In this region we see a progressive loss of ecosystem function and little regeneration (McKinney et al. 2009: Barringer et al. 2012; Fiedler and McKinney 2014). These circumstances require management intervention, and the most promising strategy is to plant seedlings from parent trees with confirmed genetic resistance to *Cronartium ribicola*. The US Forest Service has instituted programs to locate and screen potentially blister rust–resistant whitebark pine trees, collect cones from these trees, and grow and plant the seedlings (Schwandt et al. 2010; Tomback and Achuff 2010).

Quantifying Ecosystem Services: What Are Clark's Nutcrackers Worth?

Planting rust-resistant seedlings is now a major component of the restoration strategy for whitebark pine (e.g., Tomback and Achuff 2010: Keane et al. 2012). Seedlings are planted in large burns and open forests where damage to whitebark pine and mortality from both blister rust and mountain pine beetle have drastically reduced cone production. Planting whitebark pine seedlings replaces nutcracker seed dispersal services, with one major difference: the seeds collected for growing seedlings are ideally from trees with either potential or known blister rust resistance. The cones are harvested, seeds germinated, and seedlings grown following a costly and labor-intensive protocol (table 7.3). Nutcrackers under historical conditions, in contrast, would scatter-hoard seeds from trees with and without genetic resistance to blister rust.

Restoration projects are implemented by national forests using contractors to harvest cones and plant seedlings (usually at 440 seedlings/ha) and US Forest Service nursery facilities to screen for genetic resistance and grow seedlings. Seedlings from parent trees with known resistance are just now becoming available. Based on information presented in table 7.3, the costs of replacing nutcracker ecosystem services for a single hectare range from $1,980 to $2,405 (Wenny et al. 2011; Tomback et al. 2011). Within the whitebark pine's US range, the total area with a greater than 50% blister rust infection rate is estimated at about 1,450,000 ha (Keane et al. 2012). The cost of replacing nutcracker seed dispersal services for this area then ranges from $2.871 billion to $3.487 billion. The entire US range of whitebark pine encompasses about 5,770,000 ha (Keane et al. 2012), and restoration across the range would cost from $11.425 billion to $13.877 billion.

The equivalent time frame for natural regeneration is longer than that of planting all seedlings within a single field season (Tomback et al. 2011). On the basis of data collected following the 1988 Yellowstone fires, Tomback (unpublished) calculated the number of new whitebark pine seedlings that germinated from natural seed caches (Tomback et al. 2001a) and found that nutcracker seed dispersal services require a minimum of five to six years to produce 440 whitebark pine seedlings/ha. By planting seedlings with some resistance and using microsites that favor higher seedling survival rates, the replacement dispersal services are better than nutcracker services. Nutcrackers, however, spread regeneration over time, which reduces risk to seeds and seedlings because the conditions for seed germination and seedling survival vary over time. Furthermore, planting of seedlings will

TABLE 7.3 **Calculation of replacement costs for whitebark pine seed dispersal services by Clark's nutcrackers for a single hectare. Blister rust–resistant seedlings are planted as a major component of whitebark pine restoration projects. Costs vary from year to year and among national forests. Bridger-Teton National Forest, Wyoming, and Flathead National Forest, Montana, contributed information on costs and protocols, which were previously summarized in Wenny et al. 2011 and described in Tomback et al. 2011.**

A. Procedures and conservative assumptions used in calculations

Seed source trees known to be genetically resistant or potentially resistant to the blister rust pathogen must be protected from the mountain pine beetle with applications of verbenone or carbaryl, but this is not included in cost calculations.
The costs of identifying and screening potentially resistant seed source trees, of travel and transportation, and of cone storage are not included.
National forests plant whitebark pine seedlings at a density of 175 seedlings per acre, or about 440 seedlings per hectare.
Although multiple rust–resistant whitebark pine trees are used as seed sources to maintain genetic diversity, the costs here are based on obtaining seeds from one tree only, which usually produces more than enough seeds to grow 440 seedlings.
In July the whorls of maturing cones are enclosed in hardware cloth cone cages to protect them from foraging nutcrackers and pine squirrels. Typically about 30 cone cages are placed per tree by contracted teams of certified tree climbers. These calculations assume that cages for protecting cones are already available.
The trees are climbed again in September; the cages are removed and the cones collected.
At nursery facilities the seeds are extracted from the cones and are stratified and sown, and the seedlings are grown and tended.
The seedlings are usually planted by planting crews.

B. Estimated costs

Climb and cage cones: $250 to $375 per tree
Climb and collect mature cones: $250 to $425 per tree
Administrative oversight: $100 per tree
Growing seedlings: 440 seedlings @ $2 per seedling = $880
Planting one hectare: $250 to $375
Planting layout, administration: $250 per hectare

Estimated cost of replacing nutcrackers for regeneration on one hectare of forest

$1,980 to $2,405

not recreate the natural tree morphology, genetic diversity, and population genetic structure that results from nutcracker seed dispersal, and crews cannot plant in the most remote terrain.

The Importance of Corvid Seed Dispersal Services to Forested Landscapes

The corvids that disperse tree nuts and conifer seeds maintain the structure and composition of many important forest ecosystems around the world in both the temperate broad-leaved deciduous forest and needle-leaved (coniferous) forest biomes. Although the numbers of plants dispersed by corvids are fewer than those dispersed by frugivores or waterfowl (chapters 5 and 6), they are essential to the structure and function of many of these forests. A number of foundation tree species and nut-bearing shrubs depend completely or partly on the specialized corvid dispersers for regeneration—notably oaks, beeches, hazel, filbert, pinyon pines, and five-needle white pines—as well as for distributional changes in response to natural disturbance and climate change (tables 7.1 and 7.2). These dispersal services are becoming particularly important as a warming climate drives elevational and latitudinal shifts in tree and bird distribution. (Wormworth and Şekercioğlu 2011).

Corvid seed dispersal services for forests ultimately translate into ecosystem services and economic benefits for humans from forests (Costanza et al. 1997; Pimentel et al. 1997). Many corvid-dispersed forest communities provide wood products, fruits and nuts, and watershed protection (supporting services), as well as recreational opportunities (cultural services). Nuts from tree and shrub species that are dispersed by corvids—such as hazelnuts, filberts, and the seeds of Italian stone pine, Siberian stone pine, Korean pine, and the pinyon pines—are important to regional economies and are even global commodities. Moreover, whitebark pine and its disperser, Clark's nutcracker, provide an important but unsettling cautionary message: We must be ever vigilant, because no keystone or foundation tree species, nor its bird disperser, is immune to anthropogenic disturbance, no matter how remote or extensive its range.

Acknowledgments

I completed this chapter while supported by a Charles Bullard Harvard Forest fellowship.

References

Amadon, D. 1944. The genera of Corvidae and their relationships. *American Museum Novitates* 125:1–21.

American Ornithologists' Union Committee on Classification and Nomenclature. 1998. *Check-List of North American Birds*, 7th edition. Lawrence, KS: American Ornithologists' Union, Allen Press.

Andersson, M., and Krebs, J. 1978. On the evolution of hoarding behavior. *Animal Behaviour* 26:707–11.

Arno, S. F. 2001. Community types and natural disturbance processes. In *Whitebark Pine Communities: Ecology and Restoration*, ed. D. F. Tomback, S. F. Arno, and R. E. Keane, 74–88. Washington: Island Press.

Arno, S. F., and Hammerly, R. P. 1984. *Timberline: Mountain and Arctic Forest Frontiers*. Seattle: The Mountaineers.

Arno, S. F., and Hoff, R. J. 1990. *Pinus albicaulis* Engelm: Whitebark pine. In *Silvics of North America: Vol. 1, Conifers*, ed. R. P. Burns, B. H. Honkala, 268–79. Washington: USDA Forest Service, .

Balda, R. P., and Bateman, G. C. 1971. Flocking and annual cycle of the Piñon Jay, *Gymnorhinus cyanocphalus*. *Condor* 73:287–302.

Balda, R. P., and Kamil, A. C. 1989. A comparative study of cache recovery by three corvid species. *Animal Behavior* 38:486–95.

Barringer, L. E., Tomback, D. F., Wunder, M. B., and McKinney, S. T. 2012. Whitebark pine stand condition, tree abundance, and cone production as predictors of visitation by Clark's nutcracker (*Nucifraga columbiana*). *PLos ONE* 7(5): e37663.

Baud, K. S. 1993. Simulating Clark's nutcracker caching behavior: Germination and predation of seed caches. MS thesis, Department of Integrative Biology, University of Colorado Denver.

Beletsky, L. 2006. *Birds of the World*. Baltimore: John Hopkins University Press.

Birkhead, T. R. 1991. *The Magpies*. London: T & A D Poyser.

Bock, W. J., Balda, R. P., and Vander Wall, S. B. 1973. Morphology of the sublingual pouch and tongue musculature in Clark's nutcracker. *Auk* 90:491–519.

Bossema, I. 1979. Jays and oaks: An eco-ethological study of a symbiosis. *Behaviour* 70:1–117.

Brodin, A, and Ekman, J. 1994. Benefits of food hoarding. *Nature* 372:510.

Brown, J. L. 1994. Mexican Jay (*Aphelocoma ultramarina*). In *The Birds of North America*, no. 118, ed. A. Poole, and F. Gill. Philadelphia: The Birds of North America, Inc.

Bruederle, L. P., Tomback, D. F., Kelly, K. K., and Hardwick, R. C. 1998. Population genetic structure in a bird-dispersed pine, *Pinus albicaulis* (Pinaceae). *Canadian Journal of Botany* 76:83–90.

Bruederle, L. P., Rogers, D. L., Krutovskii, K. V., and Politov, D. V. 2001. Population genetics and evolutionary implications. In *Whitebark Pine Communities: Ecology and Restoration*, ed. D. F. Tomback, S. F. Arno, and R. E. Keane, 137–53. Washington: Island Press.

Bunch, K. G., and Tomback, D. F. 1986. Bolus recovery by gray jays: An experimental analysis. *Animal Behaviour* 34:754–62.

Callaway, R. M. 1998. Competition and facilitation on elevation gradients in subalpine forests of the northern Rocky Mountains, USA. *Oikos* 2:561–73.

Carsey, K. S., and Tomback, D. F. 1994. Growth form distribution and genetic relationships in tree clusters of *Pinus flexilis*, a bird-dispersed pine. *Oecologia* 98:402–11.

Clark, J. S. 1998. Why trees migrate so fast: Confronting theory with dispersal biology and the paleorecord. *American Naturalist* 152:204–24.

Clements, J. F. 2007. *The Clements Checklist of Birds of the World*, 6th edition. Ithaca, NY: Comstock Publishing Associates, Cornell University Press.

Costanza, R., d'Arge, R., de Groot, R., Farber, S., Grasso, M., Hannon, B., Limburg, K., Naeem, S., O'Neill, R. V., Paruelo, J., Raskin, R. G., Sutton, P., and van den Belt, M. 1997. The value of the world's ecosystem services and natural capital. *Nature* 387:253–60.

Cristol, D. A. 2005. Walnut-caching behavior of American Crows. *Journal of Field Ornithology* 76:27–32.

Critchfield, W. B., and Little, E. L. Jr. 1966. Geographic distribution of the pines of the world. Washington: USDA Forest Service.

Darley-Hill, S., and W. C. Johnson. 1981. Acorn dispersal by the blue jay (*Cyanocitta cristata*). *Oecologia* 50:231–32.

Davis, M. B. 1981. Quaternary history and the stability of deciduous forests. In *Forest succession: Concepts and applications*, ed. D. C. West, H. H. Shugart, and D. B. Botkin, 132–77. New York: Springer-Verlag.

DeGange, A. R., Fitzpatrick, J. W., Layne, J. N., and Woolfenden, G. E. 1989. Acorn harvesting by Florida scrub jays. *Ecology* 70:348–56.

Dittman, D. L., and Cardiff, S. W. 2009. Crows and jays: Corvidae. In *National Geographic Complete Birds of the World*, ed. T. Harris, 268–70. Washington: National Geographic Society.

Dohms, K. M., and Burg, T. M. . 2013. Molecular markers reveal limited population genetic structure in a North American corvid, Clark's Nutcracker (*Nucifraga columbiana*). PLoS ONE 8:e79621. Doi:10.1371/ journal.pone.0079621.

Ellison, A. M., Bank, M. S., Clinton, B. D., Colburn, E. A , Elliott, K., Ford, C. R., Foster, D. R., Kloeppel, B. D., Knoepp, J. D., Lovett, G. M., Mohan, J., Orwig, D. A., Rodenhouse, N. L., Sobczak, W. V., Stinson, K. A., Stone, J. K., Swan, C. M., Thompson, J., Von Holle, B., and Webster, J. R. 2005. Loss of foundation species: Consequences for the structure and dynamics of forested ecosystems. *Frontiers in Ecology and the Environment* 3:479–86.

Farley, J, and Costanza, R. 2010. Payments for ecosystem services: From local to global. *Ecological Economics* 69:2060–68.

Farnes, P. E. 1990. SNOTEL and snow course data: Describing the hydrology of whitebark pine ecosystems. In *Proceedings—Symposium on Whitebark Pine Ecosystems: Ecology and Management of a High-Mountain Resource*, ed. W. C. Schmidt and K. J. McDonald, 302–4. Ogden, UT: USDA Forest Service, Intermountain Research Station.

Fiedler, C. E., and McKinney, S. T. 2014. Forest structure, health, and mortality in two Rocky Mountain whitebark pine ecosystems: Implications for restoration. *Natural Areas Journal* 34:290–99.

Fisher, R. M., and Myres, M. T. 1980. A review of factors influencing extralimital occurrences of Clark's Nutcracker in Canada. *Canadian Field-Naturalist* 94: 43–51.

Fleck, D. C., and Tomback, D. F. 1996. Tannin and protein in the diet of a food-hoarding granivore, the western scrub-jay. *Condor* 98:474–82.

Galetti, M., Guevara, R., Côrtes, M. C., Fadini, R., Von Matter, S., Leite, A. B., Labecca, F., Ribeiro, T., Carvalho, C. S., Collevatti, R. G., Pires, M. M., Guimarães Jr., P. R., Brancalion, P. H., Ribeiro, M. C., and Jordano, P. 2013. Functional extinction of birds drives rapid evolutionary changes in seed size. *Science* 340:1086–90.

Gendron, R. P., and Reichman, O. J. 1995. Food perishability and inventory management: A comparison of three caching strategies. *American Naturalist* 145:948–68.

Gibson, K., Skov, K., Kegley, S., Jorgensen, C., Smith, S., and Witcosky, J. 2008. *Mountain Pine Beetle Impacts in High-Elevation Five-Needle Pines: Current Trends and Challenges*. Forest Health Protection R1–08–020. Missoula, MT: USDA Forest Service.

Gill, F., and D. Donsker, eds. 2014. IOC World Bird List (V. 4.2) doi: 10.14344/IOC.ML.4.2.

Giuntoli, M, and Mewaldt, L. R. 1978. Stomach contents of Clark's nutcracker collected in western Montana. *Auk* 95:595–98.

Gómez, J. M. 2003. Spatial patterns in long-distance dispersal of *Quercus ilex* acorns by jays in a heterogeneous landscape. *Ecography* 26:573–84.

Goodwin, D. 1976. *Crows of the World*. Ithaca, NY: Cornell University Press.

Government of Canada. 2012. Order amending Schedule 1 to the Species at Risk Act. *Canada Gazette* part II, vol. 146, no. 14, SOR/2012–113, 20 June 2012. Accessed 16 September 2012 at http://www.sararegistry.gc.ca/virtual_sara/files/orders/ g2–14614i_e.pdf.

Greene, E., W. Davison, and Muehter, V. R. 1998. Steller's Jay (*Cyanocitta stelleri*). 1998. In *The Birds of North America*, no. 343, ed. A. Poole and F. Gill. Philadelphia: The Birds of North America.

Grobecker, D. B., and Pietsch, T.W. 1978. Crows use automobiles as nutcrackers. *Auk* 95:760–61.

Hayashida, M. 1989. Seed dispersal and regeneration patterns of *Pinus parviflora* var. *pentaphylla* on Mt. Apoi in Hokkaido. *Research Bulletins of the College Experiment Forests, Faculty of Agriculture, Hokkaido University* 46:177–90.

Hitchcock, C. L., and Houston, A. I. 1994. The value of a hoard: Not just energy. *Behavioral Ecology* 5:202–5.

Hougner, C., Colding, J., and Söderqvist, T. 2006. Economic valuation of a seed dispersal service in the Stockholm National Urban Park, Sweden. *Ecological Economics* 59:364–74.

Howe, H. F., and Miriti, M. N. 2004. When seed dispersal matters. *BioScience* 54:651–60.

Howe, H. F., and Smallwood, J. 1982. Ecology of seed dispersal. *Annual Review of Ecology and Systematics* 13:201–28.

Hunt, G. H. 1996. Manufacture of hook-tools by New Caledonian crows. *Nature* 379:249–51.

———. 2000. Human-like, population-level specialization in the manufacture of pandanus tools by New Caledonian crows *Corvus moneduloides*. *Proceedings of the Royal Society of London* 267:403–13.

Hutchins, H. E., and Lanner, R. M. 1982. The central role of Clark's nutcracker in the dispersal and establishment of whitebark pine. *Oecologia* 55:192–201.

James, P. C., and Verbeek, N. A. M. 1983. The food storage behaviour of the northwestern crow. *Behaviour* 85:276–91.

Johnson, M., Vander Wall, S. B., and Borchert, M. 2003. A comparative analysis of seed and cone characteristics and seed-dispersal strategies of three pines in the subsection *Sabinianae*. *Plant Ecology* 168:69–84.

Johnson, W. C., and Adkisson, C. S. 1985. Dispersal of beech nuts by blue jays in fragmented landscapes. *American Midland Naturalist* 113:319–24.

Johnson, W. C., Adkisson, C. S., Crow, T. R., and Dixon, M. D. 1997. Nut caching by blue jays (*Cyanocitta cristata* L.): Implications for demography. *American Midland Naturalist* 138: 357–70.

Johnson, W. C., Thomas, L., and Adkisson, C. S. 1993. Dietary circumvention of acorn tannins by blue jays. *Oecologia* 94:159–64.

Johnson, W. C., and Webb, III, T. 1989. The role of blue jays (*Cyanocitta cristata* L.) in the postglacial dispersal of fagaceous trees in eastern North America. *Journal of Biogeography* 16:561–71.

Keane, R. E., Tomback, D. F., Aubry, C. A., Bower, A. D., Campbell, E. M., Cripps, C. L., Jenkins, M. B., Mahalovich, M. F., Manning, M., McKinney, S. T., Murray, M. P., Perkins, D. L., Reinhart, D. P., Ryan, C., Schoettle, A. W., and Smith, C. M. 2012. *A Range-Wide Restoration Strategy for Whitebark Pine (Pinus albicaulis)*. Fort Collins, CO: USDA, Forest Service, Rocky Mountain Research Station.

Kelly, D. 1994. The evolutionary ecology of mast seeding. *Trends in Ecology and Evolution* 9:465–70.

Kendall, K. C. 1983. Use of pine nuts by grizzly and black bears in the Yellowstone area. *International Conference on Bear Research and Management* 5:166–73.

Kendall, K. C., and Keane, R. E. 2001. Whitebark pine decline: Infection, mortality, and population trends. In *Whitebark Pine Communities: Ecology and Restoration*, ed. D. F. Tomback, S. F. Arno, and R. E. Keane, 221–42. Washington: Island Press.

Kilham, L. 1984. Foraging and food-storing of American crows in Florida. *Florida Field Naturalist* 12:25–48.

Koenig, W. D., and Knops, J. M. H. 2005. The mystery of masting in trees. *American Scientist* 93:340–47.

Krieger, D. J. 2001. Economic value of forest ecosystem services: A review. The Wilderness Society.

Krutovskii, K. V., Politov, D. V., and Altukov, Y. P. 1994. Genetic differentiation and phylogeny of stone pines based on isozyme loci. In *Proceedings—International Workshop on Subalpine Stone Pines and Their Environment: The Status of Our Knowledge*, ed. W. C. Schmidt and F-K Holtmeier, 19–30. Ogden, UT: USDA Forest Service, Intermountain Research Station.

Krugman, S. L., and Jenkinson, J. L. 1974. *Pinus* L. Pine. In *Seeds of Woody Plants in the United States*, 598–640. Washington: USDA Forest Service.

Kunstler, G., Thuiller, W., Curt, T., Bouchaud, M., Jouvie, R., Deruette, F., and Lepart, J. 2007. *Fagus sylvatica* L. recruitment across a fragmented Mediterranean landscape, importance of long distance effective dispersal, abiotic conditions and biotic interactions. *Diversity and Distributions* 13:799–807.

Langin, K. M., Sillett, T. S., Funk, W. C., Morrison, S. A., Desrosiers, M. A., and Ghalambor, C. K. 2015. Islands within an island: Repeated adaptive divergence in a single population. *Evolution* 69:653–65.

Lanner, R. M. 1982. Adaptations of whitebark pine for seed dispersal by Clark's nutcracker. *Canadian Journal of Forest Research* 12:391–402.

———. 1988. Dependence of Great Basin bristlecone pine on Clark's nutcracker for regeneration at high elevations. *Arctic and Alpine Research* 20:358–62.

———. 1990. Biology, taxonomy, evolution, and geography of stone pines of the world. In *Proceedings—Symposium on Whitebark Pine Ecosystems: Ecology and Management of a High-Mountain Resource*, ed. W. C. Schmidt and K. J. McDonald, 14–24. Ogden, UT: USDA Forest Service Intermountain Research Station.

———. 1998. Seed dispersal in *Pinus*. In *Ecology and Biogeography of Pinus*, ed. D. M. Richardson, 281–95. Cambridge: Cambridge University Press.

Lanner, R. M., H. E. Hutchins, and H. A. Lanner. 1984. Bristlecone pine and Clark's nutcracker: Probable interaction in the White Mountains, California. *Great Basin Naturalist* 44:357–60.

Ligon, J. D. 1974. Green cones of the Piñon pine stimulate late summer breeding in the Piñon Jay. *Nature* 250:80–82.

———. 1978. Reproductive interdependence of piñon jays and piñon pines. *Ecological Monographs* 48:111–26.

Linhart, Y. B., and Tomback, D. F. 1985. Seed dispersal by Clark's nutcrackers causes multi-trunk growth form in pines. *Oecologia* 67:107–10.

Logan, J. A, MacFarlane, W. W., and Willcox, L. 2010. Whitebark pine vulnerability to climate-driver mountain pine beetle disturbance in the Greater Yellowstone Ecosystem. *Ecological Applications* 20:895–902.

Logan, J. A., and Powell, J. A. 2001. Ghost forests, global warming, and the mountain pine beetle (Coleoptera: Scolytidae). *American Entomologist* 47:160–72.

Lorenz, T. J., and Sullivan, K. A. 2009. Seasonal differences in space use by Clark's nutcrackers in the Cascade Range. *Condor* 111:326–40.

Lorenz, T. J., Sullivan, K. A., Bakian, A. V., and Aubry, C. A. 2011. Cache-site selection in Clark's nutcracker (*Nucifraga columbiana*). *Auk* 128:237–47.

Maple, T. 1974. Do crows use automobiles as nutcrackers? *Western Birds* 5:97–98.

Martínez-Delgado, E., Mellink, E., Rogelio Aguirre-Rivera, J., and García-Moya, E. 1996. Removal of piñon seeds by birds and rodents in San Luis Potosí, México. *Southwestern Naturalist* 41:270–74.

Marzluff, J. M., and Balda, R. P. 1992. *The Pinyon Jay: Behavioral Ecology of a Colonial and Cooperative Corvid*. London: T & AD Poyser.

Mattes, H. 1978. Der Tannenhäher im Engadin: Studien zu seiner Ökologie und Funktion im Arvenwald. Münstersche Geographische Arbeiten. Paderborn, Germany: Ferdinand Schöningh.

Mattson, D. J., Blanchard, B. M., and R. R. Knight. 1992. Yellowstone grizzly bear mortality, human habituation, and whitebark pine seed crops. *Journal of Wildlife Management* 56:432–42.

McCaughey, W. W. 1993. Delayed germination and seedling emergence of *Pinus albicaulis* in a high elevation clearcut in Montana, USA. In *Dormancy and Barriers to Germination*. Proceedings International Symposium IUFRO Project Group P2.04–00 (Seed Problems), ed. D. G. W. Edwards, 67–72. Victoria, BC: Forestry Canada, Pacific Forestry Centre.

McCaughey, W. W., and Tomback, D. F. 2001. The natural regeneration process. In *Whitebark Pine Communities: Ecology and Restoration*, ed. D. F. Tomback, S. F. Arno, and R. E. Keane, 105–20. Washington: Island Press.

McKinney, S. T., Fiedler, C. E, and Tomback, D. F. 2009. Invasive pathogen threatens bird-pine mutualism: Implications for sustaining a high-elevation ecosystem. *Ecological Applications* 19:597–607.

McKinney, S. T., and Tomback, D. F. 2007. The influence of white pine blister rust on seed dispersal in whitebark pine. *Canadian Journal of Forest Research* 37:1044–57.

———. 2011. Altered community dynamics in Rocky Mountain whitebark pine forests and the potential for accelerating declines. In *Mountain Ecosystems: Dynamics, Management, and Conservation*, ed. K. E. Richards, 45–78. Hauppauge, NY: Nova Science Publishers.

Mewaldt, L. R. 1952. The incubation patch of the Clark nutcracker. *Condor* 54: 361.

———. 1956. Nesting behavior of the Clark nutcracker. *Condor* 58:3–23.

Mirov, N. T. 1967. *The Genus Pinus*. New York: Ronald Press Company.

Mitton, J. B., Kreiser, B. R., and Latta, R. G. 2000. Glacial refugia of limber pine (*Pinus flexillis* James) inferred from the population structure of mitochondrial DNA. *Molecular Ecology* 9:91–97.

Moerman, D. E. 1998. *Native American Ethnobotany*. Portland, OR: Timber Press.

Moore, J. E., and Swihart, R. K. 2006. Nut selection by captive blue jays: Importance of availability and implications for seed dispersal. *Condor* 108:377–88.

Mohatt, K. R., Cripps, C. L., and Lavin, M. 2008. Ectomycorrhizal fungi of whitebark pine (a tree in peril) revealed by sporocarps and molecular analysis of mycorrhizae from treeline forests in the Greater Yellowstone Ecosystem. *Botany* 86: 14–25.

Morris, D. 1962. The behaviour of the green acouchi (*Myoprocta pratti*) with special reference to scatter hoarding. *Proceedings of the Zoological Society of London* 139:701–32.

Pansing, E. R. 2014. The influence of cache site and rodent pilferage on whitebark pine seed germination in the northern and central Rocky Mountains. MS thesis, Department of Integrative Biology, University of Colorado Denver.

Pimentel, D., Wilson, C., McCullum, C., Huang, R., Dwen, P., Flack, J., Tran, Q., Saltman, T., Cliff, B. 1997. Economic and environmental benefits of biodiversity. *BioScience* 47:747–57.

Peterson, A. T. 1993. Adaptive geographical variation in bill shape of scrub jays (*Aphelocoma coerulescens*). *American Naturalist* 142:508–27.

Pons, J., and Pausas, J. G. 2007a. Acorn dispersal estimated by radio-tracking. *Oecologia* 153:903–11.

———. 2007b. Not only size matters: Acorn selection by the European jay (*Garrulus glandarius*). *ACTA Oecologica* 31:353–60.

Powell, J. A., and Zimmermann, N. E. 2004. Multiscale analysis of active seed dispersal contributes to resolving Reid's paradox. *Ecology* 85:490–506.

Resler, L. M., and Tomback, D. F. 2008. Blister rust prevalence in krummholz whitebark pine: Implications for treeline dynamics. *Arctic, Antarctic, and Alpine Research* 40:161–70.

Richardson, B. A., Klopfenstein, N. B., and Brunsfeld, S. J. 2002a. DNA from bird-dispersed seed and wind-disseminated pollen provides insights into postglacial colonization and population genetic structure of whitebark pine (*Pinus albicaulis*). *Molecular Ecology* 11:215–27.

———. 2002b. Assessing Clark's nutcracker seed-caching flights using maternally inherited mitochondrial DNA of whitebark pine. *Canadian Journal of Forest Research* 32:1103–7.

Richardson, D. M., and Rundel, P. W. 1998. Ecology and biogeography of *Pinus*: An introduction. In *Ecology and Biogeography of Pinus*, ed. D. M. Richardson, 3–46. Cambridge: Cambridge University Press.

Richardson, P. C., and Verbeek, N. A. M. 1986. Diet selection and optimization by northwestern crows feeding on Japanese littleneck clams. *Ecology* 67: 1219–26.

Rogers, D. L., Millar, C. I., and Westfall, R. D. 1999. Fine-scale genetic architecture of whitebark pine (*Pinus albicaulis*): Associations with watershed and growth form. *Evolution* 53:74–90.

Romme, W. H., Knight, D. H., and Yavitt, J. B. 1986. Mountain pine beetle outbreaks in the Rocky Mountains: Regulators of primary productivity? *American Naturalist* 4:484–94.

Roselaar, C. S. 1994. *Nucifraga caryocatactes* nutcracker. In *Handbook of the Birds of Europe, the Middle East, and North Africa: The Birds of the Western Palearctic. Volume VIII: Crows to Finches*, ed. S. Cramp, C. M. Perrins, and D. J. Brooks, 76–95. Oxford: Oxford University Press.

Samano, S., and Tomback, D. F. 2003. Cone opening phenology, seed dispersal, and seed predation in southwester white pine (*Pinus strobiformis*) in southern Colorado. *Écoscience* 10:319–26.

Schwandt, J. W., Lockman, I. B., Kliejunas, J. T., and Muir, J. A. 2010. Current health issues and management strategies for white pines in the western United States and Canada. *Forest Pathology* 40:226–50.

Siepielski, A. M., and Benkman, C. W. 2007. Extreme environmental variation sharpens selection that drives the evolution of a mutualism. *Proceedings of the Royal Society B* 274:1799–1805.

Smith, C. M., Wilson, B., Rasheed, S., Walker, R. C., Carolin, T., and Sheppard, B. 2008. Whitebark pine and white pine blister rust in the Rocky Mountains of Canada and northern Montana. *Canadian Journal of Forest Research* 38: 982–95.

Soons, M. B., and W. A. Ozinga. 2005. How important is long-distance seed dispersal for the regional survival of plant species? *Diversity and Distributions* 11:165–72.

Sork, V. L. 1993. Evolutionary ecology of mast-seeding in temperate and tropical oaks (*Quercus* spp.). *Vegetatio* 107/108:133–47.

Stapanian, M. C., and C. C. Smith. 1986. How fox squirrels influence the invasion of prairies by nut-bearing trees. *Journal of Mammalogy* 67:326–32.

Tarvin, K. A., and Woolfenden, G. E. 1999. Blue jay (*Cyanocitta cristata*). In *The Birds of North America*, no. 469, ed. A. Poole and F. Gill. Philadelphia: The Birds of North America.

Tillman-Sutela, E., Kauppi, A., Karppinen, K., and Tomback, D. F. 2008. Variant maturity in seed structures of *Pinus albicaulis* (Engelm.) and *Pinus sibirica* (Du Tour): Key to a soil seed bank, unusual among conifers? *Trees* 22:225–36.

Tomback, D. F. 1978. Foraging strategies of Clark's nutcracker. *The Living Bird* 16:123–61.

———. 1982. Dispersal of whitebark pine seeds by Clark's nutcracker: A mutualism hypothesis. *Journal of Animal Ecology,* 51:451–67.

———. 1983. Nutcrackers and pines: Coevolution or coadaptation? In *Coevolution*, ed. M. J. Nitecki, 179–223. Proceedings of the Fifth Annual Spring Systematics Symposium, Field Museum of Natural History, Chicago. Chicago: University of Chicago Press.

———. 1986. Post-fire regeneration of krummholz whitebark pine: A consequence of nutcracker seed caching. *Madroño* 33:100–110.

———. 1988. Nutcracker-pine mutualisms: Multi-trunk trees and seed size. In *Acta XIX Congressus Internationalis Ornithologici, Vol. 1*, ed. H. Ouellet, 518–27. Ottawa, ONT: University of Ottawa Press.

———. 1994. Ecological relationship between Clark's nutcracker and four wingless-seed *Strobus* pines of western North America. In *Proceedings—International Workshop on Subalpine Stone Pines and Their Environment: the Status of our Knowledge*, ed. W. C. Schmidt and F-K Holtmeier, 221–24. Ogden, UT: USDA Forest Service Intermountain Research Station.

———. 1998. Clark's nutcracker (*Nucifraga columbiana*). In *The Birds of North America*, no. 331, ed. A. Poole, and F. Gill. Philadelphia: The Birds of North America.

———. 2001. Clark's nutcracker: Agent of regeneration. In *Whitebark Pine Communities: Ecology and Restoration*, ed. D. F. Tomback, S. F. Arno, and R. E. Keane, 89–104. Washington: Island Press.

———. 2005. The impact of seed dispersal by Clark's nutcracker on whitebark pine: Multi-scale perspective on a high mountain mutualism. In *Mountain Ecosystems: Studies in Treeline Ecology*, ed. G. Broll, and B. Keplin, 181–201. Berlin: Springer.

Tomback, D. F., and P. Achuff. 2010. Blister rust and western forest biodiversity: Ecology, values, and outlook for white pines. *Forest Pathology* 40:186–225.

Tomback, D. F., Achuff, P., Schoettle, A.W., Schwandt, J., and Mastrogiuseppe, R. J. 2011a. The magnificent high-elevation five-needle white pines: Ecological roles and future outlook. In *Proceedings: High-Five Symposium: The Future of High-Elevation Five-Needle White Pines in Western North America*, ed. R. E. Keane, D. F. Tomback, M. P. Murray, and C. M. Smith, 2–28. Fort Collins, CO: USDA Forest Service, Rocky Mountain Research Station.

Tomback, D. F., Anderies, A. J., Carsey, K. S., Powell, M. L., and Mellmann-Brown, S. 2001a. Delayed seed germination in whitebark pine and regeneration patterns following the Yellowstone fires. *Ecology* 82:2587–2600.

Tomback, D. F., Arno, S. F., and Keane, R. E. 2001b. The compelling case for management intervention. In *Whitebark Pine Communities: Ecology and Restoration*, ed. D. F. Tomback, S. F. Arno, and R. E. Keane, 3–25. Washington: Island Press.

Tomback, D. F., Chipman, K. G. Resler, L. M., Smith-McKenna, E. K. and Smith, C. M. 2014. Relative abundance and functional role of whitebark pine at treeline in the northern Rocky Mountains. *Arctic, Antarctic, and Alpine Research* 46: 407–18.

Tomback, D. F., L. A. Hoffmann, and S. K. Sund. 1990. Coevolution of whitebark pine and nutcrackers: Implications for forest regeneration. In *Proceedings—Symposium on Whitebark Pine Ecosystems: Ecology and Management of a High-Mountain Resource*, eds. W. C. Schmidt, and K. J. McDonald, 118–29. Ogden, UT: USDA Forest Service Intermountain Research Station.

Tomback, D. F., Holtmeier, F.-K., Mattes, H., Carsey, K. S., and Powell, M. L. 1993. Tree clusters and growth from distribution in *Pinus cembra*, a bird-dispersed pine. *Arctic and Alpine Research* 25:374–81.

Tomback, D. F., and Kendall, K. C. 2001. Biodiversity losses: The downward spiral. In *Whitebark Pine Communities: Ecology and Restoration*, ed. D. F. Tomback, S. F. Arno, and R. E. Keane, 243–62. Washington: Island Press.

Tomback, D. F., and Linhart, Y. B. 1990. The evolution of bird-dispersed pines. *Evolutionary Ecology* 4:185–219.

Torick, L. L. 1995. The interaction between Clark's nutcracker and ponderosa pine, a wind-dispersed pine: Energy efficiency and multi-genet growth forms. MS thesis, Department of Biology, University of Colorado Denver.

Trakhtenbrot, A., Nathan, R., Perry, G., and Richardson, D. M. 2005. The importance of long-distance dispersal in biodiversity conservation. *Diversity and Distributions* 11:173–81.

Turcek, F. J., and Kelso, L. 1968. Ecological aspects of food transportation and storage in the Corvidae. *Communications in Behavioral Biology*, Part A, 1:277–97.

US Fish and Wildlife Service. 2011. Endangered and threatened wildlife and plants: 12-month funding on a petition to list *Pinus albicaulis* as endangered or threatened with critical habitat. *Federal Register* 76(138): 42631–54.

Vander Wall, S. B. 1988. Foraging of Clark's nutcracker on rapidly changing pine seed resources. *Condor* 90:621–31.

———. 1990. *Food Hoarding in Animals*. Chicago: University of Chicago Press.

———. 1993. A model of caching depth: Implications for scatter hoarders and plant dispersal. *American Naturalist* 141:217–32.

———. 2001. The evolutionary ecology of nut dispersal. *Botanical Review* 67:74–117.

Vander Wall, S. B., and Balda, R. P. 1977. Coadaptations of the Clark's nutcracker and the piñon pine for efficient seed harvest and dispersal. *Ecological Monographs* 47: 89–111.

———. 1981. Ecology and evolution of food storing behavior in conifer-seed-caching corvids. *Zeitschrift für Tierpsychologie* 56:217–42.

Vander Wall, S. B., Hoffman, S. W., and Potts, W. K. 1981. Emigration behavior of Clark's nutcracker. *Condor* 83:162–70.

Verbeek, N. A. M. 1997. Food cache recovery by northwestern crows (*Corvus caurinus*). *Canadian Journal of Zoology* 75:1351–56.

Waite, R. K. 1985. Food caching and recovery by farmland corvids. *Bird Study* 32:45–49.

Webster, K. L., and Johnson, E. A. 2000. The importance of regional dynamics in local populations of limber pine (*Pinus flexilis*). *Écoscience* 7:175–82.

Wenny, D. G. 2001. Advantages of seed dispersal: A re-evaluation of directed dispersal. *Evolutionary Ecology Research* 3:51–74.

Wenny, D. G., DeVault, T. L., Johnson, M. D., Kelly, D., Şekercioğlu, C. H., Tomback, D. F., and Whelan, C. J. 2011. The need to quantify ecosystem services provided by birds. *Auk* 128:1–14.

Woods, K. D., and Davis, M. B. 1989. Paleoecology of range limits: Beech in the Upper Peninsula of Michigan. *Ecology* 70:681–96.

Wormworth, J., Şekercioğlu, Ç. H. 2011. *Winged Sentinels: Birds and Climate Change*. Cambridge University Press, New York.

Zusi, R. L. 1987. A feeding adaption of the jaw articulation in New World jays (Corvidae). *Auk* 104:665–80.

CHAPTER EIGHT

Ecosystem Services Provided by Avian Scavengers

Travis L. DeVault, James C. Beasley, Zachary H. Olson, Marcos Moleón, Martina Carrete, Antoni Margalida, and José Antonio Sánchez-Zapata

Food webs developed under classical theoretical models often depict simplistic interactions among trophic levels linked by predation (Hairston et al. 1960). As a result, extensive research efforts have been devoted to studying predator-prey interactions, often ignoring the contribution of scavenging in food-web dynamics. However, recent advancements in food-web theory have recognized the widespread and critical role that scavenging plays in stabilizing food webs in ecosystems throughout the world, thus suggesting that previous models may have greatly underestimated the importance of scavenging in food web research (Wilson and Wolkovich 2011; Barton et al. 2013). Such disregard for the importance of scavenging likely stems from a number of factors, such as human aversion to decomposing matter, difficulties in identifying scavenged versus depredated materials, and the fact that most species utilize carrion opportunistically (DeVault et al. 2003). Nonetheless, recent population declines of a number of obligate scavengers (e.g., vultures) have drawn international attention to this important group of species, and have sparked a renaissance in research on scavenging (Koenig 2006; Sekercioglu 2006; Ogada et al. 2012a; Moleón and Sanchez-Zapata 2015; Buechley and Şekercioğlu 2016a, 2016b; Ogada et al. 2016).

Carrion as a Unique Food Source

From the perspective of the predator-scavenger, carrion differs from live prey in several ways. At any point in time, there is usually more live prey in an area than there is carrion, because carrion is generally assimilated very quickly after an animal dies, either by decomposition or scavenging (DeVault et al. 2003). Thus, although carrion could be harder to find (although its presence, odor, is often advertised by microbial decomposers; Janzen 1977; Putman 1983; DeVault et al. 2004; Shivik 2006), it is easier to consume than live prey, as it does not bother to hide or defend itself from predators (Moleón et al. 2014b). However, there are risks associated with consuming carrion, as microbial decomposers produce objectionable and dangerous chemicals in their attempt to sequester the resource (Janzen 1977; Burkepile et al. 2006; see Moleón et al. 2015 for further differences between predation and scavenging networks).

Even though carrion is generally scarce compared to live prey, a central question in scavenging ecology concerns carrion availability—that is, how many animals in a given area die in such a way that they become available to scavengers (DeVault et al. 2003). Houston (1979) argued that scavengers (e.g., vultures) obtain very little food from predator-killed animals, because predators either consume all of their prey or guard their kills. In such cases, scavengers must rely on carcasses from animals that die from causes other than natural predation (Pereira et al. 2014). Numerous studies of cause-specific mortality suggest that in many areas a high percentage of animals die from malnutrition, disease, exposure, collision with vehicles and anthropogenic structures, and hunting by humans (when some animal remains are left in the field), and thereby become available to scavengers, although this percentage varies widely across habitats and animal communities (see DeVault et al. 2003). In the Pyrenees, the estimated proportion of carcasses of wild and domestic ungulates available to avian scavengers ranged between 25 and 80%, depending on the habitat occupied by the prey species (forest or open landscape; Margalida et al. 2011a; Margalida and Colomer 2012).

Occasionally carcasses become available to scavengers in spatial and temporal pulses (Wilson and Wolkovich 2011). For example, several species of birds and mammals depend on salmon carcasses when they become available in fall and winter after spawning (Hewson 1995; Ben-David et al. 1997). Brown bears (*Ursus arctos*) and other carnivores extensively scavenge ungulates killed by wild fires in the western United States (Singer et al. 1989; Blanchard and Knight 1990). Also, in some areas domestic carrion (i.e., dead farm animals) provides an important source of food (Lambertucci et al. 2009). Margalida et al. (2011a) showed that in the pre-Pyrenees region of northeastern Spain, wild ungulates do not currently provide enough food to sustain avian scavengers, and domestic animal carcasses are necessary to prevent population declines. Other studies suggest that predator-killed animals are important to scavengers (e.g., Paquet 1992; Wilmers et al. 2003a, b; Selva et al. 2005). For example, Krofel et al. (2012) showed that brown bears often usurp lynx (*Lynx lynx*) kills, thus causing substantial shifts in lynx foraging patterns.

Irrespective of the cause, it is clear that upon death a sufficient number of animals become available to scavengers, which profoundly impact ecosystems (Wilson and Wolkovich 2011; Barton et al. 2013). Houston (1979) suggested that only about 30% of large ungulates in the Serengeti are killed by predators; the rest die from other causes and become available to scavengers. Moreover, Putman (1976) calculated that about 40% of the production of small mammals becomes available to scavengers (see also DeVault et al. 2003). However, these figures vary widely among seasons and regions (Pereira et al. 2014), and many questions remain regarding the availability of wildlife carrion. For example, more information is needed on how differently sized carcasses contribute to the total carrion pool (Barton et al. 2013). Similarly, it is especially difficult to determine how much available biomass within a carcass has been scavenged versus how much has been decomposed (Putman 1983). In the case of osteophagus species, such as the bearded vulture (*Gypaetus barbatus*), some bones evidently are avoided, as is suggested by their accumulation in nests and ossuaries. In this case, bone nutritive value (fat content) and handling efficiency, regardless of bone size and morphology, appear to play an important role in bone selection, because this implies an optimization of foraging time and of the increased energy gained from the food (Margalida 2008a, b).

The Consumers: Obligate and Facultative Scavengers

Defined broadly, a scavenger is any organism that feeds on a dead animal it did not kill. There is substantive evidence from the literature to suggest that most carnivorous animals will capitalize on carrion resources if given the opportunity (reviewed in DeVault et al. 2003; Pereira et al. 2014; Mateo-Tomás et al. 2015). For example, carcasses in terrestrial (Tabor et al. 2005; Selva et al. 2005) and marine ecosystems (Smith and Baco 2003)

FIGURE 8.1. African vultures on an elephant carcass. Wherever they are still present, social *Gyps* vultures are the dominant avian scavengers of medium to large vertebrate carcasses in the Old World. This photograph, taken by an automatic camera in the Hluhluwe-iMfolozi Park (South Africa), shows a group of white-backed vultures (*G. africanus*) feeding on an elephant (*Loxodonta africana*) carcass. Other avian scavengers, such as white-headed vultures (*Aegypius occipitalis*) and pied crows (*Corvus albus*), are relegated to a marginal role. Photo by Marcos Moleón.

can host several hundred species of invertebrate scavengers (Beasley et al. 2012). Thus, the diversity of organisms considered to be scavengers is large, but the extent to which each of these species uses carrion varies tremendously (DeVault et al. 2003).

Species that scavenge can be separated into two unequal groups. The first group, the obligate scavengers, relies on carrion for its survival and reproduction. The only known terrestrial vertebrates in this group are vultures (both Old- and New-World; figs. 8.1 and 8.2). In fact, due to energetic constraints, obligate vertebrate scavengers must be large soaring fliers (Ruxton and Houston 2004a).

Obligate scavengers as a group exhibit several adaptations that foster their ability to use carrion as a food source. First, they must possess efficient locomotion (Ruxton and Houston 2004a; Shivik 2006). Efficient travel allows obligate scavengers to increase their search area, effectively exchanging the spatial and temporal unpredictability of carrion at local scales for relatively predictable occurrences at much larger scales (DeVault et al. 2003; Ruxton and Houston 2004a). Second, obligate scavengers must be able to find carcasses from great distances, a feat accomplished using keenly focused senses of sight or smell (Houston 1979; DeVault et al. 2003). For example, the obligate scavenging turkey vulture (*Cathartes aura*; fig. 8.3) and congeners possess an excellent olfactory sense (Stager 1964). Finally, obligate scavengers must exhibit morphological and physiological adaptations to the problems encountered when feeding on carcasses. The obligate scavenging vultures have highly acidic stomachs (as low as pH = 1) that probably help to decrease the pathogenic risk of high microbial loads (Houston and Cooper 1975), and few to no feathers on their heads, which reduces fouling (Houston 1979). Thus, obligate scavengers have evolved to efficiently find and acquire the nutrients in carcasses.

FIGURE 8.2. Of the New World vultures, the black (*Coragyps atratus*) and turkey (*Cathartes aura*) vulture are the most widely distributed, and they overlap extensively in range. This photograph, taken by an automatic remote camera in South Carolina, shows a group of black and turkey vultures feeding on a feral pig (*Sus scrofa*) carcass. Although both species are often found using the same carrion resources in areas where they occur sympatrically, black vultures generally are more social and rely primarily on their vision to detect carrion, whereas turkey vultures are less gregarious and have well-developed olfactory capabilities. Photo by James C. Beasley.

FIGURE 8.3. Turkey vultures are abundant across much of North and South America, and are well adapted to human-dominated landscapes. Photo by Travis L. DeVault.

Facultative scavengers, on the other hand, comprise much more diversity than exists among obligate scavengers, as they include all species that scavenge when opportunities arise, but do not depend solely on carrion for survival and reproduction (figs. 8.4 and 8.5). Facultative scavengers exhibit a range in the frequency with which they are associated with scavenging activity (Pereira et al. 2014; Moreno-Opo et al. 2016). For example, carrion feeding is often listed in food-habits descriptions of natural history accounts for generalist species such as the red fox and members of the Corvidae family (McFarland et al. 1979). However, it is less frequently acknowledged that scavenging is also observed among the more specialist predators (e.g., *Buteo* hawks and buzzards, Errington and Breckenridge 1938; various snakes, DeVault and Krochmal 2002; owls, Kapfer et al. 2011), and even among herbivores (e.g., hippopotamus [*Hippopotamus amphibius*]; Dudley 1996). Many facultative scavengers (e.g., raptors; Sánchez-Zapata et al. 2010; Moreno-Opo et al. 2016) eat more carrion than is often assumed, largely because traditional pellet- and prey-remains analysis underrepresents the presence of carrion in their diets (DeVault et al. 2003). It is tempting to view the different frequencies of scavenging

FIGURE 8.4. The golden eagle (*Aquila chrysaetos*) is a facultative avian scavenger with a wide distribution in the Northern Hemisphere. Photo by Eugenio Noguera.

FIGURE 8.5. The African fish-eagle (*Haliaeetus vocifer*) is a specialized fish eater, but it does not pass by opportunities to feed on the small carcasses of both aquatic and terrestrial vertebrates, especially when it is young. Photo by David Carmona.

observed among facultative scavengers as relating directly to each species' general preferences for carrion relative to live prey. However, such differences are more often dependent on the tolerance for the by-products of microbial decomposition and, even more important, the ability to detect and acquire carrion (Janzen 1977; Houston 1979; Shivik 2006).

Implications of Avian Scavenging for Ecosystems and Humans

What Happens to Food Webs when Avian Scavengers Are Removed from Ecosystems?

Understanding the role of avian scavengers in food webs and ecosystem functioning is indispensable to adequately recognizing the regulating and cultural services that these birds provide to humanity, as well as the supporting services behind them (Moleón et al. 2014a, 2014b). As seen above, carrion represents a vast reservoir of nutrients and energy, and a huge number of bird species (among many other vertebrates, invertebrates, and microorganisms) show adaptations and/or abilities to exploit this resource. The most specialized scavengers, vultures, mobilize a large part of the nutrients and energy encapsulated in vertebrate carcasses, but many other Accipitridae (primary predators) and members of other families, like Corvidae (omnivores), also scavenge frequently (DeVault et al. 2003; Mateo-Tomás et al. 2015; Olson et al. 2016). Both the multichannel feeding characteristic of facultative scavengers and the typically high number of feeding links involving scavenging (either obligate or facultative) have been identified recently as stabilizing forces in food webs (Wilson and Wolkovich 2011).

The high connectivity of scavenging networks makes it difficult to predict the final outcome of a given perturbation to the food web, with direct bottom-up effects and indirect and complex top-down consequences taking place (Moleón et al. 2014b). Unfortunately, little is known about the ecological consequences of large-scale vulture declines and rarefaction (Ogada et al. 2012a, b), although severe cascading effects can be hypothesized and have indeed likely occurred (see below). Scavenging at the community level is not random but rather is an ordered, nested process in which carcasses visited by poorly specialized scavengers are subsets of those carcasses visited by highly specialized scavengers (Selva and Fortuna 2007). This means that the absence of vultures from the community will trigger strong effects on carrion consumption patterns and the subsequent energy and nutrient flow rates. On one hand, the most obvious effect of the loss of vultures is the decreased rate at which energy and nutrients propagate through food webs (DeVault et al. 2003; Ogada et al. 2012b). On the other hand, less specialized scavengers could potentially then fill this vacant niche and consume more carcasses, which in theory would alter the behavior and demography not only of facultative scavengers themselves, but also of other organisms directly or indirectly connected within the same food web (Moleón et al. 2014b). Processes such as hyperpredation (Courchamp et al. 2000) are therefore prone to emerge following vulture declines. For instance, populations of facultative mammalian scavengers such as feral dogs and rats seem to have increased as a consequence of the recent dramatic vulture population collapse in India, and this could have increased the predation impact of these predators on other wildlife (Pain et al. 2003).

There also might be important ecological consequences when facultative scavengers become locally rare or are extirpated (Olson et al. 2012). The nested nature of carrion consumption patterns predicts that the most specialized scavengers, vultures, are not able to functionally compensate for the extirpation of facultative avian scavengers, as the latter will normally feed on carcasses after the former are satiated. Instead, other vertebrates, invertebrates, and microorganisms are expected to use carrion at higher rates when avian facultative scavengers are absent or scarce. Because most carrion biomass is consumed by vertebrates (DeVault et al. 2003), important indirect effects related to wildlife and human health would likely result from the loss of facultative avian scavengers. Food web changes associated with the extirpation of facultative avian scavengers are expected to be more profound outside vultures' distributional ranges, where facultative scavengers normally visit more carcasses than they do in vulture-dominated environments.

Regulating Services

DISEASE AND PEST CONTROL. Consuming carcasses of wild and domestic animals has been the most enduring ecosystem service provided by scavengers to humans (Moleón et al. 2014a), as reducing the exposure to rotting matter strongly contributes to reducing the rates of transmission of infectious diseases (Ogada et al. 2012b). In the absence of obligate scavengers, facultative scavengers may increase the rate at which they ingest carrion, thus buffering ecosystem functioning to some extent (Şekercioğlu et al. 2004). But as illustrated by the increase in rats and feral dogs following severe population declines of vultures in India, drastic population declines of obligate scavengers could strongly favor some opportunistic facultative scavengers by releasing them from carcass competition (Markandya et al. 2008). Thus, the surplus of carcasses available in the absence of vultures might lead, at least partly, to a population increase wherein opportunistic scavengers could be considered as pests (i.e., as being detrimental to humans or human concerns). The Indian paradigm is a good example of that, and several authors have noted human health risks associated with elevated populations of dogs and rats, the primary reservoirs for rabies and bubonic plague respectively (Pain et al. 2003; Markandya et al. 2008). Moreover, rat and dog bites themselves are a threat to humans (Markandya et al. 2008). Similar problems are expected to emerge in other areas with severe vulture population declines, as in many African regions (Ogada et al. 2012a). In Zimbabwe, for instance, feral dogs dominate carcasses outside protected reserves but not inside them, where vultures are still present and serve as the major scavengers (Butler and du Toit 2002). Therefore, healthy vulture populations may well be key to effective pest control worldwide (Moleón et al. 2014a).

ENVIRONMENTAL AND ECONOMIC COSTS OF SUPPLANTING ECOSYSTEM SERVICES PROVIDED BY SCAVENGERS. In accordance with certain sanitary guidelines, dead livestock has been systematically removed from large regions, as was done in Europe after the bovine spongiform encephalopathy (BSE) crisis (see "Sanitary policies," below). Both government agencies and farmers have been paying for carcass transport and incineration for more than a decade, a service that vultures and other scavengers have provided cost-free for centuries. In Spain alone it is estimated that on average, vultures remove 134 to 201 t of bones and 5,551 to 8,326 t of meat each year, leading to a minimum annual savings of €0.91 to 1.49 million (€0.97–1.60 million throughout the entire European Union; Margalida and Colomer 2012). Later, Morales-Reyes et. al. (2015) estimated that supplanting the natural removal of extensive livestock carrion by scavengers with carcass collection and transport to authorized plants in Spain led to annual emissions of 77,344 metric tons of CO_2 eq. to the atmosphere and payments of about $50 million to insurance companies. In another study, Markandya et al. (2008) estimated the human health cost of the vulture decline in India. They calculated the monetary costs (i.e., medicines, doctor remuneration, and work compensation) associated with human rabies transmitted by feral dog bites, which increased dramatically following the Indian vulture crisis, at an estimated US$2.43 billion annually on average. These examples clearly illustrate the important environmental economic and social benefits that vultures can provide to humans.

INDUSTRY SERVICES. Avian scavengers can provide further economic benefits to humans through several traditional industrial activities. In India, for instance, bones from dead cattle are gathered by bone collectors and transported to supply the fertilizer industry. Vultures greatly facilitate bone collection by efficiently cleaning cattle carcasses. In the absence of vultures, the bones would be less readily available, of poorer quality, less hygienic (due to the presence of more rotten flesh), and more difficult to collect (Markandya et al. 2008).

Cultural Services

INTELLECTUAL, SPIRITUAL, AND AESTHETIC INSPIRATION. Modern humans have inherited a strong cultural benefit from the ancient and changing interspecific interactions between closely related *Homo* species and avian and other scavengers since the Plio/Pleistocene transition (Moleón et al. 2014a). Earliest Pliocene archaeological assemblages (2.5 mya) demonstrate that a major function of the earliest known human tools was meat and marrow processing of large carcasses. This human behavior extended well into the Pleistocene (Heinzelin et al. 1999). These early tools may resemble or mimic adaptations of vultures and hyenas for processing of carcasses. They may well have been essential for successful competition with better adapted natural scavengers. Thus, competition with other scavengers probably contributed to the perfection of these human tools and their use, and hence to cultural diversity. Around the same time, selective pressures associated with confrontational scavenging probably triggered perhaps the most distinctive features of *Homo* species: language and collaborative cooperation (Bickerton and Szathmáry 2011). Also, endurance running probably emerged in our lineage as a response to the foraging challenges imposed by the scattered and ephemeral resource that is carrion (Bramble and Lieberman 2011). Ultimately, improved diet quality due to increasing meat consumption, first from active scavenging and then from hunting, has been related, along with other factors, to the extraordinary brain enlargement within the human lineage (Leonard et al. 2007; Bramble and Lieberman 2011; Navarrete et al. 2011).

There are many examples of the spiritual services provided by vultures since approximately 200 kya (Moleón et al. 2014a). Recent studies suggest that Neanderthals widely exploited birds, particularly scavengers, for their feathers and claws as personal ornaments in symbolic behavior (Finlayson et al. 2012). Similarly, many cultures had closed symbolisms and myths involving vultures (Sekercioglu 2006). For example, Egyptians represented the goddess Nebjet as a vulture, and Native Americans from North America to Patagonia included condors as one of their main cultural symbols (Gordillo 2012). Also, the funeral ceremonies of numerous human cultures around the world consisted of offering the corpses of dead relatives to vultures (Donázar 1993; Eaton 2003). All these cultural and spiritual scenarios survive locally today (Moleón et al. 2014a). Humans compete for carrion in some African regions (O'Connell et al. 1988). Zoroastrianism-practicing Parsis and Tibetan Buddhists in Asia maintain the tradition of leaving human corpses to vultures for purification. Certain Native American cultures maintain traditions and festivities linked to condors. Moreover, the presence of vultures and other avian scavengers in artistic, literary, and musical expression is currently widespread (Donázar 1993; Gordillo 2002). The decline of vultures worldwide could lead to the loss of these ancient cultural and spiritual services.

RECREATIONAL SERVICES AND ECOTOURISM. Ecotourism has become a major economic resource and development tool for many regions and countries (Weaver 2008). Recreation and ecotourism activities associated with avian scavengers are flourishing worldwide (Moleón et al. 2014a). There is an increasing interest in viewing and photographing vultures and other carrion eaters such as eagles. Public agencies, nongovernmental organizations, and travel agencies are explicitly advertising and offering these opportunities (Becker et al. 2005; Piper 2005; Markandya et al. 2008; Weaver 2008; Donázar et al. 2009a). Becker et al. (2005) estimated that the potential annual value of viewing threatened griffon vultures (*Gyps fulvus*) at a nature reserve in Israel was from US$1.1 to $1.2 million, and that 85% of the visitors came to the park to view vultures. There are many other examples of the actual and potential value of ecotourism around vulture breeding areas and feeding stations as important engines for local economies (Anderson and Anthony 2005; Piper 2005; Ferrari et al. 2009).

Supporting Services

Avian scavengers consume much of the huge biomass encapsulated in vertebrate carcasses. As seen above, the absence of vultures can retard the rate at which nutrients are redistributed through the ecosystem (DeVault et al. 2003; Ogada et al. 2012b). Thus, avian scavengers play an important role in nutrient cycling.

Global Challenges to Scavenger Conservation in a Changing World

Populations of obligate scavengers have significantly declined over the last several decades across the globe, mainly due to a suite of anthropogenic factors. Scavengers are the most threatened avian functional group (chapter 12, fig. 6; Şekercioğlu et al. 2004) and 61% of the obligate avian scavengers of the world are currently threatened with extinction (BirdLife International 2014; Ogada et al. 2012a). Many of these species have therefore become high priorities for conservation. This is exemplified by the Egyptian vulture (*Neophron percnopterus*), a formerly widespread and common species, whose decline from least concern to endangered in 2007 was one of the fastest declines in conservation status of any bird species. Specific threats to scavengers vary regionally, but most scavenger species are highly susceptible to anthropogenic disturbances, owing to their unique life history traits. In particular, habitat loss and human persecution have played a prominent role in vulture declines in many regions. However, unintentional poisoning has emerged as one of the greatest threats to avian scavengers globally (Ogada et al. 2012a).

Poisoning and Other Environmental Contaminants

Vultures are particularly vulnerable to contaminants, due to their reliance on carrion. Because they often feed communally, large numbers can be poisoned at a single carcass. In particular, the deliberate poisoning of carnivores by humans is likely the most widespread cause of vulture poisoning worldwide (Donázar 1993; Margalida 2012; Ogada et al. 2012a). The chain of secondary poisoning initiated by the use of poisoned baits ultimately affects vultures and other scavengers. Although the use of poisons to manage carnivore populations has been banned in many countries, it continues to be a common illegal tool used in some regions to manage game species and protect livestock (Hernández and Margalida 2008, 2009a; Ogada et al. 2012a).

Recent links between catastrophic vulture declines and unintentional

FIGURE 8.6. Adult white-backed vulture (*Gyps africanus*) perched near a buffalo carcass in the Hluhluwe-iMfolozi Park, South Africa. Africa holds the richest diversity of vulture species in the planet. Unfortunately, the current threats to African vultures are manifold, with intentional poisoning and food shortage probably being the most widespread. As a result, vultures are increasingly disappearing from large areas and are confined to fewer and fewer secure, protected reserves. Photo by David Carmona.

poisoning through consumption of the veterinary drug diclofenac, a nonsteroidal antiinflammatory drug administered to livestock, has drawn international attention to the vulnerability of scavengers to environmental contamination. In Asia, for example, populations of *Gyps* vultures declined by more than 95% over the last two decades due to accidental poisoning through the consumption of livestock treated with diclofenac (Green et al. 2004; Oaks et al. 2004; Shultz et al. 2004). Vultures consuming livestock with diclofenac-contaminated tissues often die within days from kidney failure (Oaks et al. 2004), but susceptibility to the toxic effects of this drug is not consistent among all avian scavengers (Rattner et al. 2008; Margalida et al. 2014a). Nonetheless, this rapid and widespread crisis has captivated international attention and sparked an unprecedented scientific interest in vultures and other scavengers (Koenig 2006; Margalida et al. 2014a).

Ingestion of pellets or fragments from lead bullets poses another significant threat to some scavengers (Hunt et al. 2006; Kelly et al. 2011; Lambertucci et al. 2011). Upon impact, lead bullets often fragment and become lodged in muscle and soft tissue, where they become available to scavengers that consume viscera or muscle tissue from field-processed and unrecovered big game. Although the use of lead shot was banned for waterfowl hunting in the United States in 1991, lead-based ammunition is still used legally to harvest upland birds and big game throughout most of the United States. Today, lead poisoning remains the leading cause of death for the California condor (*Gymnogyps californianus*), and is perhaps the primary factor threatening the recovery of this species (Cade 2007). Similarly, elevated lead exposure linked to hunting also has been documented for turkey vultures, as well as for a diversity of facultative avian scavengers such as common ravens (*Corvus corax*), great horned owls (*Bubo virginianus*), red-tailed hawks (*Buteo jamaicensis*), golden eagles (*Aquila chrysaetos*), and bald eagles (*Haliaeetus leucocephalus*; Clark and Scheuhammer 2003; Craighead and Bedrosian 2009; Kelly et al. 2011; Kelly and Johnson 2011). In Europe, lead poisoning has been identified in several avian scavengers, such as the Egyptian vulture (Gangoso et al. 2009) and bearded vulture (Hernández and Margalida 2009b).

Although diclofenac and other environmental contaminants that cause rapid mortality have received widespread attention, scavengers are also exposed to numerous other toxicants that may have sublethal effects that often go unnoticed (Kumar et al. 2003). For example, more than 50% of bald eagles admitted to wildlife rehabilitators in Iowa had ingested lead, presumably scavenged from hunter-killed white-tailed deer (*Odocoileus virginianus*) remains left in the field (Neumann 2009). Such sublethal exposure to heavy metals may affect bone mineralization (Gangoso et al. 2009), reduce muscle and fat concentrations (Carpenter et al. 2003), and cause organ damage, internal lesions (Pattee et al. 1981), and reduced hatching success (Steidl et al. 1991).

Climate Change

Scavengers, particularly obligate scavengers, are inextricably linked to the distribution and availability of carrion. Thus, any shift in the quantity or temporal stability of carrion resources profoundly affects the composition and dynamics of scavenging communities. Availability of carrion is highly modulated by climate (DeVault et al. 2004; Selva et al. 2005; Parmenter and MacMahon 2009; DeVault et al. 2011) and trophic integrity (Wilmers et al. 2003a, b; Wilmers and Post 2006). Thus, a critical research priority is to elucidate the impact of anthropogenic perturbations to ecosystems, as well as to environmental contamination, on scavenging communities (Beasley et al. 2015).

Climate change is causing phenological mismatch in plant and animal communities worldwide and already resulting in bird population declines (Wormworth and Şekercioğlu 2011). However, little is known about the effects of climate change on the phenology of carrion and the scavengers that depend on it. Altered temperature and precipitation patterns resulting from climate change may shift the spatial and temporal availability of carrion, thus impacting scavenging communities across the globe (Smith et al. 2008; Wilson and Wolkovich 2011). For example, the incidence and geographic range of many diseases is projected to increase in response to global climate change (Patz et al. 1996; Harvell et al. 2002). Rather than providing a consistent increase in animal mortality, such increases would likely produce pulses of animal death, thus disrupting the temporal availability of carrion within ecosystems. Changes in the temporal and spatial distribution of carrion, mainly its aggregation, significantly reduce the diversity and evenness of carrion consumption among scavengers (Wilmers et al. 2003a; Cortés-Avizanda et al. 2012), potentially reducing the overwinter survival or fecundity of facultative scavengers that rely on carrion through the winter (Fuglei et al. 2003). However, the impact of truncated carrion availability would probably be more severe for obligate scavengers, with widespread implications for their conservation.

Climate change may also alter competitive interactions between vertebrate scavengers and microbes, potentially reducing vertebrate biodiversity in terrestrial ecosystems. Microbial decomposition doubles with every 10 °C increase in temperature (Vass et al. 1992; Parmenter and MacMahon 2009), suggesting that carrion availability to vertebrates could decrease by 20 to 40%, based on current projections of climate change models (Beasley et al. 2012). Given that the majority of obligate scavengers are currently threatened with extinction (fig. 12.1; Ogada et al. 2012a), such reductions in carrion could contribute to further population declines through resource provisioning (see below).

Sanitary Policies

Carcasses of unstabled livestock have been historically left at their death sites, thus producing an unpredictable and dispersed collection of carcasses throughout the landscape (Donázar et al. 1997). These traditional livestock disposition practices have supported many populations of avian scavengers for centuries (Donázar et al. 1997; Tella 2001). However, anthropogenic activities also can put this beneficial and reciprocal coexistence of vultures and humans at risk. This occurred recently in Europe with the detection of variant (vCJD) and new variant (nvCJD) Creutzfeldt-Jakob disease in humans, which was acquired from cattle infected by bovine spongiform encephalopathy (BSE). The subsequent application of restrictive sanitary legislation (Regulation CE 1774/2002) greatly limited the use of animal by-products that were not intended for human consumption. This legislation required that all carcasses of domestic animals had to be collected from farms and processed or destroyed in authorized facilities. As a result of these sanitary regulations, supplementary feeding points for vultures supplied by intensive farming have greatly diminished (80%) throughout Spain since 2006. The dichotomy between sanitary and environmental policies (i.e., eliminating carcasses versus conserving scavenger species) led to several European dispositions that regulated the use of animal by-products as food for scavenging birds (Tella 2001; Donázar et al. 2009a, 2009b; Margalida et al. 2010).

As a consequence of food shortages, several demographic warning signals have been documented in avian scavengers, including halted population growth, decreased breeding success, apparent increased mortality among younger age classes, and increases in the number of aggressive interactions with live livestock (Donázar et al. 2009a; Margalida et al. 2011b; Margalida et al. 2014b). An immediate solution applied by managers and conservationists has been the implementation of artificial feeding sites and vulture restaurants to counteract the presumed food shortages. With predictable food resources, habitat quality has been modified, and the regular use of feeding stations by vultures and other scavengers could change the ecological services directly and indirectly provided by these species (Deygout et al. 2009; Dupont et al. 2011).

Power Lines and Wind Farms

Global demand for energy is increasing worldwide, leading to an increase in the production and development of both traditional and alternative power sources (Northrup and Witenmeyer 2012), some of which may pose a threat to avian scavengers. For example, electrocution from and collisions with power lines and wind farms have been documented for several vulture species (Ledger and Annegarn 1981; Donázar et al. 2002; Margalida et al. 2008), as well as for facultative avian scavengers such as eagles and ravens (Lehman et al. 2010; Guil et al. 2011). Although the demographic data needed to assess the actual risk that electrocution and collision pose to populations are unavailable, some authors consider that power line mortality is not high enough to affect long-term population size (Bevanger 1994; 1998). Others, however, have identified electrocution as a primary cause of some vulture population declines. For example, Ledger and Annegarn (1980) and Krüger et al. (2004) show that in South Africa the Cape griffon is electrocuted more than any other raptor species. Nikolaus (1984, 2006) and Angelov et al. (2012) advised that electrocutions were responsible for Egyptian vulture declines near Khartoum, Sudan. Leshem (1985) suggested that griffon vulture declines in Israel may result from electrocution and other human-caused factors.

Wind farms have received public and governmental support as alternative energy sources that do not contribute to air pollution, unlike sources associated with fossil fuel technologies (Leddy et al. 1999). During the last decade, wind farm developments have increased substantially all over the world, with the greatest increases occurring in Europe and the United States. The great success achieved by countries such as Germany and Spain in developing the wind power industry serves as an example for all countries interested in expanding wind energy production (Kenisarin et al. 2006). However, the expansion of wind farms has environmental impact (i.e., habitat removal, construction of roads and power lines, visual impact; Laiolo and Tella 2006; Kuvlesky et al. 2007) that must be evaluated and considered. Most research on the subject investigates how wind farm development impacts bird and bat populations (e.g., Langston and Pullan 2003; Baerwald et al. 2008; Garvin et al. 2011). Among all species studied, vultures are among the species most frequently killed by turbines (Carrete et al. 2012). Indeed, in some areas hundreds of individuals have been killed through collisions with turbines, leading to a stoppage of the wind farms' activity in a few cases (Carrete et al. 2010). Moreover, it has been shown that slight increases in mortality at wind farms can significantly affect population trends, thus accelerating the population extinction of sensitive or endangered species (Carrete et al. 2009).

The Future of Avian Scavenger Conservation

The sustainability of free-ranging populations of many obligate scavenging birds will undoubtedly be dependent upon our ability to recognize and mitigate existing and future threats. While management strategies developed to address the effects of climate change and trophic downgrading remain challenging, the top priorities should be reduction of exposure to harmful veterinary drugs and other toxicants, sanitary regulations consistent with wildlife conservation needs, and reduced habitat alteration (Balmford 2013).

Eliminating Intentional Poisoning

Eliminating deliberate wildlife poisoning is a complex task that must involve legal, educational, economic, and punitive measures. An important advancement in fighting against illegal poisoning is the recent Life Biodiversity project, "Innovative actions against illegal poisoning in EU Mediterranean pilot areas" (http://www.lifeagainstpoison.org). This project joins several conservation and research institutions from three European countries (Portugal, Spain, and Greece) to find practical solutions to the problem that illegal poisoning poses for many threatened species, mostly vultures. Through close cooperation with local governments, hunters, and stockbreeders, this project aims to improve the conservation status of different species by identifying high-risk areas for poisoning, thus raising awareness about the detrimental effects of poisoning and decreasing the sense of impunity among offenders. One of the most innovative and successful actions is the use of trained dogs to find poisoned baits that would otherwise go unnoticed.

Sanitary and Veterinary Regulations

In response to the catastrophic population collapse of avian scavengers in Asia, veterinary use of diclofenac in domesticated livestock was banned in India, Pakistan, and Nepal in 2006. The banning of this drug reduced population declines of several vulture species, including that of the oriental white-backed vulture (*Gyps bengalensis*), which may have even increased slightly in recent years (Balmford 2013). However, additional efforts are needed to completely eliminate diclofenac from carcasses, as the presence of diclofenac in less than 1% of carcasses is sufficient to cause severe population declines in susceptible vultures (Cuthbert et al. 2011). Recent trials evaluating the effects of an alternative nonsteroidal antiinflammatory drug, meloxicam, failed to detect any lethal or sublethal effects on captive or wild *Gyps* vultures (Swan et al. 2006; Swarup et al. 2007). Meloxicam is also of low toxicity to other birds and appears to be rapidly metabolized; this suggests that repeated long-term exposure is unlikely to negatively impact scavengers (Swarup et al. 2007). However, additional trials evaluating the impact of meloxicam and other veterinary drugs on scavengers, as well as reduced presence of diclofenac in livestock carcasses, are needed to ensure the recovery of critically endangered *Gyps* vultures. Veterinary treatments should also be carefully applied to both stabled and unstabled livestock to reduce the prevalence of other antiinflammatory, antibiotic, and antiparasitic agents (Donázar et al. 2009a).

Legislation to minimize wildlife exposure to lead also should be implemented or improved, depending on the country, to reduce the effects of lead on large predatory and scavenging birds, including emblematic species such as the California condor. For this reason, lead ammunition was banned from use for big-game hunting within portions of the condor's range in 2008. After this ban, blood lead concentrations decreased 2.5 to 3 times in both golden eagles and turkey vultures, suggesting that a complete ban of lead ammunition would decrease lead exposure for many scavengers (Kelly et al. 2011).

The BSE crisis in Europe required new sanitary regulations for both human health and ecological reality. Fortunately, recommendations made by scientists, conservationists, and managers have recently led to new European guidelines allowing farmers to abandon dead animals in the field and/or feeding stations (Margalida et al. 2012). This illustrates how scientific arguments can trigger positive political action and help to reconcile conservation challenges and human activities (Sutherland et al. 2004; Margalida et al. 2012).

Improving Power Lines and Wind Farms

Measures to mitigate electrocution and collisions include reviewing the placement of new electric lines, removing earth wires or fitting them with markers, and changing pylon design. Such measures have been used in several countries, particularly in Europe, North America, and South Africa (Lehman et al. 2007). For example, low-utility and medium-voltage distribution lines have been placed underground in the Netherlands, Belgium, the United Kingdom, Norway, Denmark, and Germany. Also, most countries in southern Europe require all poles and technical components of power lines to be manufactured and constructed in a way that is safe for birds and protects them from electrocution (Schürenberg et al. 2010). In Spain, several efforts have identified mortality hotspots and modified pylons and lines (Tintó et al 2010). In South Africa, the collision rates of some species such as cranes and bustards were partially reduced after bird-flight diverters were attached to ground wires (Anderson 2002). Regrettably, the actual effectiveness of these measures for vultures is unknown, although for other raptor species some mortality reductions have been reported (Benson 1981).

In the case of wind-farm mortality, our understanding of the problem is less advanced, and mitigation often fails (Drewitt and Langston 2006). Hence the most effective guideline for wind farms is to place them far from sensitive species (Carrete et al. 2012). Recent research found that vultures possess large visual fields that provide comprehensive coverage of the ground ahead and the sky on either side, but which leave large blind spots above and below their heads. Thus, when vultures fly, they tilt their heads downwards so that the space directly in front of them becomes a blind area (Martin et al. 2012).

Demographic studies indicate that fecundity and survival of vultures are negatively influenced by mortality at wind turbines (Carrete et al. 2009; Martínez-Abraín et al. 2011; no data available for power lines), and that small reductions in the survival of territorial and nonterritorial birds associated with wind farms can strongly impact the population viability of these long-lived species (Carrete et al. 2009). Altogether, existing data highlight the need to examine the long-term impact of power lines and wind farms rather than focusing on short-term mortality, as is often promoted by power companies and some wildlife agencies. Unlike other non-natural causes of mortality which are difficult to eradicate or control, power line and wind farm fatalities can be lowered by powering down or removing risky poles (or entire lines) or turbines (or entire farms) and, in certain cases, by placing them outside areas critical for endangered birds.

Vulture Restaurants

Some countries have created feeding stations where carcass piles are maintained to provide supplemental food to threatened or at-risk avian scavengers. Vulture restaurants, vulture feeding stations, or *muladares* were first created in the late 1960s in both the French and the Spanish slopes of the Pyrenees (Donázar 1993). The impetus was the unfounded but pervasive idea that food shortage resulting from the progressive transformation of traditional systems of animal husbandry was a major cause of the widespread decline of vultures in Europe. Since that time, vulture feeding stations have played an essential role in many avian scavenger conservation programs worldwide, including reintroductions, recoveries of small populations, and range expansions (Donázar et al. 2009a). Several advantages have been traditionally attributed to vulture restaurants (see reviews in Anderson and Anthony 2005; Piper 2005; Donázar et al. 2009a). The most obvious benefit is the provision of food in situations of presumed temporal or, above all, spatial food scarcity (Moreno-Opo et al. 2015a). Vulture restaurants are also generally safe places for birds to feed; they provide poison-free food, keep avian scavengers out of poisoning areas (Oro et al. 2008), and supplement vulture diets with rare nutrients like calcium. As indirect benefits, vulture restaurants can be important for raising awareness among landowners, farmers, and the general public, and for ecotourism and science (e.g., as places for scientists to read the bands or wing tags of marked birds).

Despite these benefits, vulture restaurants should not be conceived as the panacea for vulture conservation in modern times. Numerous objections to these artificial feeding schemes have been identified or suggested (Anderson and Anthony 2005; Piper 2005; Donázar et al. 2009a). First and most important, it remains unclear whether food provisioning improves the long-term viability of the target populations, although some local population recoveries have been partially attributed to such programs (Donázar et al. 2009a). Benefits to some demographic parameters, such as an increase in pre-adult survival (Oro et al. 2008), can be counteracted by density-dependent processes (Carrete et al. 2006a) and behavioral changes (Carrete et al. 2006b) that compromise the long-term success of these management actions (Donázar et al. 2009a). Second, vulture restaurants are single places in which food is supplied at a fairly constant rate. This differs from the ecological context in which vultures evolved, with food available only randomly in space and time and located by searching strategies based on long-distance movements and using behavioral processes such as social facilitation and information sharing (Cortés-Avizanda et al. 2012; Kane et al. 2014). The repercussions of this scenario on individuals, populations, and communities of avian scavengers are largely unknown, but are undoubtedly significant (Donázar et al. 2009a). The accumulation of carrion in a single place favors the most numerous and dominant species in the guild, so that other species, which often are more threatened, are nonetheless excluded from this conservation-motivated food source (Cortés-Avizanda et al. 2010, 2012; Moreno-Opo et al. 2015b).

Societal Involvement and Ecotourism

Societal involvement is essential in any modern conservation strategy. Any initiative aimed at spreading awareness of the ecological value of avian scavengers should be welcomed, especially in educational and academic arenas. Interdisciplinary and international cooperation is also highly desirable for global awareness about avian scavengers. A good example of this is the annual International Vulture Awareness Day (September 1; www.vultureday.org), which was established recently as a way to bring the public closer to the problems affecting vultures and the actions of scientists and conservationists to address them. This initiative, consisting of internationally coordinated and media-covered activities including talks and censuses, resulted from cooperation by the South African Endangered Wildlife Trust (Birds of Prey Programme) and the English Hawk Conservancy Trust. It is now joined by both public and private conservation organizations, as well as by many other people around the world who are concerned with vultures.

The growing source of income associated with vulture and eagle viewing and photography is emerging as a powerful conservation tool, provided that part of the profits are allocated to endangered avian scavenger management programs and also to the benefit of local human communities (Becker et al. 2005; Piper 2005; Donázar et al. 2009a). However, further research is needed to widely evaluate the current economic value and future potential of avian scavenger tourism (Becker et al. 2005). Care should be taken that human presence does not induce potential selective pressures on those wild populations (Carrete and Tella 2010).

Ecosystem Disservices

Vultures are resistant to bacterial toxins in decomposing carcasses (Houston and Cooper 1975), and they decrease the propagation of some diseases. Ogada et al. (2012b) showed that, in the absence of vultures, the time necessary for complete depletion of carcasses nearly tripled, and that the number of carnivorous mammals (which are known to spread some diseases) using carcasses increased threefold. Also, the collapse of vulture populations in Asia apparently led to increases in the number of feral dogs and rats (Pain et al. 2003), coinciding with substantial increases in the prevalence of human rabies (Markandya et al. 2008). Thus, vultures and other avian scavengers play an important role in limiting disease spread overall. Even so, some research suggests that in certain circumstances, scavengers might propagate disease (see Jennelle et al. 2009). For example, VerCauteren et al. (2012) demonstrated that infectious scrapie prions survived passage through the digestive systems of American crows (*Corvus brachyrhynchos*), and speculated that the crows could proliferate

FIGURE 8.7. The black vulture (*Coragyps atratus*) is a species of New World vulture whose range extends from the Midwestern United States to central Chile. This individual has been marked with a uniquely numbered patagial tag to allow researchers to collect data on the population dynamics and behavior of the species. Photo by James C. Beasley.

prion diseases. However, relatively little is known about how avian scavengers influence disease ecology. Additional research in this area would be beneficial to humans and wildlife (fig. 8.7).

Acknowledgments

We thank our respective institutions for funding support during the preparation of this chapter, and Elizabeth Poggiali for assistance in proofing and formatting. This work was partially funded by project CGL2012-40013-C02-02 Spanish Government. Bradley F. Blackwell, Thomas W. Seamans, and Mark E. Tobin provided helpful comments on the manuscript. The input of James C. Beasley was supported through funding provided by the US Department of Energy under Award Number DE-FC09-07SR22506 to the University of Georgia Research Foundation. This work has partially been funded by the Spanish Ministry of Economy and Competitiveness through grant no. CGL2012-40013-C02-01/02 MIMECO and FEDER funds.

References

Álvarez, F., de Reyna, A., and Hiraldo, F. 1976. Interactions among avian scavengers in southern Spain. *Ornis Scandinavica* 2:215–26.

Anderson, M. D. 2002. Karoo large terrestrial bird powerline project, report no. 1. Johannesburg: Eskom (unpublished report).

Anderson, M. D., and Anthony, A. 2005. The advantages and disadvantages of vulture restaurants versus simply leaving livestock (and game) carcasses in the veldt. *Vulture News* 53:42–45.

Balmford, A. 2013. Pollution, politics, and vultures. *Science* 339:653–54.

Barton, P. S., Cunningham, S. A., Lindenmayer, D. B., and Manning, A. D. 2013. The role of carrion in maintaining biodiversity and ecological processes in terrestrial ecosystems. *Oecologia* 171:761–72.

Beasley, J. C., Olson, Z. H., and DeVault, T. L. 2012. Carrion cycling in food webs: Comparisons among terrestrial and marine ecosystems. *Oikos* 121:1021–26.

Beasley, J. C., Olson, Z. H., and DeVault, T. L., 2015. Ecological role of vertebrate scavengers. In *Carrion Ecology, Evolution and Their Applications*, E. M. Benhow, J. Tomberlin, and A. Tarone, eds., 107–27. CRC Press, Boca Raton, FL.

Becker, N., Inbar, M., Bahat, O., Choresh, Y., Ben-Noon, G., and Yaffe, O. 2005. Estimating the economic value of viewing griffon vultures *Gyps fulvus*: A travel cost model study at Gamla Nature Reserve, Israel. *Oryx* 39:429–34.

Ben-David, M., Hanley, T. A., Klein, D. R., and Schell, D. M. 1997. Seasonal changes in diets of coastal and riverine mink: The role of spawning Pacific salmon. *Canadian Journal of Zoology* 75:803–11.

Bevanger, K. 1994. Bird interactions with utility structures: Collision and electrocution, causes and mitigation measures. *Ibis* 136:412–25.

———. 1998. Biological and conservation aspects of bird mortality caused by electricity power lines: A review. *Biological Conservation* 86:67–76.

Bickerton, D., and Szathmáry, E. 2011. Confrontational scavenging as a possible source for language and cooperation. *BMC Evolutionary Biology* 11:261.

BirdLife International. 2014. IUCN Red List for Birds. Accessed September 5, 2014. http://www.birdlife.org.

Blanchard, B. M., and Knight, R. R. 1990. Reactions of grizzly bears, *Ursus arctos horribilis*, to wildfire in Yellowstone National Park, Wyoming. *Canadian Field-Naturalist* 104:592–94.

Bramble, D. M., and Lieberman, D. E. 2011. Endurance running and the evolution of *Homo*. *Nature* 432:345–52.

Buechley, E. R., Şekercioğlu, Ç. H. 2016a. The avian scavenger crisis: Looming extinctions, trophic cascades, and loss of critical ecosystem function. *Biological Conservation*. In press.

Buechley, E. R., Şekercioğlu, Ç. H. 2016b. Vanishing vultures: The collapse of critical scavengers. *Current Biology*. In press.

Bump, J. K., Peterson, R. O, and Vucetich, J.A. 2009. Wolves modulate soil nutrient heterogeneity and foliar nitrogen by configuring the distribution of ungulate carcasses. *Ecology* 90:3159–67.

Burkepile, D. E., Parker, J. D., Woodson, C. B., Mills, H. J., Kubanek, J., Sobecky, P. A., and Hay, M. E. 2006. Chemically mediated competition between microbes and animals: Microbes as consumers in food webs. *Ecology* 87:2821–31.

Butler, J. R. A., and du Toit, J. T. 2002. Diet of free-ranging domestic dogs (*Canis familiaris*) in rural Zimbabwue: Implications for wild scavengers on the periphery of wildlife reserves. *Animal Conservation* 5:29–37.

Cade, T. J. 2007. Exposure of California condors to lead from spent ammunition. *Journal of Wildlife Management* 71:2125–33.

Carpenter, J. W., Pattee, O. H, Fritts, S. H, Rattner, B. A., Wiemeyer, S. N., Royle, J. A., and Smith, M. R. 2003. Experimental lead poisoning in turkey vultures (*Cathartes aura*). *Journal of Wildlife Diseases* 39:96–104.

Carrete, M., Donázar, J. A., and Margalida, A. 2006a. Density-dependent productivity depression in Pyrenean bearded vultures: Implications for conservation. *Ecological Applications* 16:1674–82.

———. 2006b. Linking ecology, behaviour and conservation: Does habitat saturation change the mating system of bearded vultures? *Biology Letters* 2: 624–27.

Carrete, M., Sánchez-Zapata, J. A, Benítez, J. R, Lobón, M., and Donázar, J. A. 2009. Large scale risk-assessment of wind-farms on population viability of a globally endangered long-lived raptor. *Biological Conservation* 142:2954–61.

Carrete, M., Sánchez-Zapata, J. A, Benítez, J. R., Lobón, M., Montoya, F., and Donázar, J. A. 2012. Mortality at wind-farms is positively correlated to large-scale distribution and aggregation in griffon vultures. *Biological Conservation* 145:102–8.

Carrete, M., and Tella, J. L. 2010. Individual consistency in flight initiation distances in burrowing owls: A new hypothesis on disturbance-induced habitat selection. *Biology Letters* 6:167–70.

Chaudhry, M., Arshad, S., Ali Mahmood, A., and Khan, A. A. 2004. Diclofenac residues as the cause of vulture population decline in Pakistan. *Nature* 427: 630–33.

Clark, A. J., and Scheuhammer, A. M. 2003. Lead poisoning in upland-foraging birds of prey in Canada. *Ecotoxicology* 12:23–30.

Cooper, S. M., Holekamp, K. E., and Smale, L. 1999. A seasonal feast: Long-term analysis of feeding behaviour in the spotted hyaena (*Crocuta crocuta*). *African Journal of Ecology* 37:149–60.

Cortés-Avizanda, A., Carrete, M., and Donázar, J. A. 2010. Managing supplementary feeding for avian scavengers: Guidelines for optimal design using ecological criteria. *Biological Conservation* 143:1707–15.

Cortés-Avizanda, A., Jovani, R., Carrete, M., and Donázar, J. A. 2012. Resource unpredictability promotes species diversity and coexistence in an avian scavenger guild: A field experiment. *Ecology* 93:2570–79.

Courchamp, F., Langlais, M., and Sugihara, G. 2000. Rabbits killing birds: Modelling the hyperpredation process. *Journal of Animal Ecology* 69:154–64.

Craighead, D. and Bedrosian, B. 2009. A relationship between blood lead levels of common ravens and the hunting season in the southern Yellowstone Ecosystem. In *Ingestion of Lead from Spent Ammunition: Implications for Wildlife and Humans*, ed. R. T. Watson, M. Fuller, M. Pokras, and W. G. Hunt. Boise, ID: Peregrine Fund.

Cuthbert, R., Taggart, M. A., Prakash, V., Saini, M., Swarup, D., Upreti, S., Mateo, R., Sunder Chakraborty, S., Deori, P., and Green, R. E. 2011. Effectiveness of action in India to reduce exposure of *Gyps* vultures to the toxic veterinary drug diclofenac. *PLoS ONE* 6(5):e19069.

Danell, K., Berteaux, D., and Brathen, K. A. 2002. Effect of Muskox Carcasses on Nitrogen Concentration in Tundra Vegetation. *Arctic* 55:389–92.

DeVault, T. L., Brisbin, I. L. Jr., and Rhodes, O. E. Jr. 2004. Factors influencing the acquisition of rodent carrion by vertebrate scavengers and decomposers. *Canadian Journal of Zoology* 82:502–9.

DeVault, T. L., and Krochmal, A. R. 2002. Scavenging by snakes: An examination of the literature. *Herpetologica* 58:429–36.

DeVault, T. L., Olson, Z. H., Beasley, J. C., and Rhodes, O. E. Jr. 2011. Mesopredators dominate competition for carrion in an agricultural landscape. *Basic and Applied Ecology* 12:268–74.

DeVault, T. L., Rhodes, O. E. Jr., and Shivik, J. A. 2003. Scavenging by vertebrates: Behavioral, ecological, and evolutionary perspectives on an important energy transfer pathway in terrestrial ecosystems. *Oikos* 102:225–34.

Donázar, J. A. 1993. *Los buitres ibéricos*. Madrid: Ed. Quercus.

Donázar, J. A., Margalida, A., and Campión, D., eds. 2009a. *Vultures, Feeding Stations and Sanitary Legislation: A Conflict and Its Consequences from the Perspective of Conservation Biology*. Spain: Sociedad de Ciencias Aranzadi.

Donázar, J. A., Margalida, A., Carrete, M., and Sánchez-Zapata, J. A. 2009b. Too sanitary for vultures. *Science* 326:664.

Donázar, J. A., Naveso, M. A., Tella, J. L., and Campión, D. 1997. Extensive grazing and raptors in Spain. In *Farming and Birds in Europe*, ed. D. Pain and M. W. Pienkowsky, 117–49. London: Academic Press.

Drewitt, A. L., and Langston, R. H. W. 2006. Assessing the impacts of wind farms on birds. *Ibis* 148:29–42.

Dudley, J. P. 1996. Record of carnivory, scavenging and predation for *Hippopotamus amphibius* in Hwange National Park, Zimbabwe. *Mammalia* 60:486–88.

Dunne, J. A., Williams, R. J., and Martinez, N. D. 2002. Network structure and biodiversity loss in food webs: Robustness increases with connectance. *Ecology Letters* 5:558–67.

Dupont, H., Mihoub, J.-B., Bobbe, S., and Sarrazin, F. 2012. Modelling carcass disposal practices: Implications for the management of an ecological service provided by vultures. *Journal of Applied Ecology* 49:404–11.

Duriez, O., Herman, S., and Sarrazin, F. 2012. Intra-specific competition in foraging Griffon Vultures *Gyps Fulvus*: 2. The influence of supplementary feeding management. *Bird Study* 59:193–206.

Eaton, J. 2003. Silent towers, empty skies. *Earth Island Journal* 18(4): 30–33.

Errington, P., and Breckenridge, W. 1938. Food habits of buteo hawks in north-central United States. *Wilson Bulletin* 50:113–21.

Ferrari, S., MacNamara, M., Albrieu, C., Asueta, R., and Alarcón, S. 2009. The use of wild fauna for the promotion of ecotourism activities: The case of the Andean condor (*Vultur gryphus*) in the Río Turbio coal basin. *Ambientalmente Sustentable* 8:173–84.

Finlayson, C., Brown, K., Blasco, R., Rosell, J., Negro, J. J., et al. 2012. Birds of a feather: Neanderthal exploitation of raptors and corvids. *PLoS ONE* 7:e45927.

Fuglei, E., Oritsland, N. A., and Prestrud, P. 2003. Local variation in arctic fox abundance on Svalbard, Norway. *Polar Biology* 26:93–98.

Gangoso, L., Agudo, R., Anadón, J. D., de la Riva, M., Suleyman, A. S., Porter, R., and Donázar, J. A. 2013. Reinventing mutualism between humans and wild fauna: Insight from cultures as ecosystem services providers. *Conservation Letters* 6:172–79.

Gangoso, L., Álvarez-Lloret, P., Rodríguez-Navarro, A. A. B., Mateo, R., Hiraldo, F., and Donázar, J. A. 2009. Long-term effects of lead poisoning on bone mineralization in vultures exposed to ammunition sources. *Environmental Pollution* 157:569–74.

Gennard, D. 2007. *Forensic Entomology: An Introduction*. West Sussex, UK: John Wiley & Sons.

Gibbs, J. P., and Stanton, E. J. 2001. Habitat fragmentation and arthropod community change: Carrion beetles, phoretic mites, and flies. *Ecological Applications* 11: 79–85.

Gordillo, S. 2002. El cóndor andino como patrimonio natural y cultural de Sudamérica. . Córdoba, Argentina: Cuadernos de Taller, 5.

Green, G. I., Mattson, D. J., and Peek, J. M. 1997. Spring feeding on ungulate carcasses by grizzly bears in Yellowstone National Park. *Journal of Wildlife Management* 61:1040–55.

Green, R. E., Taggart, M. A., Das, D., Pain, D. J., Kumar, C. S., Cunningham, A. A., and Cuthbert, R. 2004. Collapse of Asian vulture populations: Risk of mortality from residues of the veterinary drug diclofenac in carcasses of treated cattle. *Journal of Applied Ecology* 43:949–56.

Guil, F., Fernández-Olalla, M., Moreno-Opo, R., Mosqueda, I., García, M. E., Aranda, A., Arredondo, A., Guzmán, J., Oria, J., González, L. M., and Margalida, A. 2011. Minimising mortality in endangered raptors due to power lines: The importance of spatial aggregation to optimize the application of mitigation measures. *PLoS ONE* 6:e28212.

Hairston, N. G., Smith, F. E., and Slobodkin, L. B. 1960. Community structure, population control, and competition. *American Naturalist* 94:421–25.

Hanski, I. 1987. Carrion fly community dynamics: Patchiness, seasonality and coexistence. *Ecological Entomology* 12:257–66.

Hanski, I., and Kuusela, S. 1980. The structure of carrion fly communities: Differences in breeding seasons. *Annales Zoologici Fennici* 17:185–90.

Harvell, C. D., Mitchell, C. E., Ward, J. R., Altizer, S., Dobson, A. P., Ostfeld, R. S., and Samuel, M. D. 2002. Climate warming and disease risks for terrestrial and marine biota. *Science* 296:2158–62.

Heinzelin, J. D., Desmond Clark, J., White, T., Hart, W., Renne, P., Wolde-Gabriel, G., Beyene, Y., and Vrba, E. 1999. Environment and behaviour of 2.5-Million-Year-Old Buri Hominids. *Science* 284:625–29.

Hernández, M., and Margalida, A. 2008. Pesticide abuse in Europe: Effects on the Cinereous vulture (*Aegypius monachus*) population in Spain. *Ecotoxicology* 17:264–72.

———. 2009a. Poison-related mortality effects in the endangered Egyptian vulture (*Neophron percnopterus*) population in Spain. *European Journal of Wildlife Research* 55:415–23.

———. 2009b. Assessing the risk of lead exposure for the conservation of the endangered Pyrenean bearded vulture (*Gypaetus barbatus*) population. *Environmental Research* 109:837–42.

Hewson, R. 1984. Scavenging and predation upon sheep and lambs in west Scotland. *Journal of Applied Ecology* 21:843–68.

Houston, D. C. 1979. The adaptations of scavengers. In *Serengeti, Dynamics of an Ecosystem*, ed. A. R. E. Sinclair and M. Norton-Griffiths. Chicago: University of Chicago Press.

Houston, D. C., and Cooper, J. 1975. The digestive tract of the whiteback griffon vulture and its role in disease transmission among wild ungulates. *Journal of Wildlife Diseases* 11:306–13.

Hunt, W. G., Burnham, W., Parish, C. N., Burnham, K. K., Mutch, B., and Oaks, J. L. 2006. Bullet fragments in deer remains: Implications for lead exposure in avian scavengers. *Wildlife Society Bulletin* 34:167–70.

Janzen, D. 1977. Why fruits rot, seeds mold, and meat spoils. *American Naturalist* 111:691–713.

Jennelle, C., Samuel, M. D., Nolden, C. A., and Berkley, E. A. 2009. Deer carcass decomposition and potential scavenger exposure to chronic wasting disease. *Journal of Wildlife Management* 73:655–62.

Jinová, B., Podskalská, H., Růzicka, J., and Hoskovec, M. 2009. Irresistible bouquet of death: How are burying beetles (Coleoptera: Silphidae: Nicrophorus) attracted by carcasses. *Naturwissenschaften* 96:889–99.

Johnson, M. D. 1975. Seasonal and microseral variations in the insect populations on carrion. *American Midland Naturalist* 93:79–90.

Kane, A., Jackson, A. L., Ogada, D. L., Monadjem, A., and McNally, L. 2014. Vultures acquire information on carcass location from scavenging eagles. *Proc. R. Soc. B.* 281:20141072.

Kapfer, J., Gammon, D., and Groves, J. 2011. Carrion–feeding by barred owls (Strixvaria). *Wilson Journal of Ornithology* 123:646–49.

Kelly, T. R., Bloom, P. H., Torres, S. G., Hernandez, Y. Z., Poppenga, R. H., Boyce, W. M., and Johnson, C. K. 2011. Impact of the California lead ammunition ban on reducing lead exposure in golden eagles and turkey vultures. *PLoS ONE* 6: e17656.

Kelly, T. R., and Johnson, C. K. 2011. Lead exposure in free-flying turkey vultures is associated with big game hunting in California. *PLoS ONE* 6:e15350.

Kenisarina, M., V. M. Karsli, and M. Caglar. 2006. Wind power engineering in the world and perspectives of its development in Turkey. *Renewable and Sustainable Energy Reviews* 10:341–69.

Koenig, R. 2006. Vulture research soars as the scavengers' numbers decline. *Science* 312:1591–92.

Krofel, M., Kos, I., and Jerina, K. 2012. The noble cats and the big bad scavengers: Effects of dominant scavengers on solitary predators. *Behavioral Ecology and Sociobiology* 66:1297–1304.

Krüger, R., Maritz, A., van Rooyen, C. 2004. Vulture electrocutions on vertically configured medium voltage structures in the Northern Cape Province, South Africa. In Chancellor, R. D., and Meyburg, B.–U. (Eds.), *Raptors Worldwide*. World Working Group on Birds of Prey and Owls, Berlin, Germany, and MME/BirdLife Hungary, Budepest, 437–41.

Kumar, K. S., Bowerman, W. W., DeVault, T. L., Takasuga, T., Rhodes, O. E. Jr., Brisbin, I. L. Jr., and Masunaga, S. 2003. Chlorinated hydrocarbon contaminants in blood of black and turkey vultures from Savannah River Site, South Carolina, USA. *Chemosphere* 53:173–82.

Laiolo, P., and Tella, J. L. 2006. Fate of unproductive and unattractive habitats: Recent changes in Iberian steppes and their effects on endangered avifauna. *Environmental Conservation* 33:223–32.

Lambertucci, S. A., Donázar, J. A., Delgado, A., Jiménez, B., Sáez, M., Sánchez-Zapata, J. A., and Hiraldo, F. 2011. Widening the problem of lead poisoning to a South-American top scavenger: Lead concentrations in feathers of wild Andean condors. *Biological Conservation* 144:1464–71.

Lambertucci, S. A., Trejo, A., Di Martino, S., Sánchez-Zapata, J. A., Donázar, J. A., and Hiraldo, F. 2009. Spatial and temporal patterns in the diet of the Andean condor: Ecological replacement of native fauna by exotic species. *Animal Conservation* 12:338–45.

Ledger, J. A., and Annegarn, H. J. 1981. Electrocution hazards to the cape vulture *Gyps coprotheres* in South Africa. *Biological Conservation* 20:15–24.

Lehman, R. N., Savidge, J. A., Kennedy, P. L., and Harness, R. E. 2012. Raptor electrocution rates for a utility in the intermountain Western United States. *Journal of Wildlife Management* 74:459–70.

Leonard, W. R., Snodgrass, J. J., and Robertson, M. L. 2007. Effects of brain evolution on human nutrition and metabolism. *Annual Review of Nutrition* 27: 311–27.

Leshem, Y. 1985. Griffon Vultures in Israel: Electrocution and other reasons for a declining population. *Vulture News* 13:14–20.

McFarland, W. N., Pough, F. H., Cade, T. J., and Heiser, J. B. 1979. *Vertebrate Life*. New York: Macmillan.

Margalida, A. 2008a. Bearded vultures (*Gypaetus barbatus*) prefer fatty bones. *Behavioural Ecology and Sociobiology* 63:187–93.

———. 2008b. Presence of bone remains in the ossuaries of bearded vultures *Gypaetus barbatus*: Storage or nutritive rejection? *Auk* 125:560–64.

——. 2012. Baits, budget cuts: A deadly mix. *Science* 338:192.

Margalida, A., Bogliani, G., Bowden, C., Donázar, J. A., Genero, F., Gilbert, M., Karesh, W., Kock, R., Lubroth, J., Manteca, X., Naidoo, V., Neimanis, A., Sánchez-Zapata, J. A., Taggart, M., Vaarten, J., Yon, L., Kuiken, T., and Green, R. E. 2014. One health approach to use of pharmaceuticals. *Science* 346:1296–98.

Margalida, A., Campión, D., and Donázar, J. A. 2011b. European vultures' altered behaviour. *Nature* 480:457.

Margalida, A., Carrete, M., Sánchez-Zapata, J. A., and Donázar, J. A. 2012. Good news for European vultures. *Science* 335:284.

Margalida, A., and Colomer, M. A. 2012. Modelling the effects of sanitary policies on European vulture conservation. *Scientific Reports* 2:753.

Margalida, A., Colomer, M. A., and Sanuy, D. 2011a. Can wild ungulate carcasses provide enough biomass to maintain avian scavenger populations? An empirical assessment using a bio-inspired computational model. *PLoS ONE* 6: e20248.

Margalida, A., Donázar, J. A., Carrete, M., and Sánchez-Zapata, J. A. 2010. Sanitary versus environmental policies: Fitting together two pieces of the puzzle of European vulture conservation. *Journal of Applied Ecology* 47:931–35.

Margalida, A., Heredia, R., Razin, M., and Hernández, M. 2008. Sources of variation in mortality of the bearded vulture *Gypaetus barbatus* in Europe. *Bird Conservation International* 18:1–10.

Markandya, A., Taylor, T., Longo, A., Murty, M. N., Murty, S., and Dhavala, K. 2008. Counting the cost of vulture decline: An appraisal of the human health and other benefits of vultures in India. *Ecological Economics* 67:194–204.

Mateo-Tomás, P., Olea, P., Moleón, M., Vicente, J., Botella, F., Selva, N., Viñuela, J., and Sánchez-Zapata, J. A. 2015. From regional to global patterns in vertebrate scavenger communities subsidized by big game hunting. *Diversity and Distributions* 21:913–24.

Matuszewski, S., Bajerlein, D., Konwerski, S., and Szpila, K. 2011. Insect succession and carrion decomposition in selected forests of Central Europe. Part 3: Succession of carrion fauna. *Forensic Science International* 207:150–63.

McCann, K. 2000. The diversity-stability debate. *Nature* 405:228–33.

McCann, K., Hastings, A., and Huxel, G. R. 1998. Weak trophic interactions and the balance of nature. *Nature* 395:794–98.

Melis, C., Selva, N., Teurlings, L., Skarpe, C., Linnell, J. D. C., and Andersen, R. 2007. Soil and vegetation nutrient response to bison carcasses in Białowieża Primeval Forest, Poland. *Ecological Research* 22:807–13.

Moleón, M., and Sánchez-Zapata, J. A. 2015. The living dead: Time to integrate scavenging into ecological teaching. *BioScience* 65:1003–10.

Moleón, M., Sánchez-Zapata, J. A., Margalida, A., Carrete, M., Donázar, J. A., and Owen-Smith, N. 2014a. Humans and scavengers: The evolution of interactions and ecosystem services. *BioScience* 64: 394–403.

Moleón, M., Sánchez-Zapata, J. A., Sebastián-González, E., and Owen-Smith, N. 2015. Carcass size shapes the structure and functioning of an African scavenging assemblage. *Oikos* 124:1391–1403.

Moleón, M., Sánchez-Zapata, J. A., Selva, N., Donázar, J. A., and Owen-Smith, N. 2014b. Inter-specific interactions linking predation and scavenging in terrestrial vertebrate assemblages. *Biological Reviews* 89:1042–1054, doi: 10.1111/brv.12097.

Morales-Reyes, Z., Pérez-Garcia, J. M., Moleón, M., Botella, F., Carrete, M., Lazcano, C., Moreno-Opo, R., Margalida, A., Donázar, J. A., and Sánchez-Zapata, J. A. 2015. Supplanting ecosystem services provided by scavengers raises greenhouse gas emissions. *Scientific Reports* 5:7811.

Moreno-Opo, R., A. Trujillano, and A. Margalida. 2015a. Optimization of supplementary feeding programs for European vultures depends on environmental and management factors. *Ecosphere* 6(7):127.

Moreno-Opo, R., Trujillano, A., Arredondo, A., González, L. M., and Margalida, A. 2015b. Manipulating size, amount and appearance of food inputs to optimize supplementary feeding programs for European vultures. *Biological Conservation* 181:27–35.

Moreno-Opo, R., Trujillano, A., and Margalida, A. 2016. Behavioural coexistence and feeding efficiency drive niche partitioning at carcasses within the guild of European avian scavengers. *Behavioral Ecology* doi:10.1093/beheco/arw010.

Navarrete A., van Schaik, C. P., and Isler, K. 2011. Energetics and the evolution of human brain size. *Nature* 480:91–93.

Neumann, K. 2009. Bald eagle lead poisoning in winter. In *Ingestion of Lead from Spent Ammunition: Implications for Wildlife and Humans*, ed. R. T. Watson, M. Fuller, M. Pokras, and W. G. Hunt. Boise, ID: Peregrine Fund.

Nikolaus, G. 1984. Large numbers of birds killed by electric power line. *Scopus* 8:42.

———. 2006. Where have all the African vultures gone? *Vulture News* 55:65–67.

Northrup, J. M., and Wittenmeyer, G. 2012. Characterising the impacts of emerging energy development on wildlife, with an eye towards mitigation. *Ecology Letters*, doi: 10.1111/ele.12009.

Oaks, J. L., Gilbert, M., Virani, M. Z., Watson, R. T., Meteyer, C. U., Rideout, B. A., Shivaprasad, H. L., Ahmed, S., Chaudhry, M. J., Arshad, M., Mahmood, S., Ali, A. and Khan, A. A. 2004. Diclofenac residues as the cause of vulture population decline in Pakistan. *Nature* 427:630–33.

O'Connell, J. F., Hawkes, K., Blurton-Jones, N. 1988. Hadza scavenging: Implications for Plio/Pleistocene hominid subsistence. *Current Anthropology* 29: 356–63.

Ogada, D. L., Keesing, F., and Virani, M. Z. 2012a. Dropping dead: Causes and consequences of vulture population declines worldwide. *Annals of the New York Academy of Sciences* 1249:57–71.

Ogada, D., Shaw, P., Byers, R. L., Buij, R., Murn, C., Thiollay, J. M., Beale, C. M., Holdo, R. M., Pomeroy, D., Baker, N., Kruger, S., Botha, A., Virani, M. Z., Monadjem, A. and Sinclair, A. R. E. 2016. Another continental vulture crisis: Africa's vultures collapsing toward extinction. *Conservation Letters* doi: 10.1111/conl.12182.

Ogada, D. L., Torchin, M. E., Kinnaird, M. F., and Ezenwa, V. O. 2012b. Effects of vulture declines on facultative scavengers and potential implications for mammalian disease transmission. *Conservation Biology* 26:453–60.

Olson, Z. H., Beasley, J. C., DeVault, T. L., and Rhodes, Jr., O. E. 2011. Scavenger community response to the removal of a dominant scavenger. *Oikos* 121:77–84.

Olson, Z. H., Beasley, J. C., and Rhodes, Jr., O. E., 2016. Carcass type affects local scavenger guilds more than habitat connectivity. *PlosOne* 11(2) e0147798.

Oro, D., Margalida, A., Carrete, M., Heredia, R., and Donázar, J. A. 2008. Testing the goodness of supplementary feeding to enhance population viability in an endangered vulture. *PLoS ONE* 3:e4084.

Pain, D. J., Cunningham, A. A., Donald, P. F., Duckworth, J. W., Houston, D. C., Katzner, T., Parry-Jones, J., Poole, C., Prakash, V., Round, P., and Timmins, R. 2003. Causes and effects of temporospatial declines of Gyps vultures in Asia. *Conservation Biology* 17:661–71.

Paquet, P. C. 1992. Prey use strategies of sympatric wolves and coyotes in Riding Mountain National Park, Manitoba. *Journal of Mammalogy* 73:337–43.

Parmenter, R. R., and MacMahon, J. A. 2009. Carrion decomposition and nutrient cycling in a semiarid shrub-steppe ecosystem. *Ecological Monographs* 79:637–61.

Patz, J. A., Epstein, P. R., Burke, T. A., and Balbus, J. M. 1996. Global climate change and emerging infectious diseases. *Journal of the American Medical Association* 275:217–23.

Payne, J. A. 1965. A summer carrion study of the baby pig *Sus scrofa* Linnaeus. *Ecology* 592–602.

Payne, L. X., and Moore, J. W. 2006. Mobile scavengers create hotspots of freshwater productivity. *Oikos* 115:69–80.

Pearson, O. P. 1964. Carnivore-mouse predation: An example of its intensity and bioenergetics. *Journal of Mammalogy* 45:177–88.

Pereira, L. M., Owen-Smith, N., and Moleón, M. 2014. Facultative predation and scavenging by mammalian carnivores: Seasonal, regional and intra–guild comparisons. *Mammal Review*: 44:44-55.

Piper, S. E. 2005. Supplementary feeding programs: How necessary are they for the maintenance of numerous and healthy vultures populations? In *Proceedings of the International Conference on Conservation and Management of Vulture Populations*, ed. D. C. Houston and S. E. Piper, 41–50. Thessaloniki, Greece: Natural History Museum of Crete–WWF Greece.

Putman, R. J. 1976. Energetics of the decomposition of animal carrion. PhD diss., University of Oxford.

———. 1983. *Carrion and Dung: The Decomposition of Animal Wastes*. London: Edward Arnold.

Ragg, J. R., Mackintosh, C. G., and Moller, H. 2000. The scavenging behavior of ferrets (*Mustela furo*), feral cats (*Felis domesticus*), possums (*Trichosurus vulpecula*), hedgehogs (*Erinaceus europaeus*), and harrier hawks (*Circus approximans*) on pastoral farmlands in New Zealand: Implications for bovine tuberculosis transmission. *New Zealand Veterinary Journal* 48:166–75.

Rattner, B. A., Whitehead, M. A., Gasper, G., Meteyer, C. U., Link, W. A., Taggart, M. A., Meharg, A. A., Pattee, O. H., and Pain, D. J. 2008. Apparent tolerance of turkey vultures (*Cathartes aura*) to the non-steroidal anti-inflammatory drug diclofenac. *Environmental Toxicology and Chemistry* 27:2341–45.

Remonti, L., Balestrieri, A., Domensis, L., Banchi, C., Lo Valo, T., Robetto, S., and Orusa, R. 2005. Red fox (*Vulpes vulpes*) cannibalistic behaviour and the prevalence of *Trichinella britovi* in NW Italian Alps. *Parasitology Research* 97:431–35.

Rozen, D. E., Engelmoer, D. J. P., and Smiseth, P. T. 2008. Antimicrobial strategies in burying beetles breeding on carrion. *Proceedings of the National Academy of Sciences* 105:17890–95.

Ruxton, G. D., and Houston, D. C. 2004a. Obligate vertebrate scavengers must be large soaring fliers. *J. Theor. Biol.* 228, 431–36.

Schürenberg, B., Schneider, R., and Jerrentrup, H. 2010. Implementation of recommendation No.110/2004 on minimising adverse effects of above ground electricity transmission facilities (power lines) on birds. Report by the NGOs to the 30th meeting of the Standing Committee of the Bern Convention, Strassbourg. T–PVS/Files (2010) 21. Council of Europe.

Şekercioğlu, C.H. 2006. Ecological significance of bird populations. In *Handbook of the Birds of the World*, vol. 11, ed. J. del Hoyo, A. Elliott, and D. A. Christie, 15–51. . Barcelona and Cambridge: Lynx Press and BirdLife International.

Şekercioğlu, C. H., Daily, G. C., and Ehrlich, P. R. 2004. Ecosystem consequences of bird declines. *Proceedings of the National Academy of Sciences* 101:18042–47.

Selva, N., and Fortuna, M. A. 2007. The nested structure of a scavenger community. *Proceedings of the Royal Society of London, Series B* 274:1101–8.

Selva, N., Jędrzejewska, B., Jędrzejewski, W., and Wajrak, A. 2005. Factors affecting carcass use by a guild of scavengers in European temperate woodland. *Canadian Journal of Zoology* 83:1590–1601.

Shivik, J. A. 2006. Are vultures birds, and do snakes have venom, because of macro- and microscavenger conflict? *BioScience* 56:819–23.

Shultz, S., Baral, H. S., Charman, S., Cunningham, A. A., Das, D., Ghalsasi, G. R., Goudar, M. S., Green, R. E., Jones, A., Nighot, P., Pain, D. J., and Prakash, V. 2004. Diclofenac poisoning is widespread in declining vulture populations across the Indian subcontinent. *Proceedings of the Royal Society London B* 271:S458–S460.

Singer, F. J., Schreier, W., Oppenheim, J., and Garton, E. O. 1989. Drought, fires, and large mammals. *BioScience* 39:716–22.

Smith, C. R., and Baco, A. R. 2003. Ecology of whale falls at the deep-sea floor. *Oceanography and Marine Biology* 41:311–54.

Smith, C. R., De Leo, F. C., Bernardino, A. F., Sweetman, A. K., and Martinez Arbizu, P. 2008. Abyssal food limitation, ecosystem structure, and climate change. *Trends in Ecology and Evolution* 23:518–28.

Smith, S., and Paselk, R. 1986. Olfactory sensitivity of the turkey vulture (*Cathartes aura*) to three carrion-associated odorants. *Auk* 103:586–92.

Songserm, T., Amonsin, A., Jam–on, R., Sae-Heng, N., Meemak, N., Pariyothorn, N., Payungporn, S., Theamboonlers, A., and Poovorawan, Y. 2006. Avian influenza H5N1 in naturally infected domestic cat. *Emerging Infectious Diseases* 12:681–83.

Steidl, R. J., Griffin, C. R., and Niles, L. J. 1991. Contaminant levels of osprey eggs and prey reflect regional differences in reproductive success. *Journal of Wildlife Management* 55:601–8.

Swan, G., Naidoo, V., Cuthbert, R., Green, R. E., Pain, D. J., Swarup, D., Prakash, V., Taggart, M., Bekker, L., Das, D., Diekmann, J., Diekmann, M., Killian, E., Meharg, A., Patra, R. C., Saini, M., and Wolter, K. 2006. Removing the threat of diclofenac to critically endangered Asian vultures. *PLoS Biology* 4:396–402.

Swarup, D., Patra, R. C., Prakash, V., Cuthbert, R., Das, D., Avari, P., Pain, D. J., Green, R. E., Sharma, A. K., Saini, M., Das, D., and Taggart, M. 2007. Safety of meloxicam to critically endangered Gyps vultures and other scavenging birds in India. *Animal Conservation* 10: 192–98.

Tabor, K. L., Fell, R. D., and Brewster, C. C. 2005. Insect fauna visiting carrion in Southwest Virginia. *Forensic Science International* 150:73–80.

Tella, J. L. 2001. Action is needed now, or BSE crisis could wipe out endangered birds of prey. *Nature* 410:408.

Tintó, A., Real, J., and Mañosa, S. 2010. Predicting and correcting electrocution of birds in Mediterranean areas. *Journal of Wildlife Management* 74:1852–62.

Towne, E.G. 2000. Prairie vegetation and soil nutrient responses to ungulate carcasses. *Oecologia* 122:232–39.

Vass, A. A. 2001. Beyond the grave: Understanding human decomposition. *Microbiology Today* 28:190–93.

Vass, A. A, Bass, W. M., Wolt, J. D., Foss, J. E., and Ammons, J. T. 1992. Time since death determinations of human cadavers using soil solution. *Journal of Forensic Science* 37: 1236–53.

VerCauteren, K. C., Pilon, J. L., Nash, P. B., Phillips, G. E., and Fischer, J. W. 2012. Prion remains infectious after passage through digestive system of American crows (*Corvus brachyrhynchos*). *PLoS ONE* 7:e45774.

Weaver, D. B. 2008. *Ecotourism*. Milton, Australia: John Wiley & Sons.

Wilmers, C. C., Crabtree, R. L., Smith, D. W., Murphy, K. M., and Getz, W. M. 2003b. Trophic facilitation by introduced top predators: Grey wolf subsidies to scavengers in Yellowstone National Park. *Journal of Animal Ecology* 72:909–16.

Wilmers, C. C., and Post, E. 2006. Predicting the influence of wolf-provided carrion on scavenger community dynamics under climate change scenarios. *Global Change Biology* 12:403–9.

Wilmers, C. C., Stahler, D. R., Crabtree, R. L., Smith, D. W., and Getz, W. M. 2003a. Resource dispersion and consumer dominance: Scavenging at wolf- and hunter-killed carcasses in Greater Yellowstone, USA. *Ecology Letters* 6:996–1003.

Wilson, E. E., and Wolkovich, E. M. 2011. Scavenging: How carnivores and carrion structure communities. *Trends in Ecology and Evolution* 26:129–35.

Wormworth, J., Şekercioğlu, C. H. 2011. *Winged Sentinels: Birds and Climate Change*. Cambridge University Press, New York.

CHAPTER NINE

Nutrient Dynamics and Nutrient Cycling by Birds

Motoko S. Fujita and Kayoko O. Kameda

Flight allows birds to use various habitats of various spatial scales over both short and long periods of time. Seabirds, for instance, establish colonies and roosts on oceanic islands or coastal areas, forage in pelagic waters, and transport captured prey back to the colonies. Seabirds thus transport nutrients from pelagic regions to land areas. Avian life cycles encompass the use of different habitats for foraging, breeding, and resting. These characteristics give birds a special function in ecosystems and landscapes. For example, birds link distant ecosystems by transporting nutrients from aquatic to terrestrial habitats. Thus, they transport nutrients that otherwise would remain in a certain place, in ways that few other animals can.

In this chapter, we explore birds as drivers of nutrient dynamics across ecosystems. We first explain why nutrient transport by birds is important and how the characteristics of birds are especially effective for nutrient transport. We present case studies that show the direct and indirect ecological effects of avian nutrient transport. We then describe provisioning services provided by those ecological interactions. Lastly, we discuss some negative effects of bird nutrient transport on people and environments, underlining the importance of assessing the costs and benefits of bird-mediated nutrient dynamics in human-dominated ecosystems.

Birds as Drivers of Nutrient Dynamics

Why Is Nutrient Cycling Important?

Nutrients are critical elements for organisms, providing the building blocks for growth, maintenance, metabolism, and reproduction. Among bio-elements (elements needed to maintain living organisms), some are considered especially important as nutrients for controlling growth and reproduction. Quantitatively important elements for organisms include carbon, nitrogen, and phosphorus. Carbon is the most abundant bio-element in living tissue, and it can be easily gained from the environment by plants in the form of CO_2. Nitrogen is an essential component of amino acids and nucleotides, the building blocks of proteins and nucleic acids (DNA and RNA), respectively. Phosphorus is an essential component of intracellular energy transfer (ATP), DNA, RNA, and cell membranes. Both nitrogen and phosphorus availability can limit primary production in aquatic and terrestrial ecosystems (Hecky and Kilham 1988; Vitousek and Howarth 1991; Elser et al. 2007). Therefore, sufficient nitrogen and phosphorus through nutrient cycling within an ecosystem or from outside the system is vital for populations of both aquatic and terrestrial organisms.

Global cycles for nitrogen and phosphorus differ (fig. 9.1). Both elements have water-soluble forms that move from terrestrial to aquatic areas through water runoff. However, the pathways of nitrogen and phosphorus from aquatic to terrestrial areas differ. In aquatic systems, microbial denitrification converts nitrates (NO_3^-) to atmospheric molecular nitrogen (N_2). N_2 cycles back into terrestrial systems primarily through bacterial N_2 fixation (fig. 9.1a). Phosphorus, which has no gaseous form at ambient temperatures, does not move from aquatic to terrestrial areas through the atmosphere, but instead is deposited into ocean sediments (fig. 9.1b), or is absorbed by aquatic plants, whereby it moves through the aquatic food web. Long-term geological phenomena such as subduction, eruptions, or uplifts (the rock cycle), and short-term biological transportations by mobile animals are the only pathways for phosphorus to move from sea to land (Schlesinger 1997).

Allochthonous Nutrient Transport

When considering nutrient cycling, ecosystem boundaries must be defined because the impacts of nutrient transport on ecosystems may differ

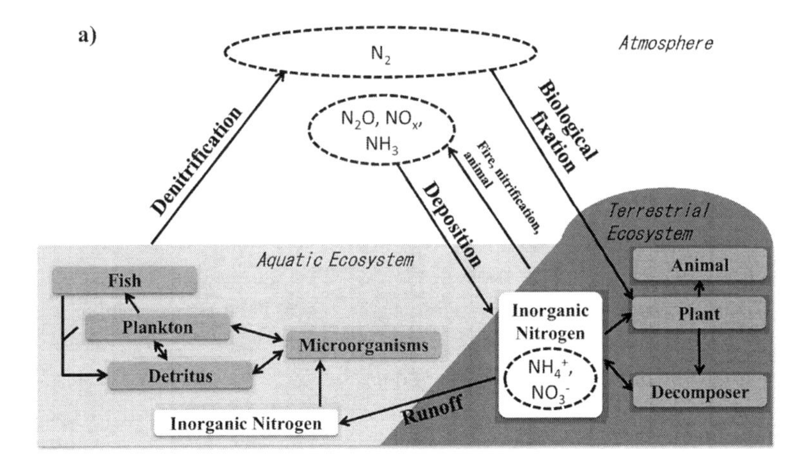

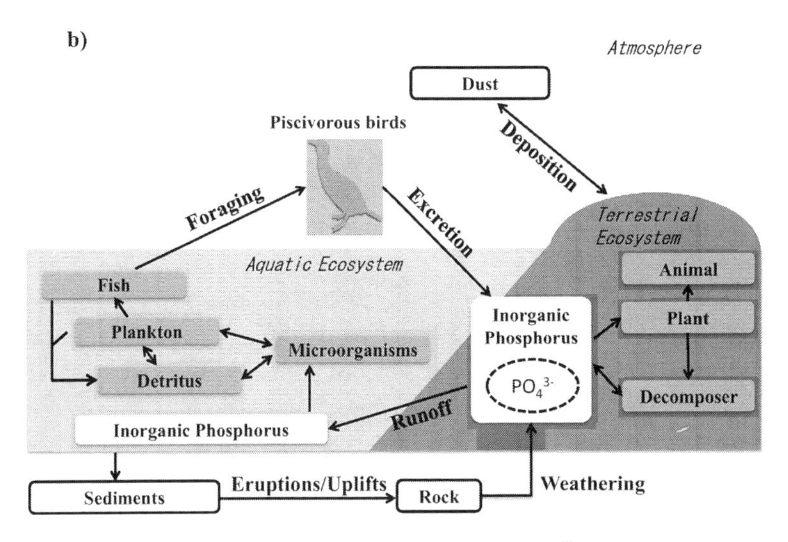

FIGURE 9.1. Global nutrient cycles: (a) nitrogen cycle, (b) phosphorous cycle. Artificial factors (N_2 fixation, NO_x, etc.) are not included.

among ecosystems. For example, a lake ecosystem has a border with the land and any rivers that flow into it. *Allochthonous nutrient transport* is the nutrient input from outside the ecosystem, whereas *autochthonous nutrient flow* comes from inside the ecosystem, and is sometimes referred to as the *internal cycle*. In the case of a lake ecosystem, nutrients that come from the river flow are considered allochthonous input, whereas the nutrients that are recycled from lake sediments are considered autochthonous input.

Elements that arise from both autochthonous and allochthonous fluxes are important for organisms. Autochthonous nutrient flux of nutrients is usually quite large in comparison to allochthonous flux. Litter fall in terrestrial forest ecosystems, for example, the main source of autochthonous N, accounts for up to 25 times more N than total allochthonous N input by deposition and nitrogen fixation (Schlesinger 1997). Nevertheless, allochthonous input is important. Massive quantities of some elements received from outside sources, including the atmosphere, lithosphere, and biosphere, may be retained within an ecosystem over long time frames (millions of years) by internal cycling.

Ecosystems cannot rely solely on autochthonous flow without becoming deficient in certain nutrients. Despite its relatively low total flux, allochthonous input increases primary productivity, with subsequent productivity increases in higher trophic levels, in many ecosystems (Polis et al. 1997). For elements like phosphorus that cannot cycle among ecosystems as gases, allochthonous input via biological transportation is critical. Although weathered rock and dust transported by wind constitute major sources of phosphorus over long time horizons (Schlesinger 1997), phosphorus transport by animals is the only mechanism over short time frames.

Nutrient Transport by Birds and Other Animals

Many animal species transport resources across ecosystems. Other than birds, marine mammals, sea turtles, anadromous fish, and insects all transport nutrients. Some well-known examples of animal nutrient transport between aquatic and terrestrial ecosystems include waterbirds and some mammals. Ibises (Oliver and Legović 1988), geese (Kitchell et al. 1999), cormorants (Hobara et al. 2005), penguins (Lindeboom 1984), and shearwaters (Fukami et al. 2006), as well as coyotes (Rose and Polis 1998) and bears (Hilderbrand et al. 1999) that feed on fish and distribute feces on land, transport lake- or sea-derived nutrients to terrestrial ecosystems. Transported in this way, the nutrients are thus allochthonous in origin. Such nutrient loading produces complex effects involving primary producers and various levels of consumers in ecosystems (Oliver and Legović 1988; Fukami et al. 2006).

Species that transport nutrients across ecosystems share two key traits:

they are highly mobile and they use multiple habitats. Seabirds, anadromous fish, and sea turtles live and forage in the sea and lay eggs on land or in inland waters far from the pelagic environments. Terrestrial predators such as bears or coyotes forage for fish in rivers and then go back to the land. On the other hand, emergent aquatic insects have an opposite life history: they live in aquatic systems as larvae and emerge as adults from the water into terrestrial systems after they pupate. Migratory birds like geese often use different habitats during their breeding and nonbreeding seasons. Other birds, such as crows and swiftlets, forage during the day and return to roosts at night (Koon and Cranbrook 2002; Fujita and Koike 2009). Clearly, birds should be considered among the most important nutrient transporters at various spatiotemporal scales (table 9.1).

Bird droppings, especially from piscivores, contain high concentrations of nitrogen and phosphorus (Sekercioglu 2006). Hutchinson (1950) carried out several chemical analyses of fresh seabird guano, finding that nitrogen comprised from 8.41 to 20.09% and P_2O_5 from 5.53 to 17.40% (equivalent to 2.43 to 7.66% phosphorus) of its total mass. These amounts are large compared to those found in fresh bat droppings, which contain from 8.25 to 11.73% of total nitrogen and 2.25 to 7.42% of total P_2O_5, equivalent to 0.99 to 3.26% phosphorus by mass (Hutchinson 1950). On the other hand, fresh droppings of large-billed crows, *Corvus macrorhynchos*, a non-seabird, contain only 6.01% of total nitrogen and 0.57% of total phosphorus by mass in winter (Fujita and Koike 2009). In addition to the high concentrations of nitrogen and phosphorus in bird droppings, the colonial behavior of some piscivorous birds such as penguins, pelicans, boobies, cormorants, and gulls substantially increases the magnitude of nutrient input in nesting and roosting areas.

Aquatic bird species transport nutrients from water to land when they consume fish, krill, or other "seafood" and defecate or regurgitate in roosts or colonies (Gillham 1956; Lindeboom 1984; Anderson and Polis 1999; Croll et al. 2005). Such nutrient transport counters the movement from land to water via runoff and erosion. In fact, in the absence of animal nutrient transport, the only mechanisms for returning phosphorus from water to land is through the long-term geological phenomena of eruption or uplift. Therefore, nutrient transport from water to land mediated by birds and other animals, particularly of phosphorus, represents an extremely important supporting ecosystem service. Globally, seabirds transfer an estimated 10,000 to 100,000 tons of phosphorus from sea to land annually (Murphy 1981).

TABLE 9.1 **Bird species that play a role in transporting nutrients**

Donor ecosystem	Recipient ecosystem	Species	Scientific name	References
Sea	Island / coastal land	Buller's shearwater	*Puffinus bulleri*	Fukami et al. 2006
		Cape cormorant	*Phalacrocorax capensis*	Bosman et al. 1986
		Cape gannet	*Sula capensis*	Bosman et al. 1986
		Double-crested cormorant	*Phalacrocorax auritus*	Ellis 2005
		Gentoo penguin	*Pygoscelis papua*	Smith 1978
		Glaucous-winged gull	*Larus glaucescens*	Wooton 1991
		Great black-backed gull	*Larus marinus*	Ellis 2005
		Great cormorant	*Phalacrocorax carbo*	Breuning-Madsen et al. 2010
		Jackass penguin	*Spheniscus demersus*	Bosman et al. 1986
		King penguin	*Aptenodytes patagonica*	Smith 1978; Lindeboom 1984
		Macaroni penguin	*Eudyptes chrysolophus*	Lindeboom 1984
		Northern giant petrel	*Macronectes halli*	Smith 1978
		Pelagic cormorant	*Phalacrocorax pelagicus*	Wooton 1991
		Pigeon guillemot	*Cepphus columba*	Wooton 1991
		Southern giant petrel	*Macronectes giganteus*	Smith 1978
		Wandering albatross	*Diomedea exulans*	Smith 1978
		Wedge-tailed shearwater	*Puffinus pacificus*	Bancroft et al. 2005
		Unspecified seabirds		Polis and Hurd 1996; Anderson and Polis 1999; Sanchez-Pinero and Polis 2000; Croll et al. 2005; Maron et al. 2006
Lake	Forest	Great cormorant	*Phalacrocorax carbo*	Ishida 1996; Hobara et al. 2001, 2005; Kameda et al. 2006; Osono et al. 2002, 2006
Agricultural field	Lake	Lesser snow goose	*Chen caerulescens*	Kitchell et al. 1999
		Ross's goose	*Chen rossii*	Kitchell et al. 1999
		Tundra swan	*Cygnus columbianus*	Nakamura et al. 2010a
		Pochard	*Aythya ferina*	Nakamura et al. 2010a

TABLE 9.1 ***(continued)***

Donor ecosystem	Recipient ecosystem	Species	Scientific name	References
		Tufted duck	*Aythya fuligula*	Nakamura et al. 2010a
		Northen pintail	*Anas acuta*	Nakamura et al. 2010a
Residential area	Forest	Carrion crow	*Corvus corone*	Fujita and Koike 2009; Fujita and Koike 2007
		Jungle crow	*Corvus macrorhynchos*	Fujita and Koike 2009; Fujita and Koike 2007
Land	Marsh	Red-winged blackbirds	*Agelaius phoeniceus*	Hayes and Caslick 1984
		Common grackles	*Quiscalus quiscula*	Hayes and Caslick 1984
		Brown-headed cowbirds	*Molothrus ater*	Hayes and Caslick 1984
		European starlings	*Sturnus vulgaris*	Hayes and Caslick 1984

Characteristics of Birds as Nutrient Transporters

BIRD BEHAVIOR AND HABITAT USE. Anadromous fish and aquatic birds are the animals most typically responsible for transporting resources from aquatic to terrestrial areas. A common feature of both fish and bird species is that they migrate from marine to inland areas, greatly influencing inland ecosystems. Colonial-nesting waterbirds also impact terrestrial areas surrounding their colonies.

However, birds possess specific characteristics that set them apart from fish with respect to nutrient transport. Many bird species have a “spot-to-spot” migration pattern (fig. 9.2) and can fly over inhospitable environments, much like an airplane flying from one airport to another. Moreover, birds usually deposit nutrients directly on land in their droppings. Therefore, the two main characteristics of cross-ecosystem resource subsidies transported by birds are (1) bottom-up effects through the nutrient supply from excreta and (2) widespread distribution of the subsidies in terrestrial areas.

In contrast, for a fish to transport nutrients from one area to another (say, from a lake to a river tributary), the two bodies of water must be connected. Anadromous fish transport nutrients hundreds or even thousands

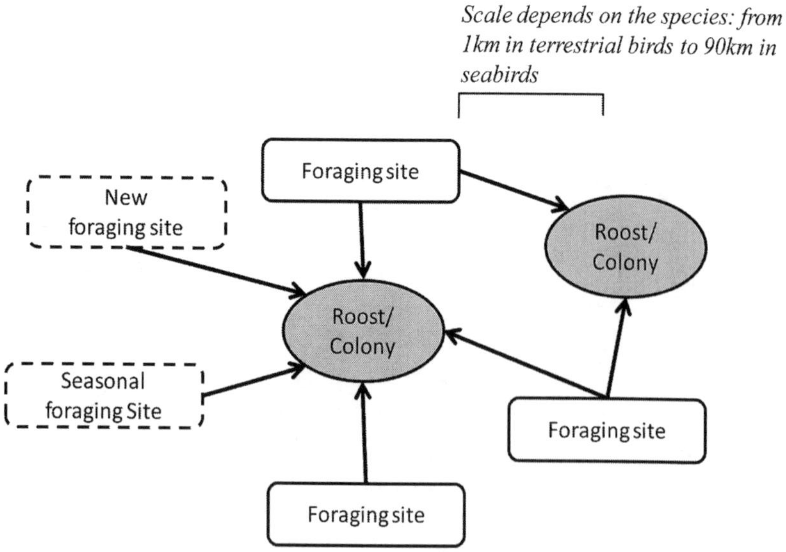

FIGURE 9.2. Spot-to-spot migration and nutrient flow by birds.

of kilometers inland (Hilderbrand et al. 1999), but they need continuous water systems to do so. Many resources supplied by fish arise from their excreta and carcasses, often at the conclusion of spawning. Thus, nutrients derived from fish primarily remain in the rivers in which they spawn. Although fish-derived nutrients might be assimilated by riverine plants, they can be transported to the land basically only when terrestrial predators eat fish. From this perspective, their movement, unlike that of birds, is somewhat like a train running along railroad tracks.

SPATIAL SCALE AND LANDSCAPE HETEROGENEITY. Urban landscape mosaics may consist of several patches of small forests, parks, grasslands, and farmlands scattered in a matrix of residential areas. Rural mosaics may consist of remnant forest patches within an agricultural matrix. Landscape heterogeneity drives diurnal bird foraging and roosting patterns. Many bird species, such as crows, herons, egrets, and cormorants, forage in various habitats including residential areas, rivers, lakes, and farmlands during the daytime, then congregate in night roosts, usually in forest remnants without human disturbance. Roosting leads to the accumulation of droppings at the roosting site. Because the elements that comprise the droppings derive from the food eaten, roosting behavior drives nutrient transport from foraging grounds to roosting sites. Landscape heterogeneity thus influences diurnal movement and, ultimately, heterogeneous nutrient deposition by birds.

The spatial scale of nutrient transport depends on the home range of birds, notably the distance from foraging grounds to the roosting sites. Large-billed crows in urban settings in Japan, for example, feed on garbage in residential areas located 1 to 5 km away from their roosting sites (Morishita et al. 2003). Similar spatial scales were observed in red-legged cormorants (*Phalacrocorax gaimardi*) in Argentina; mean foraging range was 1.9 ± 0.9 km from their nesting site (Gandini et al. 2005). In the United States, double-crested cormorants (*Phalacrocorax auritus*) were observed foraging 2.9 km (SE ± 180 m, max = 14.2 km) from their nesting location (Coleman et al. 2005). Similarly, the foraging range of great cormorants (*Phalacrocorax carbo*) was 2 to 11 km in Japan (Hino and Ishida 2012). Seabirds forage over much longer distances. Atlantic puffins (*Fratercula arctica*) in Britain made two types of foraging trips; long absences that included an overnight stay at distant (38–66 km) foraging areas and short daytime excursions to areas much nearer (9–17 km) the colony (Harris et al. 2012). As reviewed by Thaxter et al. (2012), northern gannets (*Morus bassanus*), lesser black-backed gulls (*Larus fuscus*), and northern fulmars (*Fulmarus glacialis*) had the largest mean foraging ranges, with distances of 92.5, 71.9, and 47.5 km respectively. In the case of seabirds, their foraging behavior also depends on the spatiotemporal distribution of their prey.

AVIFAUNAL SHIFTS AND THEIR IMPACTS. Nutrient transport can differ among bird species. Therefore, avifaunal changes alter nutrient transport. Bird community composition changes along the urban-rural landscape gradient (Clergeau et al. 1998; Marzluff 2001; Natuhara and Imai 1996) and urban bird communities have higher homogeneity and lower species richness than do rural communities. In Yokohama, Japan, for example, many urban forest fragments are characterized by brown-eared bulbuls (*Hypsipetes amaurotis*), Japanese white-eyes (*Zosterops japonicus*), and great tits (*Parus major*). In forest fragments with winter crow roosts, jungle crows (*Corvus macrorhynchos*) and carrion crows (*C. corone*) accounted for more than 80% of the abundance (Fujita and Koike 2009). Bird biomass in these crow roosts was 10 times greater than in other urban fragmented forests (Fujita 2010), and led to 50 times more nitrogen and phosphorus input in crow roosts than in other urban fragmented forests (Fujita and Koike 2009). These results illustrate substantial differences in nutrient transport driven by different avian communities.

Effects of Nutrient Transport on Ecosystems

Direct Effects on Chemical Properties and Primary Production

AQUATIC ECOSYSTEMS. Nutrient transport into aquatic ecosystems involves two processes (Young et al. 2011): (1) direct supply from the droppings of waterbirds or seabirds; and (2) indirect supplies from terrestrial to aquatic environments by runoff, by soil infiltration and percolation into water, and by movement from land to water through volatilization and precipitation.

Direct effects of bird droppings have been investigated frequently for ducks, geese, and swans (Anseriformes; Manny et al. 1994; Post et al. 1998; Kitchell et al. 1999; Nakamura et al. 2010a). Many species in this group are herbivores with large home ranges, which often forage in grasslands or wetlands (including rice fields). Their foraging sites may be far from their roosting lakes, to which they return after foraging. Because they congregate in large flocks of hundreds to millions of individuals, the amount and the distance of nutrients transported by Anseriformes may be quite large. Oversupply of nutrients transported by waterfowl typically leads to the eutrophication of aquatic ecosystems (Kitchell et al. 1999; Nakamura et al. 2010a). Nakamura et al. (2010a) found that nitrogen concentration in a shallow pond increased rapidly following the arrival of migrant waterfowl. In contrast, concentration of total phosphorus and chlorophyll *a*, indicators of phytoplankton abundance, peaked about one month after the birds' departure. The dominant primary producers in the pond switched from phytoplankton to submerged macrophytes in summer, about five months after the phytoplankton peak. In this case, nutrient supply derived at least in part from waterfowl correlates with eutrophication of a shallow pond.

Indirect supply of nutrients from terrestrial colonies and roosts may alter surrounding aquatic environments as well (Bosman and Hockey 1986; Wootton 1991; Staunton Smith and Johnson 1995; Schmidt et al. 2004; Nakamura et al. 2010b). Nutrients or algae tend to increase around the colonies and roosts, especially in small freshwater bodies (Klimaszyk et al 2008; Nakamura et al. 2010b), while the responses of consumers are different among species or environments, resulting in the change of community structures (Wootton 1991).

TERRESTRIAL ECOSYSTEMS. Deposition of a large quantity of droppings has two primary effects on terrestrial vegetation. First, droppings directly adhere to and damage leaves (Ishida 1997). This negative effect is most pronounced among shrubs and herbaceous plants. Second, nutrients from droppings increase soil fertility, a potentially large bottom-up effect. Anderson and Polis (1999) showed that nitrogen and phosphorus contents of leaves on islands with seabird colonies or roosts was higher than on seabird-free islands, and increased biomass by as much as 11.8 times in wet years. On the other hand, an overload of nutrients generally affects plant growth negatively, especially in the areas with extremely high bird density. Many studies found that plant species richness decreases in bird colonies (Ishizuka 1966; Sobey and Kenworthy 1979; Ishida 1996). Instead, some cosmopolitan species, annual species, or species preferring nitrogen-rich soils, increase in bird colonies (Gillham 1963; Hogg and Morton 1983; Ishida 1996). This variation of the bird-mediated effects on plant communities depends on bird density, temperature, and precipitation, and on proximity to anthropogenic seed sources (Ellis 2005).

Birds in large roosts or breeding colonies supply large amounts of nutrients to those sites and the surrounding areas (Lindeboom 1984; Staunton Smith and Johnson 1995; Bancroft et al. 2005). As nitrogen and phosphorus are abundant in birds' excreta (Staunton Smith and Johnson 1995; Sekercioglu 2006) but are often limited in ecosystems, they affect nutrient dynamics in the soils under colonies and roosts. Nitrogen is supplied as uric acid (Bird et al. 2008), which rapidly decomposes to ammonia in soil. In many seabird colonies, ammonia volatilizes (Lindeboom 1984; Mizutani and Wada 1988) and reaches large atmospheric concentrations around the colonies, up to about one kilometer away from the rookeries, as determined by low nitrogen stable isotope values of leaves (Erskine et al. 1998). Nitrification converts soil ammonium to nitrates, which run off from the soil and may enter surrounding lands and water bodies. In some cases, large amounts of nitrogen from both bird droppings and leaf and twig litter used for nest building activities become immobilized, and accumulate (Hobara et al. 2001; Osono et al. 2006). Under these circumstances, nitrates increase in soils and pH declines, causing nitrogen saturation in some cases.

Bird-mediated transport of phosphorus from water to land is extremely important, but less understood than that of nitrogen. The phosphorus content of soils and plants have been investigated in the colonies or roosts of many birds such as penguins (Smith 1979), gulls (Ellis et al. 2006), cormorants (Hobara et al. 2005; Ellis et al. 2006; Breuning-Madsen et al. 2010), prions, shearwaters (Mulder and Keall 2001; Fukami et al.

2006), and puffins (Maron et al. 2006). Phosphorus content of soil tends to increase with bird density (Anderson and Polis 1999), although the relationship between bird densities and phosphorus content differs among species and environments (Ellis et al. 2006; Mulder et al. 2011). Phosphorus supplied as dissolved or available phosphate accumulates in soil (Staunton Smith and Johnson 1995), whereas nitrates run off from the soil. Consequently, N:P ratio decreases in soils under colonies and roosts (Hobara et al. 2005; Mulder et al. 2011). Hence, increased soil phosphorus and $\delta^{15}N$ content (Mizutani et al. 1991) may represent a long-term effect of the nutrient transportation by birds (Morita et al. 2010).

Indirect Effects on Trophic Systems and Community Composition

Allochthonous nutrient resources not only affect ecosystems directly, but also alter trophic structure indirectly. The changes are induced by increased nutrient availability and subsequent increases in primary production, or by changes in the composition of the primary producer community. This bottom-up effect is one of the major changes that occurs with allochthonous nutrient input. The importance of allochthonous resources on food webs is well studied (Polis and Strong 1996; Polis et al. 1997), and spatial subsidies are one of the primary forces of top-down and bottom-up effects in any ecosystem.

The input of guano alters community structure in intertidal ecosystems by enhancing competition for nutrients among lichens, barnacles, and algae (Wootton 1991). Great cormorant excreta that falls to the forest floor under roosting trees changes the abundance, diversity, and community structure of fungal communities (Osono et al. 2002). Comparative studies in geographically similar but ecologically different islands yielded many insights into the effects of allochthonous nutrient inputs on community dynamics. Maron et al. (2006) showed that Aleutian island populations of seabirds provided nutrients that resulted in high biomass plant communities dominated by graminoids, whereas islands with fewer seabirds featured plant communities dominated by forbs and dwarf shrubs. A more complex indirect effect is known from islands on which seabirds roost and leave abundant guano. Such islands exhibit enhanced plant productivity, which in turn increases the density of beetles that eat plant detritus by approximately a factor of five (Sãnchez-Piñero and Polis 2000). The wide-ranging cascading effects are known also in New Zealand islands, where invasive rat populations reduced seabird populations and therefore reduced sea-to-land nutrient transport. This reduction in soil fertility reduced primary production, which in turn caused declines in herbivorous invertebrates (Fukami et al. 2006).

Indirect effects of guano may be verified by analyzing sources of the nitrogen that becomes incorporated into organisms. Croll et al. (2005) showed that all organisms on islands with seabirds used marine-derived nitrogen. Analysis of $\delta^{13}C$ and $\delta^{15}N$ stable isotope ratios allowed identification of the sources and the movement of elements in the food webs. Kolb et al. (2010), for example, used $\delta^{15}N$ analyses to demonstrate that bird-derived nitrogen provided a significant nitrogen source for algae and invertebrate consumers near islands with high seabird densities. The key to the analysis is that $\delta^{15}N$ values differ significantly between bird droppings and other nitrogen sources available in the ecosystem. The $\delta^{15}N$ values of the algae and invertebrates were close to the value of bird droppings, indicating that they partially used nitrogen from bird droppings. In a simple ecosystem with only two nitrogen sources, referred to as "end-members," the relative proportions of the sources contributing to the organisms is easily calculated.

Case Studies of Nutrient Transport by Birds

Seabirds: From the Sea to Islands

Owing to their volcanic origin, the soils of many isolated marine islands are nutrient-poor. Moreover, nutrient transport from continents to isolated islands is limited. Consequently, island ecosystems depend critically on nutrient input from the surrounding ocean. As seabirds often establish colonies on such islands, their transport of nutrients from ocean to land often becomes essential for maintaining island ecosystems.

Many examples are provided by the research on hyperarid and nutrient-poor islands of Baja California, Mexico. Seabirds feeding on marine organisms living in the nutrient-rich waters around these islands subsidize the nutrient budgets on these islands (Polis et al. 2004). For example, Sãnchez-Piñero and Polis (2000) showed that density of tenebrionid beetles was high on islands with seabird nesting colonies or roost. On nesting islands, beetles directly ate seabird carrion. On the other hand, on roosting islands, beetles were affected by seabirds indirectly through eating the plant detritus enhanced by fertilization by seabird droppings.

The importance of avian nutrient transport was demonstrated on islands in the Aleutian archipelago. Seabird colonies on islands on which arctic foxes (*Vulpes lagopus*) were introduced declined from predation by foxes. Unintentionally, artificial introduction of the fox became a field experiment on the seabird effects on this island ecosystem. Seabird density, total soil phosphorus, grass biomass, and nitrogen content of grass and forbs were greater on islands without foxes than on those with foxes (Croll et al. 2005). The $\delta^{15}N$ of soils, plants, and some animal consumers showed that marine-derived nutrients subsidized island ecosystems even at higher trophic levels on islands with abundant seabirds. Marine-derived nutrients thus decreased with the decline of the seabird density, and nutrient-poor soil caused the shift from grasslands to dwarf shrub / forb-dominated vegetation.

Cormorants: From Lakes to Forests

Great cormorants are colonial piscivorous birds inhabiting both coastal and inland waters. This species transports nutrients from freshwater habitats to forests (Hobara et al. 2001, 2005; Kameda et al. 2006; Mizota et al. 2007). The large input of nutrients transported to the colonized forest changed nutrient cycling dynamics. In forests, autochthonous nutrient cycling usually exceeds allochthonous cycling. This is no longer the case when extremely large amounts of nutrient are transported to the forest from elsewhere. This example demonstrates both direct and indirect consequences of bird nutrient transport.

Nitrogen and phosphorus content in soils and dominant trees increased following cormorant colonization of Lake Biwa in Japan (Hobara et al. 2005). Nutrients were abundant both around an occupied colony and in abandoned areas, although the status and pools of both nitrogen and phosphorus differed between the sites. Organic nitrogen from cormorant excreta rapidly decomposes to ammonia and nitrates. As mentioned above, ammonia volatilization is often an important pathway for marine-derived nitrogen at penguin rookeries (Lindeboom 1984; Mizutani and Wada 1988) but not in the Lake Biwa system, owing to low soil pH (ca. 3–4). Although nitrates were still abundant in the abandoned areas, they can easily be lost from the system via runoff. Compared with nitrates, phosphorus tends to accumulate in the soil in the form of phosphates. Consequently, the N/P ratio in soils of this forest declined following establishment of the cormorant colony (Hobara et al. 2005).

Vegetation also changed drastically under cormorant colonies. Ishida (1996) found that vegetation undergoes a four-phase succession as cormorant nests increase. Increased cormorant density increases nutrient input, eventually damaging trees to such an extent that the colony is abandoned. Therefore, decreased tree layer coverage resulted in increased herb layer coverage. Nitrophilous herbs such as *Phytolacca americana* became dominant in an abandoned colony in response to a combination of increased light on the forest floor (owing to canopy dieback) and the increase in soil nitrogen. Fujiwara and Takayanagi (2001) reported a rapid physiological decline of trees damaged by colony formation. Tree damage is caused not only by increased nutrients, but also from cormorants collecting twigs and leaves for nest material (Ishida 2002). Thus, the effects of cormorants on forest vegetation occur in two pathways: direct, physical damage from collecting nest materials, and indirect physiological damage resulting from increased soil fertility.

Waterfowl: From Fields to Lakes

The largest natural lake in France, Grand-Lieu, has experienced high nutrient input from breeding and roosting birds. European starlings (*Sturnus vulgaris*), mallards (*Anas platyrhynchos*), and grey herons (*Ardea cinerea*) played significant roles in transporting nitrogen and phosphorus from outside the lake. According to Marion et al. (1994), about 1,600 to 2,000 breeding herons and great cormorants, 20,000 to 33,000 wintering ducks, gulls, and cormorants, and 1 to 2.4 million starlings deposited about 5,800 kg total N in 1981–82 and 7,640 kg total N in 1990–91. About 2,000 and 2,530 kg total P was deposited over the same time periods, respectively. During the breeding season, up to 37% of the total phosphorus input to the lake was attributable to birds. Because human sewage and agricultural runoff contributed more than did birds, the relative importance of birds was low in this particular example. However, bird input may be higher in other locations with less artificial input.

Kitchell et al. (1999) reported similar levels of nutrient input by waterfowl in a New Mexican wetland. These authors found that geese (mainly snow geese, *Chen caerulescens*) increased the nutrient loading rates in some wetland ponds by up to 40% for total nitrogen and 75% for total phosphorus, in comparison to ponds uninhabited by waterfowl. They also found that fish and crayfish in the ponds showed relatively low nitrogen stable isotope ratios, indicating that fish and crayfish used nitrogen from grains from the surrounding fields deposited by birds, rather than the nitrogen from the river. Primary production and food web structure of the wetland were dominated by avian nutrient inputs.

Crows: From Cities to Forests

In urban landscapes, birds contribute to allochthonous nutrient flow by consuming food in residential areas and depositing feces in forest fragments. Allochthonous P input (kg P ha^{-1} y^{1}) is estimated at 0.0307 in urban fragmented forests and 2.31 in forests with crow roosts, whereas N input (kg N ha^{-1} y^{1}) is 0.397 in urban fragmented forests and 23.2 in forests with crow roosts (Fujita and Koike 2009). Thus, in urban forests with roosts, birds contribute 2.7 times the amount of allochthonous P contributed via other pathways, including rock weathering and precipitation. For nitrogen, birds contribute 0.66 times the amount of allochthonous N input compared to other pathways such as N_2 fixation and precipitation of HNO_3^-. Stable nitrogen and carbon isotope ratios showed high $\delta^{15}N$ and $\delta^{13}C$ values in crow roosts, which in turn indicates that these species eat foods such as livestock meat, C_4 maize, or fish found in garbage. The large nutrient input in urban forests with crow roosts is the result of high crow abundance in urban settings.

Use of Bird Excreta by People

Bird nutrient transport represents supporting services in the form of allochthonous input into an ecosystem. When used by humans as fertilizer, bird excreta represent provisioning services.

Seabirds congregate to nest colonially on many oceanic islands, typically within regions of rich fisheries caused by oceanic upwelling. Here their excreta, or guano, accumulates on the land around their nesting areas over many years and can attain great depths. Guano became extremely valuable in the 19th century because of its rich concentration of nitrates and phosphates and its importance for fertilizer, gunpowder, other explosives, and various chemical industries. Some phosphate ores, a diminishing mineral resource, derive from bird guano as well.

Global Trade of Guano (Peru)

Droppings of piscivorous birds contain large amount of nitrogen and phosphorus, which are important constituents of plant fertilizers. Guano, the organic fertilizer made from the excreta of seabirds or other animals, has been used by Peruvian farmers for more than 1,000 years (Skaggs 1994).

Upwelling of the Pacific Ocean along the coast of Peru creates an extremely productive marine environment, and colonies of guanay cormorants (*Phalacrocorax bougainvillii*), Peruvian boobies (*Sula variegata*), and Peruvian pelicans (*Pelecanus thagas*) inhabit the Peruvian coast and offshore islands. Guano accumulates on the ground and covers it with little runoff, due to the dry climate. Environmental conditions in this area are ideal for guano formation.

Guano has also caused some territorial disputes. In the early 19th century, guano was introduced to European countries and became valuable for its high nitrogen and phosphorus content (Skaggs 1994). The amount of guano exported from Peru to European countries increased rapidly in the 19th century. Later, many countries, including the United States, went in search of guano, leading to a "guano rush." The United States passed the Guano Islands Act in 1856, allowing US citizens to claim uninhabited marine islands for the mining of guano. As a result, guano was exhausted in Peru and on many marine islands by the end of the 19th century. After invention of the Haber-Bosch process for making ammonia from atmospheric nitrogen, the trade in guano for fertilizer and other uses gradually declined.

Peruvians still collect and use guano domestically, despite drastic declines in seabird populations. In the 1950s, about 20 million seabirds inhabited the coast of Peru (Duffy 1983). By 2009, however, seabird numbers had decreased by more than 80% (to 3.4 million) due to the impact of commercial fishing operations on anchovies, which are a primary food source for the birds (Duffy 1994). The Peruvian national government later established protected marine reserves to maintain seabird populations for guano collection.

Local Use of Guano (Unoyama in Japan, Cave Swiftlets in Malaysia)

Guano was also used at a local level in some Asian countries. A typical example comes from Unoyama, Japan. Great cormorants established a nesting colony in Unoyama forest more than 100 years ago. The forest has been used and managed by local residents of Kaminoma village. Local people valued the cormorants' excreta as a fertilizer, and established a unique guano collecting technique (Fujii 2010; fig. 9.3). The local people scattered sand on the forest floor to absorb liquid guano. After five to seven days, the guano-sand mixture was then collected and stored until it was used as fertilizer. This practice continued until the 1960s.

To ensure a long-term supply of guano, the local residents managed

FIGURE 9.3. Guano collecting in Unoyama, Aichi, Japan: (a) Collecting the guano-sand mixture with a *joren*—an agricultural tool used to sift sand—and putting the guano into straw baskets. (b) Lunchtime of the local guano collectors. Permission from the Education Board of Mihama Town.

FIGURE 9.4. Use of the great cormorant as a symbol of Mihama Town: (a) as an emblem of Kaminoma Elementary School; (b) as mascots on the signboard of a tourist farm shop near Unoyama. Photos by Kayoko O. Kameda.

the Japanese black pine (*Pinus thunbergii*) forest. Forests damaged by guano deposition often experience dieback, after which the breeding cormorants abandon their colonies and move to other forests. To prevent the cormorants from moving away, people planted new trees and maintained the colony at Unoyama (Fujii 2010). Weeding the forest floor enhanced guano collection. With these management techniques, the local community maintained the colony and use of guano for more than a century.

In the 1960s, the introduction of chemical fertilizers changed agricultural practices in the Kaminoma area, and guano collection in Unoyama ceased at about this time. However, local people had developed a cultural and sentimental familiarity with cormorants, which are now regarded as a symbol of the town. The great cormorant is used as the emblem of the local elementary school, and as the mascot of a tourist farm shop near Unoyama (fig. 9.4). Great cormorants, which once supplied guano as a provisioning service, thus evolved to represent a new value in the form of a cultural service.

Nests of cave swiftlets (*Aerodramus fuciphagus* and *A. maximus*) in Southeast Asia have long been used by local people and Chinese traders for a delicacy used to make "birds-nest-soup." Besides the birds' nests, the guano of cave swiftlets (including other species as *Collocalia esculenta* and *A. salangana*) and bats has been used by local people as fertilizer. Most of the old phosphate-rich guano in Niah Cave in Sarawak, Malaysia, was removed by guano collectors despite the establishment of a licensing system of guano collection in the 1950s. Nearly 35 tons of guano was collected in 1986 (Leh and Hall 1996). Mean annual wet fresh guano production in Niah Cave was estimated as 8.75 tons in 1986 (Leh and Hall 1996), thus suggesting an unsustainable guano harvest. Moreover, guano production decreased to 1.15 tons in 1992, possibly due to the population decline of swiftlets and bats. Due to unsustainable nest harvesting in caves and large-scale forest decline and its subsequent conversion to cropland, swiftlet populations have declined, some as much as 80% to 90% in the past 30 to 40 years (Koon and Cranbrook 2002). Developing ways to sustainably use guano and bird nests is essential to maintaining both the cave swiftlets and their vital provisioning services.

Conclusion: Nutrient Cycling and Dynamics as an Ecosystem Service

Birds provide significant supporting services as nutrient transporters among ecosystems. From sea to islands, from lakes to forests, from fields to lakes, and from cities to forests, birds transport nutrients and enrich the recipient ecosystems. Allochthonous nutrient transport supports nutrient cycling in many ecosystems. Where birds aggregate in roosts and colonies, people have collected guano as organic fertilizer, a provisioning service. Both of these bird-mediated services are influenced by birds' "spot-to-spot migration" behavior, by landscape heterogeneity, and by bird community composition.

Hutchinson (1950) made rough estimates of the annual guano-derived phosphorus deposition on land by seabirds. By assuming the P_2O_5 content to be about 30%, he estimated that 10,000 metric tons of phosphorus form permanent guano deposits that have been transported from the ocean to the land. He also estimated the present total quantity of guano-derived phosphorus from deposits formed during the post-Pleistocene period at 2,800,000 metric tons, while the amount formed during the Pleistocene period comprises 32,500,000 metric tons, for a total of 35,300,000 metric tons.

World phosphorus stocks are widely expected to run out in the near future, due to excessive use by humans. The world's phosphate rock reserves are estimated to last another 125 years (Gilbert 2009), or 60 to 130 years (Steen 1998). Renewal of phosphate reserves will be limited by the tendency of phosphorus to follow gravity and accumulate on the ocean floor. A shortage of this element could limit food and plant-based energy production. Therefore, bird transport of phosphorus against gravity is tremendously important.

Under some circumstances, however, bird nutrient transport may lead to effects that humans could perceive negatively as disservices. Extremely large amounts of nutrients such as nitrogen and phosphorus can damage the plants or habitats that people want to conserve. A very fine line sometimes separates positive from negative nutrient effects. For example, local use of guano gave local residents not only fertilizer but also a deep sentimental attachment to the great cormorants. The residents in Kaminoma village value the cormorants not only as providers of fertilizer but also as cultural icons.

The social meaning of the bird habitat is also important for ecosystem services. Unoyama is originally a "Satoyama" forest, used by local residents for collecting firewood and charcoal. The social good of guano collecting extends to its ability to facilitate forest recovery and decrease the likelihood of abandonment of the Unoyama forest by local people (Kameda 2010). As a result, cormorant-human interactions at Unoyama may even decrease conflicts with people of the neighboring village. At present, some rice fields may even possibly get benefits from nutrient-rich water from the irrigation pond next to a cormorant colony (Kazama et al. 2013). In contrast, Chikubu Island in Lake Biwa is a "sacred" forest with historical, religious, and sightseeing values, where damage to the vegetation by cormorants was not accepted. The people there were eager to decrease the cormorant population to avoid damage to the forest (Kameda and Tsuboi 2013).

Management of guano collection and of harmful bird impact is needed to maintain not only bird populations, but also bird habitats and the social attitudes of people. Doing so ensures that we can all continue to enjoy the many ecosystem services gained through bird impact on ecosystem nutrient dynamics. We must consider the diversity of ways in which humans value habitats and ecosystem services of birds, in order to find ways to balance the various ecological functions of birds.

References

Anderson, W. B., and Polis, G.A. 1999. Nutrient fluxes from water to land: Seabirds affect plant nutrient status on Gulf of California islands. *Oecologia* 118: 324–32.

Bancroft, W. J., Garkaklisb, M. J., and Roberts, D. J. 2005. Burrow building in seabird colonies: A soil-forming process in island ecosystems. *Pedobiologia* 49:149–65.

Bird, M. I., Tait, E., Wurster, C. M., Furness, R. W. 2008. Stable carbon and nitrogen isotope analysis of avian uric acid. *Rapid Communications in Mass Spectrometry* 22:3393–3400.

Bosman, A. L. and Hockey, P. A. R. 1986. Seabird guano as a determinant of rocky intertidal community structure. *Marine Ecology Progress Series* 32:247–57.

Breuning-Madsen, H., Ehlers-Koch, C., Gregersen, J., and Løjtnant, C. L. 2010. Influence of perennial colonies of piscivorous birds on soil nutrient contents in a temperate humid climate. *Geografisk Tidsskrift-Danish Journal of Geography* 110:25–35.

Clergeau, P., Savard, J. L., Mennechez, G., and Falardeau, G. 1998. Bird abundance and diversity along an urban-rural gradient: A comparative study between two cities on different continents. *Condor* 100:413–25.

Coleman, J. T. H., Richmond, M. E., Rudstam, L. G., and Mattison, P. M. 2005. Foraging location and site fidelity of the double-crested cormorant on Oneida Lake, New York. *Waterbirds* 28:498–510.

Croll, D. A., Maron, J. L., Estes, J. A., Danner, E. M., and Byrd, G. V. 2005. Introduced predators transform subarctic islands from grassland to tundra. *Science* 307:1959–61.

Duffy, D. C. 1983. Competition for nesting space among Peruvian guano birds. *Auk* 100:680–88.

———. 1994. The guano islands of Peru: The once and future management of a renewable resource. In *Seabirds on Islands Threats, Case Studies and Action*

Plans, BirdLife Conservation Series 1., ed. D. N. Nettleship, J. Burger, and M. Gochfeld, 68–76. Cambridge: BirdLife International.

Ellis, J. C. 2005. Marine birds on land: A review of plant biomass, species richness, and community composition in seabird colonies. *Plant Ecology* 181:227–41.

Ellis, J. C., Fariña, J. M., and Witman, J. D. 2006. Nutrient transfer from sea to land: The case of gulls and cormorants in the Gulf of Maine. *Journal of Animal Ecology* 75:565–74.

Elser, J. J., Bracken, M. E. S., Cleland, E. E., Gruner, D. S., Harpole, W. S., Hillebrand, H., Ngai, J. T., Seabloom, E. W., Shurin, J. B., and Smith, J. E. 2007. Global analysis of nitrogen and phosphorus limitation of primary producers in freshwater, marine and terrestrial ecosystems. *Ecology Letters* 10: 1135–42.

Erskine, P. D., Bergstrom, D. M., Schmidt, S., Stewart, G. R., Tweedie, C. E., Shaw, J. D. 1998. Subantarctic Macquarie Island: A model ecosystem for studying animal-derived nitrogen sources using ^{15}N natural abundance. *Oecologia* 117: 187–93.

Fujii, H. 2010. The folk techniques to live with Great Cormorants: A historical folkloristic study of "Unoyama" in Kaminoma district of Mihama, Aich Prefecture. *Annual Bulletin of Rural Studies* 46:45–72.

Fujita, M. 2010. Role of *Corvus macrorhynchos* in cities–from the perspective of nutrient cycling. In *Natural History of Crows* ed. H. Higuchi and R. Kurosawa, 83–94. Sapporo: Hokkaido University Press.

Fujita, M. and Koike, F. 2007. Birds transport nutrients to fragmented forests in an urban landscape. *Ecological Applications* 17:648–54.

———. 2009. Landscape effects on ecosystems: Birds as active vectors of nutrient transport to fragmented urban forests versus forest-dominated landscapes. *Ecosystems* 12:391–400.

Fujiwara, S., and Takayanagi, A. 2001. The influence of the common cormorant (*Phalacrocorax carbo* Kuroda) on forest decline. *Applied Forest Science* 10: 85–90.

Fukami, T., Wardle, D. A., Bellingham, P. J., Mulder, C. P. H., Towns, D. R., Yeates, G. W., Bonner, K. I., Durrett, M. S., Grant-Hoffman, M. N., and Williamson, W. M. 2006. Above- and below-ground impacts of introduced predators in seabird-dominated island ecosystems. *Ecology Letters* 9:1299–1307.

Gandini, P., Frere, E., Quintana, F. 2005. Feeding performance and foraging area of the red-legged cormorant. *Waterbirds* 28:41–45.

Gilbert, N. 2009. Environment: The disappearing nutrient. *Nature* 461:716–18.

Gillham, M. E. 1956. The ecology of the Pembrokeshire Islands V: Manuring by the colonial seabirds and mammals, with a note on seed distribution by gulls. *Journal of Ecology* 44:428–54.

———. 1963. Some interactions of plants, rabbits and sea-birds on South African islands. *Journal of Ecology* 51:275–94.

Harris M. P., Bogdanova, M. I., Daunt, F., and Wanless, S. 2012. Using GPS technology to assess feeding areas of Atlantic Puffins *Fratercula arctica*. *Ringing and Migration* 27:43–49.

Hayes, J. P. and Caslick, J. W. 1984. Nutrient deposition in cattail stands by communally roosting blackbirds and starlings. *American Midland Naturalist* 112: 320–31.

Hecky, R. E. and Kilham, P. 1988. Nutrient limitation of phytoplankton in freshwater and marine environments: A review of recent evidence on the effects of enrichment. *Limnology and Oceanography* 33:796–822.

Hilderbrand, G. V., Hanley, T. A., Robbins, C. T., and Schwartz, C. C. 1999. Role of brown bears (*Ursus arctos*) in the flow of marine nitrogen into a terrestrial ecosystem. *Oecologia* 121:546–50.

Hino, T., and Ishida, A. 2012. Home ranges and seasonal movements of great cormorants *Phalacrocorax carbo* in the Tokai area, based on GPS-Argos tracking. *Japanese Journal of Ornithology* 61:17–28.

Hobara, S., Koba, K., Osono, T., Tokuchi, N., Ishida, A., and Kameda, K. 2005. Nitrogen and phosphorus enrichment and balance in forests colonized by cormorants: Implications of the influence of soil adsorption. *Plant and Soil* 268: 89–101.

Hobara, S., Osono, T., Koba, K., Tokuchi, N., Fujiwara, S., and Kameda, K. 2001. Forest floor quality and N transformations in a temperate forest, affected by avian-derived N deposition. *Water, Air, and Soil Pollution* 130:679–84.

Hogg, E. H. and Morton, J. K. 1983. The effects of nesting gulls on the vegetation and soil of islands in the Great Lakes. *Canadian Journal of Botany* 61: 3240–54.

Hutchinson, G. E. 1950. Survey of contemporary knowledge of biogeochemistry. 3. The biogeochemistry of vertebrate excretion. *Bulletin of the American Museum of Natural History* 96:1–554.

Ishida, A. 1996. Effects of the common cormorant, *Phalacrocorax carbo*, on evergreen forests in two nest sites at Lake Biwa, Japan. *Ecological Research* 11: 193–200.

———. 1997. Seed germination and seedling survival in a colony of the common cormorant, *Phalacrocorax carbo*. *Ecological Research* 12:249–256.

——. 2002. A review of studies on effects of the great cormorant (*Phalacrocorax carbo hanedae*) colonies and roosts on forest ecosystem. *Japanese Journal of Ornithology* 51:29–36.

Ishizuka, K. 1966. Ecology of the ornithocoprophilous plant communities on breeding places of the black-tailed gull, *Larus crassirostris*, along the coast of Japan. *Ecological Review* 16:229–44.

Kameda, K., Koba, K., Hobara, S., Osono, T., and Terai, M. 2006. Pattern of natural ^{15}N abundance in lakeside forest ecosystem affected by cormorant-derived nitrogen. *Hydrobiologia* 567:69–86.

Kameda, K. 2010. *Study of the Measurements of Conservation Management for Forest Decline Caused by the Great Cormorant.* The Report of the River Fund 21-1215-021. Tokyo: Foundation of River and Watershed Environment Management (FOREM).

Kameda, K. and Tsuboi, J. 2013. Cormorants in Japan: Population development, conflicts and management. *EU Cormorant Platform*, 1–6. *http://ec.europa.eu/environment/nature/cormorants/files/Cormorants_in_Japan.pdf.*

Kazama, K., Murano, H., Tsuzuki, K., Fujii, H., Niizuma, Y., and Mizota, C. 2013. Input of seabird-derived nitrogen into rice-paddy fields near a breeding/roosting colony of the great cormorant (*Phalacrocorax carbo*), and its effects on wild grass. *Applied Geochemistry* 28:128–34.

Kitchell, J. F., Schindler, D. E., Herwig, B. R., Post, D. M., Olson, M. H., and Oldham, M. 1999. Nutrient cycling at the landscape scale: The role of diel foraging migrations by geese at the Bosque del Apache National Wildlife Refuge, New Mexico. *Limnology and Oceanography* 44:828–36.

Klimaszyk, P., Joniak, T., Sobczyński, T. and Andrzejewski, W. 2008. Can a cormorant (*Phalacrocora carbo* L.) colony be a factor of Ostrowieckie Lake eutrophication? *Proceedings of Taal2007: The 12th World Lake Conference*, 861–63. Jaipur, India: Ministry of Environment and Forests, Government of India.

Kolb, G. S., Jerling, L., and Hamback, P. A. 2010. The impact of cormorants on plant-arthropod food webs on their nesting Islands. *Ecosystem* 13:353–66.

Koon, L. C., and Cranbrook, Earl of. 2002. *Swiftlets of Borneo: Builders of Edible Nests*. Kota Kinabalu, Malaysia: Natural History Publications.

Leh, C., and Hall, L. S. 1996. Preliminary studies on the production of guano and the socioeconomics of guano collection in Niah Cave, Sarawak. *Sarawak Museum Journal* 71:25–38.

Lindeboom, H. J. 1984. The nitrogen pathway in a penguin rookery. *Ecology* 65:269–77.

Manny A. B., Johnson C. W., and Wetzel G. R. 1994. Nutrient additions by waterfowl to lake and reservoirs: Predicting their effects on productivity and water quality. *Hydrobiologia* 279/280:121–32.

Marion, L., Clergeau, P., Brient, L., and Bertru, G. 1994. The importance of avian-contributed nitrogen (N) and phosphorus (P) to Lake Grand-Lieu, France. *Hydrobiologia* 279/280:133–47.

Maron, J. L., Estes, J. A., Croll, D. A., Danner, E. M., Elmendorf, S. C., and Buckelew, S. L. 2006. An introduced predator alters Aleutian island plant communities by thwarting nutrient subsidies. *Ecological Monographs* 76:3–24.

Marzluff, J. M. 2001. Worldwide urbanization and its effects on birds. In *Avian Ecology and Conservation in an Urbanizing World*, ed. J. M. Marzluff, R. Bowman, and R. Donnelly, 19–47. Norwell, MA: Kluwer Academic Publishers.

Mizota, C., Sasaki, M., and Yamanaka, T. 2007. Temporal variation in the concentration and nitrogen isotopic ratios of inorganic nitrogen from soils under cormorant and heron colonies. *Japanese Journal of Ornithology* 56:115–30.

Mizutani, H., Kabaya, Y., Moors, P. J., Speir, T. W., and Lyon, G. L. 1991. Nitrogen isotope ratios identify deserted seabird colonies. *Auk* 108:960–64.

Mizutani, H., and Wada, E. 1988. Nitrogen and carbon isotope ratios in seabird rookeries and their ecological implications. *Ecology* 69:340–49.

Morishita, E., Itao, K., Sasaki, K., and Higuchi, H. 2003. Movements of crows in urban areas, based on PHS Tracking. *Global Environmental Research* 7:181–91.

Morita, S., Kato, H., Iwasaki, N., Kusumoto, Y., Yoshida, K., and Hiradate, S. 2010. Unusually high levels of bio-available phosphate in the soils of Ogasawara Islands, Japan: Putative influence of seabirds. *Geoderma* 160:155–64.

Mulder, C. P. H., Jones, H. P., Kameda, K., Palmborg, C., Schmidt, S., Ellis, J. C., Orrock, J. L., Wait, D. A., Wardle, D. A., Yang, L., Young, H., Croll D. A., and Vidal, E. 2011. Impacts of seabirds on plant and soil properties. In *Seabird Islands: Ecology, Invasion, and Restoration*, ed. C. P. H. Mulder, W.B. Anderson, D.R. Towns, and P.J. Bellingham, 135–76. Oxford: Oxford University Press.

Mulder, C. P. H., and Keall, S. N. 2001. Burrowing seabirds and reptiles: Impacts on seeds, seedlings and soils in an island forest in New Zealand. *Oecologia* 127:350–60.

Murphy, G. I. 1981. Guano and the anchovetta fishery. *Research and Management of Environmental Uncertainty* 11:81–106.

Nakamura, M., Yabe, T., Ishii, Y., Kamiya, K., and Aizaki, M. 2010a. Seasonal changes of shallow aquatic ecosystems in a bird sanctuary pond. *Journal of Water and Environment Technology* 8:393–401.

Nakamura, M., Yabe, T., Ishii, Y., Kido, K., and Aizaki, M. 2010b. Extreme eutrophication in a small pond adjacent to a forest colonized by great cormorants (*Phalacrocorax carbo*). *Japanese Journal of Limnology* 71:19–26.

Natuhara, Y., and Imai, C. 1996. Spatial structure of avifauna along urban-rural gradients. *Ecological Research* 11:1–9.

Oliver, J. D., and Legović, T. 1988. Okefenokee marshland before, during and after nutrient enrichment by a bird rookery. *Ecological Modelling* 43:195–223.

Osono, T., Hobara, S., Fujiwara, S., Koba, K., and Kameda, K. 2002. Abundance, diversity, and species composition of fungal communities in a temperate forest affected by excreta of the great cormorant *Phalacrocorax carbo*. *Soil Biology and Biochemistry* 34:1537–47.

Osono T., Hobara, S., Koba, K., Kameda, K., and Takeda, H. 2006. Immobilization of avian excreta-derived nutrients and reduced lignin decomposition in needle and twig litter in a temperate coniferous forest. *Soil Biology and Biochemistry* 38:517–25.

Polis, G. A., Anderson, W. B., and Holt, R. D. 1997. Toward an integration of landscape and food web ecology: The dynamics of spatially subsidized food webs. *Annual Review of Ecology and Systematics* 28:289–316.

Polis, G. A., and Hurd, S. D. 1996. Linking marine and terrestrial food webs: Allochthonous input from the ocean supports high secondary productivity on small islands and coastal land communities. *American Naturalist* 147:396–423.

Polis, G. A., Sãnchez-Piñero, F., Stapp, P. T., Anderson, W. B., and Rose, M. D. 2004. Trophic flows from water to land: Marine input affects food webs of islands and coastal ecosystems worldwide. In *Food Webs at the Landscape Level*, ed. G. A. Polis, M. E. Power, and G. R. Huxel, 200–216. Chicago: University of Chicago Press.

Polis, G. A., and Strong, D. R. 1996. Food web complexity and community dynamics. *American Naturalist* 147:813–46.

Post, D. M., Taylor, J. P., Kitchell, J. F., Olson, M. H., Schindler, D. E., and Herwig, B. R. 1998. The role of migratory waterfowl as nutrient vectors in managed wetlands. *Conservation Biology* 12:910–20.

Rose, M. D., and Polis, G. A. 1998. The distribution and abundance of coyotes: The effects of allochthonous food subsidies from the sea. *Ecology* 79:998–1007.

Sãnchez-Piñero, F., and Polis, G. A. 2000. Bottom-up dynamics of allochthonous input: Direct and indirect effects of seabirds on islands. *Ecology* 81:3117–32.

Schlesinger, W. H. 1997. *Biogeochemistry: an analysis of global change.* San Diego: Academic Press.

Schmidt, S., Dennison, W. C., Moss, G. J., and Stewart, G. R. 2004. Nitrogen ecophysiology of Heron Island, a subtropical coral cay of the Great Barrier Reef, Australia. *Functional Plant Biology* 31:517–28.

Şekercioğlu, Ç. H. 2006. Ecological significance of bird populations. In *Handbook of the Birds of the World*, vol. 11, ed. J. Del Hoyo, A. Elliott, and D. A. Christie, 15–51. Barcelona and Cambridge: Lynx Press and Birdlife International.

Skaggs, J. M. 1994. *The Great Guano Rush: Entrepreneurs and American Overseas Expansion.* London: Palgrave Macmillan.

Smith, V. R. 1978. Animal-plant-soil nutrient relationships on Marion Island (subantarctic). *Oecologia* 32:239–53.

———. 1979. The influence of seabird manuring on the phosphorus status of Marion Island (subantarctic) soils. *Oecologia* 41:123–26.

Sobey, D. G., and Kenworthy, J. B. 1979. The relationship between herring gulls and the vegetation of their breeding colonies. *Journal of Ecology* 67:469–96.

Staunton Smith, J., and Johnson, C. R. 1995. Nutrient inputs from seabirds and humans on a populated coral cay. *Marine Ecology Progress Series* 124:189–200.

Steen, I. 1998. Phosphorus availability in the 21st century: Management of a nonrenewable resource, phosphorus and potassium. *British Sulphur Publishing* 217:25–31.

Thaxter, C. B., Lascelles, B., Sugar, K., Cook, A. S. C. P., Roos, S., Bolton, M., Langston, R. H. W., and Burton, N. H. K. 2012. Seabird foraging ranges as a preliminary tool for identifying candidate marine protected areas. *Biological Conservation* 156:53–61.

Vitousek, P. M. and Howarth, R.W. 1991. Nitrogen limitation on land and in the sea: How can it occur? *Biogeochemistry* 13:87–115.

Wootton, J. T. 1991. Direct and indirect effects of nutrients on intertidal community structure: Variable consequences of seabird guano. *Journal of Experimental Marine Biology and Ecology* 151:139–53.

Young, H. S., Hurrey, L., and Kolb, G. S. 2011. Effects of seabird-derived nutrients on aquatic systems. In *Seabird Islands: Ecology, Invasion, and Restoration*, ed. C. P. H. Mulder, W. B. Anderson, D. R. Towns, and P. J. Bellingham, 242–60. New York: Oxford University Press.

CHAPTER TEN

Avian Ecosystem Engineers

Birds That Excavate Cavities

Chris Floyd and Kathy Martin

Many species of birds excavate cavities as part of their nesting, roosting, or feeding activities. Cavity-excavating birds are considered ecosystem engineers because they transform the physical environment in ways that create resources for other species (Jones et al. 1994; Robles and Martin 2013). While interest in assessing the value of ecosystem services provided by birds is growing, most of the literature on this topic comes from studies of avian contributions to seed dispersal, pollination, and pest control (Wenny et al. 2011). Far fewer studies examine the potential ecosystem services contributed by avian cavity excavation (Şekercioğlu 2006), even though about 10% of all birds and many other vertebrates use excavated cavities in trees for nesting, and many species use cavities excavated in other substrates (Cockle et al. 2011). The avian ecosystem engineers that have received the most attention are the woodpeckers, best known for their habit of drilling holes into trees and other woody plant tissues. Usually used only once and then abandoned, the nest cavities of woodpeckers can provide shelter for other cavity-dwelling species for up to two decades (Blanc and Martin 2012).

Other avian taxa provide plentiful examples of species that burrow into soil and other nonwoody substrates, but relatively little is known about the extent to which this activity contributes critically needed cavities for nesting and roosting. Even less well understood are the broader impacts of feeding excavations (Drapeau et al. 2009). Woodpeckers create abundant holes and associated wood fragments when they forage for wood-dwelling invertebrates (Winkler et al. 1995). In addition, some woodpeckers drill sap wells that are visited by other sap feeders (Montellano et al. 2013). Woodpeckers may thus contribute significantly to trophic structure and decomposition cycles in forests (Bednarz et al. 2004; Fayt et al. 2005; Drapeau et al. 2009). In this chapter, we describe the different forms of cavity excavation, review the ecological effects of this activity, discuss ecosystem services that potentially flow from cavity excavation, and suggest research needed to quantify the importance of these services.

Excavated Cavities and Their Use by Nonexcavating Species

Cavities Used for Nesting and Roosting

NEST CAVITIES EXCAVATED INTO TREES. Tree cavities used by cavity-dwelling species are categorized as either nonexcavated ("natural") or excavated (Aitken and Martin 2007). Natural cavities are generated by a combination of decay and mechanical injury (e.g., stem breakage, insect feeding, or fire), while excavated cavities are created by vertebrates (Remm and Lõhmus 2011). Multiple taxa are represented among the species that excavate tree cavities (e.g., Sittidae, Paridae, and Trogonidae), but the most rapid, powerful excavators are in the family Picidae, a group that is collectively known as the woodpeckers but also includes piculets and wrynecks (Winkler et al. 1995). Woodpeckers have a nearly cosmopolitan distribution and are morphologically and taxonomically diverse, with 214 species, ranging in size from the 7.5-cm-long bar-breasted piculet (*Picumnus aurifrons*) to the 50-cm great slaty woodpecker (*Mulleripicus pulverulentus*; Winkler et al. 1995).

The frequency of nest cavity reuse varies among woodpecker species and populations (Wiebe et al. 2007), but in most cases woodpeckers excavate a new cavity for each nesting attempt (Short 1979; Blanc and Martin 2012). These fresh excavations ensure a continuous supply of abandoned, reusable shelters for species that require cavities but cannot construct their own ("secondary cavity-nesters"; Aitken and Martin 2004; Cockle et al. 2011; Remm and Lõhmus 2011). Without woodpecker holes, secondary cavity nesters are largely dependent on human-provided nest boxes or the relatively slow formation of natural cavities. The literature abounds with documentation of secondary cavity-nesters using former woodpecker nest holes (Newton 1994). For example, in a survey of cavity nests in interior British Columbia, Martin et al. (2004) found that most were excavated by red-naped sapsuckers (*Sphyrapicus nuchalis*) and northern flickers (*Colaptes auratus*) into quaking aspen (*Populus tremuloides*), and that these holes provided nesting/roosting habitat for at least seven other bird species and five mammal species. In aspen woodlands of the Colorado Rocky Mountains, violet-green swallows (*Tachycineta thalassina*) and tree swallows (*T. bicolor*) nested almost exclusively in cavities excavated by red-naped sapsuckers (Daily et al. 1993). Studies of cavity-nesting bird communities in forests of central interior British Columbia, western Florida, and central Estonia revealed that the great majority of suitable cavities were drilled by woodpeckers (Martin et al. 2004; Remm et al. 2006; Blanc and Walters 2008). The importance of woodpeckers is especially striking in the case of large-bodied secondary cavity nesters. Aubry and Raley (2002) reported that at least five species of ducks, five species of owls, and nine species of mammals nested in cavities produced by North America's largest extant woodpecker, the pileated woodpecker (*Dryocopus pileatus*). Similarly, nest holes of the black woodpecker (*D. martius*), Europe's largest avian excavator, were used for nesting by the jackdaw (*Corvus monedula*), Tengmalm's owl (*Aegolius funereus*), and stock dove (*Columba oenas*; Mikusiński 1995).

Use of tree cavities by mammals is not as commonly reported as it is for birds. Tree cavities provide shelter for hundreds of species of bats (Kunz and Lumsden 2003; Ruczyński and Bogdanowicz 2005), but the extent to which bat populations are limited by the availability of woodpecker holes is unknown (Kunz and Lumsden 2003; Miller et al. 2003). In the Cascade Range of southern Oregon, nests of female fishers (*Martes pennanti*) were primarily in cavities excavated by pileated woodpeckers (Aubry and Raley 2006). Multiple studies have documented use of woodpecker-excavated cavities by flying squirrels (*Glaucomys spp.*) and red squirrels (*Tamiasciurus hudsonicus*) and by bushy-tailed woodrats (*Neotoma cinerea*; Aubry and Raley 2002; Martin et al. 2004).

NEST CAVITIES EXCAVATED INTO ARBORESCENT SUCCULENT PLANTS. In deserts of North America, several bird species nest or roost in cavities excavated into large columnar cacti, including the saguaro (*Carnegiea gigantean*; Hardy and Morrison 2001) and cardón (*Pachycereus pringlei*; Zwartjes and Nordell 1998). The primary cavity nesters in these cases are primarily the Gila woodpecker (*Melanerpes uropygialis*) and the gilded flicker (*C. chrysoides*; Hardy and Morrison 2001). These cavities are later used by several species of secondary cavity nesters, such as the elf owl (*Micrathene whitneyi*;

Hardy and Morrison 2001), ash-throated flycatcher (*Myiarchus cinerascens*; Cardiff and Dittmann 2002), and purple martin (*Progne subis*; Tarof and Brown 2013). At least two species of bats, the Underwood's mastiff bat (*Eumops underwoodi*; Tibbitts et al. 2002) and the big brown bat (*Eptesicus fuscus*; Cross and Huibregtse 1964), were reported to roost in saguaro cavities. In central Mexico, ladder-backed woodpeckers (*Picoides scalaris*) excavate nest cavities into stems of *Yucca* spp. and *Agave* spp. (Acosta-Perez et al. 2013). Outside of North America there are few published reports of avian excavations in arborescent succulent plants, though Beals (1970) noted that at least one woodpecker species in the Great Rift Valley of Ethiopia excavated its nests into *Euphorbia candelabrum*.

NEST CAVITIES EXCAVATED INTO NONPLANT SUBSTRATES. Many bird species excavate nesting or roosting burrows in the ground (e.g., cliff faces and river banks). This behavior is particularly common among the Coraciiformes (e.g., kingfishers, motmots, and bee eaters; Fry et al. 1992), Psittaciformes (parrots; Brightsmith 2005), and seabirds (Procellariiformes; Schumann et al. 2013), but is also found in species of megapodes (Craciformes; Dekker and Brom 1992), kiwis (*Apteryx* spp.; Jolly 1989), and swallows (Hirundinidae; Winkler and Sheldon 1993). Several species of trogons and parrots excavate nest cavities in arboreal termitaria (Brightsmith 2005; Valdivia-Hoeflich et al. 2005), but there is relatively little literature on secondary use of these cavities. Casas-Crivillé and Valera (2005) found that holes excavated by European bee eaters (*Merops apiaster*) into sandy cliffs in southeast Spain supplied shelter for numerous species of birds, mammals, and invertebrates. In western Mexico, Valdivia-Hoeflich et al. (2005) examined abandoned cavity nests of the citreoline trogon (*Trogon citreolus*) in arboreal termitaria, and documented the reuse of the cavities by several species of arthropods and mammals. Throughout its range in Central and North America, the northern rough-winged swallow (*Stelgidopteryx serripennis*) most commonly nests in burrows excavated into banks by kingfishers and other primary burrow excavators (De Jong 1996).

Cavities Produced during Feeding

EXCAVATION OF FEEDING HOLES IN TREES. Woodpeckers exhibit a great diversity of feeding strategies (e.g., hawking for flying insects, drinking sap, probing ant colonies, gleaning on tree bark, collecting mast, and visiting bird feeders), but they are best known for noisily drilling into trees in search of wood-feeding insects, such as the larvae of bark beetles (Winkler et al. 1995). These feeding excavations often leave conspicuous holes in stems. A few investigators have proposed that woodpecker feeding activity might contribute significantly to wood decomposition and the spread of wood-decaying fungi (discussed below; Farris et al. 2004), but the literature on the ecological effects of woodpecker activity has focused largely on nest cavities, not on feeding excavations. Observations of other birds, such as chickadees and nuthatches, foraging among the feeding excavations of the pileated woodpecker (Bull and Jackson 2011) and American three-toed woodpecker (*P. dorsalis*; Leonard 2001), suggest that the feeding activity of woodpeckers might open up a source of food normally available only to species with sturdy beaks capable of digging into bark or heartwood.

EXCAVATION OF SAP WELLS. Several species of woodpeckers are known to feed at sap wells that they excavate into the sapwood of shrubs and trees. This behavior is particularly advanced in the sapsuckers (*Sphyrapicus* spp.), which possess specialized morphological and behavioral adaptations for creating wells and sipping sap (Walters et al. 2002; Winkler et al. 1995). The sap-feeding woodpecker most familiar to people in eastern North America is the yellow-bellied sapsucker (*S. varius*), which spends much of its time constructing, maintaining, and/or feeding at sap wells (Walters et al. 2002). Wells excavated in xylem are circular and arranged in horizontal rows, while those in phloem are rectangular and arrayed vertically (with the newest well drilled out above the previous one; Walters et al. 2002). Sapsucker wells have been documented in approximately 1,000 species of perennial woody plants (Tate 1973; Eberhardt 2000; Walters et al. 2002). Other species reported to excavate sap wells include the white-fronted woodpecker (*M. cactorum*; Montellano et al. 2013), the acorn woodpecker (*M. formicivorus*; Kattan 1988), the Magellanic woodpecker (*Campephilus magellanicus*; Schlatter and Vergara 2005), the white-headed woodpecker (*P. albolarvatus*; Kozma 2010), and the American three-toed woodpecker; Leonard 2001).

The ecological value of excavated sap wells lies primarily in the high concentration of sucrose in the sap (measurements ranging from 2 to 49%; Tate 1973; Ehrlich and Daily 1988; Walters et al. 2002), which provides energy not only for the excavators but also for many other species that visit the wells. Dozens of vertebrate and invertebrate species have been documented to feed on sap from sapsucker wells (Ehrlich and Daily 1988; Rissler et al. 1995). Sap wells created by white-fronted woodpeckers in the

Chaco region and the Monte desert of Argentina were fed upon by over a dozen species of birds, especially during times of food scarcity (Blendinger 1999; Montellano et al. 2013). In the forests of Tierra del Fuego Island, the presence of sap wells created by the Magellanic woodpecker boosted the local abundance of other sap-feeding bird species (Schlatter and Vergara 2005).

Hummingbirds are among the most frequently documented visitors to sap wells. Several lines of evidence suggest that sap from excavated wells provides an essential energy source for hummingbirds in some regions, and that hummingbird migration in spring is timed to coincide with the availability of sap wells. In a study of rufous hummingbirds (*Selasphorus rufus*) migrating through northern California, Sutherland et al. (1982) found that some hummingbirds spent several days feeding exclusively on tree sap from sapsucker wells and defending the wells against other sap feeders. In northern Michigan, ruby-throated hummingbirds (*Archilochus colubris*) nesting within 300 meters of sap wells excavated by sapsuckers fed almost entirely on tree sap rather than flower nectar (Southwick and Southwick 1980). In Canada, ruby-throated hummingbirds arrive at their breeding sites before their food plants have started flowering (Miller and Nero 1983). The link between sap well availability and spring migration by ruby-throated hummingbirds is also evidenced by the observations that the hummingbird's breeding range extends no further north than that of the sapsuckers, and that the hummingbirds typically do not arrive in their northern summer range until after the arrival of the sapsuckers (Southwick and Southwick 1980; Miller and Nero 1983).

Ecosystem Services Potentially Provided by Cavity Excavation

The Millennium Ecosystem Assessment (2003) recognized four categories of ecosystem services: provisioning, supporting, regulating, and cultural. The potential services provided by cavity excavators fall into the last three of these categories, all of which are indirect and thus have not yet been included in ecosystem valuation studies (Wenny et al. 2011).

Supporting Services: Contributions to Biodiversity and Ecosystem Function

If cavity excavation contributes to the flow of supporting services, it does so through its influence on biodiversity or ecosystem functioning (Wenny et al. 2011). Multiple studies have found positive effects of enhanced biodiversity on ecosystem stability (Tilman et al. 2006) or resilience (Fischer et al. 2006). In addition to supplying shelter to secondary cavity nesters, the drilling of holes in trees contributes to the decomposition processes that characterize mature forests (Lonsdale et al. 2008). Birds that excavate burrows in the ground can influence soil dynamics such as nutrient flux and erosion (Meysman et al. 2006). Below we discuss the contribution of cavity-excavating birds to biodiversity and the potential effects of cavity excavation on ecosystem function.

ENHANCEMENT OF BIODIVERSITY: CAVITY EXCAVATORS AS KEY ELEMENTS OF NEST WEBS. Cavity excavators, particularly woodpeckers, are frequently referred to in the literature as keystone species because they can strongly influence the abundance and diversity of secondary cavity-nesting species (e.g., Daily et al. 1991; Aubry and Raley 2002). According to various interpretations of the keystone species concept, the importance of a keystone species ranges from being literally keystone-like—in that its removal would precipitate a collapse of the associated community—to being disproportionately influential, in that its impact is large compared to its biomass or abundance (Cottee-Jones and Whittaker 2012). Determining where along the keystone gradient a cavity-excavating species ranks, and thus its contribution to biodiversity, requires studies that quantify the relative strengths of the links comprising the associated community. Such studies have recently been carried out using the theoretical framework of the nest web (Martin and Eadie 1999). Nest webs are analogous to food webs in that the interactions among members of a community are hierarchically arranged and depicted with arrows showing the direction in which resources are flowing. In nest webs, the top consumers are the secondary cavity nesters, which nest in cavities produced by cavity excavators. The strength of the relationship between producer and consumer is quantified by measuring either the correlation between their abundances or the proportion of the total relationships at that web level accounted for by that relationship. For example, in their study of cavity-nesting vertebrate communities in forests of interior British Columbia, Martin et. al. (2004) found that cavities excavated by northern flickers were used disproportionately more by most of the secondary cavity-nesting species than the cavities provided by the other five primary excavators. However, the relative importance of flickers as cavity providers decreased in later years after an outbreak of mountain pine beetle resulted in increased densities of the less common woodpeckers, such as the hairy

(*P. villosus*), the downy (*P. pubescens*) and the American three-toed woodpeckers (Cockle and Martin 2015).

While nest webs based on observational data (e.g., from nest counts) have proven useful for documenting indirect interactions in cavity-nesting communities, they cannot conclusively assess the extent to which a cavity excavator is a keystone species. Such causal relationships must be probed using controlled experiments (Newton 1994). Some recent studies have used manipulative methods to assess the importance of nest cavity availability in cavity-nesting communities (e.g., Aitken and Martin 2008; Cockle et al. 2010). For example, Aitken and Martin (2008) reduced quality cavities by blocking more than 50% of the highly used cavities, and found a 50% reduction in nesting densities during the period of blocking; the densities returned to normal levels when the cavities were unblocked. Aitken and Martin (2012) also performed cavity-addition experiments and found a large increase in the nesting densities of birds and mammals. In a review of studies that investigated the effect of nest cavity addition versus subtraction (nest box provisioning versus cavity blocking or snag removal) on breeding densities of cavity-nesting birds, Newton (1994) found consistently positive effects of nest cavity addition and negative effects of subtraction.

PROMOTION OF DECOMPOSITION IN FORESTS. In contrast to the considerable literature on the importance of woodpecker nest cavities, very little has been published on the ecological implications of their feeding excavations. Almost all of the studies involving woodpeckers and wood decomposition are focused on fungal decay that softens heartwood, thus enabling or promoting cavity excavation (Drapeau et al. 2009). Aubrey and Raley (2002) suggested that feeding holes created by pileated woodpeckers might significantly accelerate forest decomposition—both directly, via the bird's bill removing fragmented wood, and indirectly, by providing openings for invasion by fungi and saproxylic insects. Farris et al. (2004) found a higher frequency of saproxylic fungi on the bills of woodpeckers than on those of a group of non–cavity-nesting species, suggesting that woodpeckers might be important vectors for fungal invasion.

BIOTURBATION. Species of birds that excavate burrows in the ground can be important agents of bioturbation, defined as the biological reworking (e.g., the erosion, turning over, or mixing) of soils and sediments (Meysman et al. 2006). This form of ecosystem engineering can play an important role in subsurface ecosystems (Jones et al. 1994). For example,

Casas-Crivillí and Valera (2005) found that the excavation of 67 burrows by European bee eaters nesting in sandy banks over three breeding seasons directly removed approximately 867 kg of sand, and triggered the loss of an additional 4,500 kg when portions of a bank collapsed. More powerful sources of bioturbation are found among burrowing seabirds in the order Procellariformes, many of which congregate in enormous colonies and nest in burrows dug into the ground (Schumann et al. 2013). While these birds are well known to play an important role in nutrient flow in some marine ecosystems (Polis and Hurd 1996), their contributions to bioturbation are not as well documented (Mulder and Keall 2001). McKechnie (2006) studied colonies of sooty shearwaters (*Puffinus griseus*) on islands off southern New Zealand, and found that approximately 18 to 34% of ground surface area was undermined by shearwater burrows, and that the burrowing activity of the birds transported an average of 33 to 96 g/m^{-2} of vegetation below ground. Bancroft et al. (2005) found that burrowing by wedge-tailed shearwaters (*Puffinus pacificus*) nesting on an island off the coast of Western Australia significantly altered the physical and chemical properties of the soil (e.g., bioturbated soils were drier and denser and had higher nitrate levels).

Another form of bioturbation potentially contributed by birds is the erosion produced when birds intentionally feed on soil, an activity known as geophagy (Meysman et al. 2006). Geophagy has been documented in at least six avian orders, but it is particularly common in the parrots, some species of which gather in enormous flocks on cliff faces to feed on clay (Brightsmith 2004). Published studies on avian geophagy, however, have focused largely on the hypothesized nutritional benefits of the behavior (e.g., Diamond 1999), while little or nothing has been published on the importance of geophagy as an agent of bioturbation.

Regulating Services: Removal of Invertebrate Pests

Woodpeckers, especially those in the genus *Picoides*, are probably unsurpassed among the vertebrates in their ability to remove and consume woodboring beetles and other insects that invade tree tissues. These invertebrates can have a major impact on forest structure, a notorious recent example being the massive kills of coniferous forests in western North America due to infestations of bark beetles (*Dendroctonus spp.*; Bentz et al. 2010). Given that wood-feeding insects, particularly bark beetle larvae, are the frequent prey of many woodpeckers (Winkler et al. 1995; Martin et al.

2006), it is reasonable to expect these birds to be significant players in forest food webs, and possibly important regulators of invertebrate pests. It is therefore surprising how few studies have been conducted on the effects that woodpeckers have on populations of their prey (Fayt et al. 2005). The reverse relationship—the importance of invertebrate prey to woodpecker populations—has been shown in many studies (e.g., Fayt 2004; Horn and Hanula 2008). However, the evidence that woodpeckers exert important top-down effects in trophic webs is largely circumstantial, coming primarily from studies showing a positive response of woodpeckers to outbreaks of invertebrate pests (Fayt et al. 2005). A recent example comes from forests of interior British Columbia, where woodpecker densities increased following the massive outbreak of mountain pine beetles during 2003–4 (Edworthy et al. 2011). Other positive responses of woodpecker densities to beetle outbreaks have been observed, but those were mostly in recently burned forests (e.g., Kreisel and Stein 1999; Saab et al. 2007).

An encouraging finding from a study of avian responses to the invasion of the emerald ash borer (*Agrilus planipennis*) in five Midwestern cities of the United States was an increase in the abundance of red-bellied woodpeckers (*M. carolinus*; Koenig et al. 2013), a known predator of the ash borer (Lindell et al. 2008). Similarly, Koenig et al. (2011) found a general increase in the abundance of woodpeckers in the decades following the invasion of the gypsy moth (*Lymantria dispar*) and the subsequent, massive defoliation in the northeastern United States. In a review of published studies involving trophic interactions between *Picoides* woodpeckers and bark beetles invading spruce (*Picea* spp.) trees, Fayt et al. (2005) found that the woodpeckers tended to aggregate at sites where there were bark beetle outbreaks, and were, under some circumstances, physically capable of locally depleting beetle populations.

Cultural Services

CONTRIBUTIONS TO BIRD WATCHING. By creating shelter for other cavity-nesting or sap-feeding organisms, woodpeckers potentially boost local species richness; these species include birds valued by people who enjoy watching them in the wild or at feeders. In 2011 approximately 52 million Americans over 16 years of age spent an estimated US$4.07 billion ($110/spender) on wild bird feed, and US$970 million ($51/spender) on nest boxes, birdhouses, feeders, and baths (USFWS 2011). A recent survey found that 12.6 million (48%) of households in the United Kingdom feed wild birds (Davies et al. 2009). Woodpeckers and other cavity nesters are among the most common visitors to bird feeders in North America (Project Feederwatch). In a recent survey of 1,291 Canadians and Americans who fed wild birds, Horn and Johansen (2013) found that approximately 75% of respondents reported observing the downy woodpecker and black-capped chickadee (*Poecile atricapillus*), and that those two species—along with three other cavity nesters, the red-bellied woodpecker, tufted titmouse (*Baeolophus bicolor*), and eastern bluebird (*Sialia sialis*)—were among the top 13 species that the respondents wanted to attract to the feeders. The first 40 results yielded by a search on the Google Shopping website using the keywords "wild bird seed" had packages that featured North American birds on the front labels; 50% of those labels included images of secondary cavity nesters, and 35% included images of woodpeckers (C. Floyd 2014, unpublished data). Of these birds, the most commonly included species were the chickadee (Carolina [*P. carolinensis*] or black-capped), at 32.5%; the downy woodpecker, at 22.5%; the white-breasted nuthatch (*Sitta carolinensis*), at 15%; and the red-bellied woodpecker, at 12.5%. A similar survey of labels on suet cake feeders sold online revealed the vast majority of featured birds were cavity nesters (C. Floyd, unpublished data); many people who feed birds provide suet as well as seed (Horn and Johansen 2013). To the extent that bird feed labels represent accurate marketing to the interests of people who feed wild birds, it appears that Americans who provision wild birds are willing to pay for the pleasure of observing cavity nesters.

A taxon of cavity-nesting birds that rivals if not surpasses the woodpeckers in popularity among bird-watchers is the Psittaciformes. Diamond (1999) reported that clay-rich riverbanks in the Peruvian Amazon rain forest attracted not only 21 species of parrots but also thousands of bird-watching tourists per year. The economic benefits locally produced by this example of ecotourism included hundreds of jobs and thousands of US dollars (Diamond 1999). Other cavity-excavating species that are highly popular with ecotourists include trogons (e.g., the resplendent quetzal, *Pharomachrus mocinno*; Vivanco 2001), toucans (Jennings 2008), and burrowing seabirds (Yorio et al. 2001).

CAVITY EXCAVATORS AS SURROGATE SPECIES IN FOREST MANAGEMENT. Forest managers in Europe and North America are under increasing pressure to protect the complex multispecies interactions and decay processes that characterize natural forest ecosystems, including links with disturbance, pathogens, saproxylic insects, wood-decaying fungi, and cavity nesters (Hansen and Goheen 2000; Mikusiński et al. 2001; Messier and Puettmann 2011). Because woodpeckers are closely associated with these characteristics, these birds are widely considered to be ideal surrogate species for guiding forest management (Mikusiński 2006; Martin et al. 2015). The US Forest Service, for example, uses the black-backed woodpecker (*P. arcticus*) as an indicator of early successional burned forests and associated snags (Odion and Hanson 2013). Martikainen et al. (1998) argued for using the declining white-backed woodpecker (*Dendrocopos leucotos*) as an umbrella species for threatened saproxylic beetles, since both groups depend strongly on forests rich in decaying wood. Surrogates or not, woodpeckers are popular targets of forest management efforts in North America and Europe (e.g., see Carlson 2000; Wesolowski et al. 2005; Koivula and Schmiegelow 2007), where it is generally assumed that high-quality habitat for woodpeckers is synonymous with high ecological value of forests (Virkkala 2006).

Using woodpeckers to guide sustainable forest management assumes that measures of bird species richness or abundance are in fact accurate indicators of the management goals, or are reliable proxies for a target group of species. These assumptions have considerable support in the literature. Multiple studies have shown that woodpeckers (species and assemblages) were good indicators of species richness across a range of spatial scales (landscape scale, Mikusiński et al. 2001; stand scale, Roberge and Angelstam 2006; Drever et al. 2008) and forest conditions (different habitats, cutting regimes, and forest health, Drever et al. 2008; Martin et al. 2015). Further evidence that some woodpeckers depend on healthy forests comes from studies of forest restorations. A commonly cited example is the restoration of pine-grassland habitats for red-cockaded woodpeckers (*Picoides borealis*) in Mississippi, which boosted the local populations of threatened bird species (Wood et al. 2004). In northeastern Spain, the ranges of great spotted woodpeckers (*Dendrocopos major*) and black woodpeckers have expanded over the last few decades, probably as a result of increased forest maturation and connectivity after a reduction in intensive forestry (Gil-Tena et al. 2013).

CAVITY EXCAVATORS AS CHARISMATIC OR FLAGSHIP SPECIES IN CONSERVATION EFFORTS. Some cavity-excavating birds are so admired that they have been proposed as symbols or communication tools for conserving forest habitat (Roberge et al. 2008). Arango et al. (2007) surveyed residents in the Cape Horn Biosphere Preserve and found that the most charismatic bird was the Magellanic woodpecker; this led the authors to propose making that bird a symbol for conservation of the preserve's threatened austral forests. Public attraction to woodpeckers—or at least to extremely rare iconic birds—was prominently displayed during the quest to find the ivory-billed woodpecker (*Campephilus principalis*) in the Big Woods region of eastern Arkansas during 2004–9 (Jackson 2006). Although the search failed to produce a confirmed sighting, it spawned a tremendous amount of enthusiasm for the birds, generated economic activity in towns near the search area, and spurred efforts to conserve significant tracts of hardwood swamp and pine forest (Jackson 2006).

Conclusions and Recommendations

In sum, the rich literature on birds that nest in cavities shows that cavity-excavating birds generally add to biodiversity wherever they exist, as they provide essential resources of nesting sites and foraging opportunities. A more limited body of research indicates that cavity excavation contributes significantly (at least at local scales) to wood decomposition, invertebrate pest regulation, and bioturbation. Also, millions of people are willing to pay for the enjoyment of watching cavity-nesting birds. Thus, the potential exists to draw causal connections between cavity-excavating birds and ecosystem services that are valued by humans. Quantitatively assessing these connections, however, will not be easy. For one thing, there is the difficult task of valuing the ecosystem service in general (de Groot et al. 2010), which is beyond the scope of this chapter. Another challenge is that of quantitatively measuring how the flow of ecosystem services is augmented by enhanced biodiversity and ecosystem function (Balvanera et al. 2006). Throughout this chapter we have assumed that activities that contribute to biodiversity (e.g., violet-green and tree swallows using abandoned sapsucker nest cavities) or to ecosystem function (e.g., enhanced recycling of nutrients in soils excavated by seabirds) are likely to contribute positively to ecosystem services. The challenges are to confirm these assumptions and quantify their specific contributions. This requires that we first measure the importance of the cavity-excavation activity relative to that of other processes that contribute to the service. For example, many species that use woodpecker-excavated cavities also use natural cavities, which in some regions are used more frequently than woodpecker-excavated holes (Cockle et al. 2011). Measuring the extent to which populations of secondary cavity nesters are limited by the availability of cavities (whether woodpecker-excavated or natural) requires experimental manipulations. Given the logistical and ethical problems with directly excluding or removing woodpeckers from an area, more studies employing cavity blocking and nest box addition and removal are necessary. There may also be geographic variation in the strength of these relationships and, thus, in the valuation of the ecosystem services.

Rigorously measuring the value of woodpeckers as regulators of invertebrate pests in forests will require controlled, replicated experiments conducted across a variety of forest types (Fayt et al. 2005). The most manipulative investigations of top-down effects of woodpeckers on saproxylic prey were studies that used exclosure experiments (e.g., covering bark with hardware cloth to exclude woodpeckers) to measure the extent to which woodpeckers could deplete spruce bark beetles (reviewed by Fayt et al. 2005). Results from these studies indicated that *Picoides* woodpeckers (downy, hairy, and three-toed) can have a strong impact on beetle populations, with mortality rates attributed to the woodpeckers ranging from 19 to 98% in experimental trials. A more powerful way of testing for pest regulation by woodpeckers would be to experimentally remove them from or add them to an area, though it is difficult to imagine how this could be accomplished at a meaningfully large scale. It is clear that to measure the top-down influences of woodpeckers on a large scale, creative methods are needed. The future outbreaks of bark beetles in the western United States could be used as natural experiments for studying the potential of woodpeckers to regulate pests, especially if comparative data are collected before the outbreaks (Drever et al. 2009).

Another question that needs experimental study is the extent to which woodpeckers contribute to fungal decay, as opposed to merely being dependent on the process. The dearth of research on the contribution of woodpeckers to fungal transmission or wood decomposition in forests reflects a limited understanding of the ecology of wood-decaying fungi (Lonsdale et al. 2008). For example, little is known about the reproductive, dispersal, and colonization dynamics of heartrot fungi (Jackson and Jackson 2004). The growing field of deadwoodology—the ecology of wood-decaying fungi and their role in forest ecosystems—has the potential to address gaps in knowledge of the importance of woodpeckers as vectors of fungi (Grove 2002; Lonsdale et al. 2008; Drapeau et al. 2009). Some of the recent work in this field has focused on studying the pathways of snag recruitment, persistence, and degradation, including links with woodpeckers and saproxylic organisms (Lonsdale et al. 2008; Drapeau et al. 2009). A better understanding of the ecology of wood-decaying fungi might also come from active management techniques, such as fungal inoculation to promote the creation of dead and senescent wood. In managed forests of western Washington, for example, Bednarz et al. (2013) examined trees that had been inoculated with *Fomitopsis pinicola* (red-belted conk) approximately ten years earlier, and they found a greater prevalence of *F. pinicola* and a higher rate of woodpecker excavations (including sapsucker wells) in inoculated trees than in control trees.

It is clear that many knowledge gaps must be filled in order to quantify the contributions of cavity-excavating species to ecosystem services. Some questions could be addressed using observational or consultative methods; for example, the question of how much money people are willing to pay to see cavity nesters (de Groot et al. 2010). But most other research will require manipulative experiments; for example, measuring the contribution of sap to the energy budgets of birds that visit sap wells (McWhorter and López-Calleja 2000). Such work is intricately connected to the conservation needs of cavity-nesting birds and their habitats. For example, a common take-home message from ecological studies of woodpeckers is that certain characteristics of natural forest ecosystems (e.g., an abundance of deadwood) must be maintained in order to conserve woodpecker populations and the associated secondary cavity-nesting community (Mikusiński 2006; Virkkala 2006). An alternative approach to gaining support for woodpecker conservation would be to treat the taxon as a noncommodity resource supplied by natural forest ecosystems. Ascribing positive economic benefits to cavity excavators would help alleviate the financial bias against maintaining ecological quality that is often inherent in harvest-based forestry (Rohweder et al. 2000)

Acknowledgments

We thank Tomás Ibarra and two anonymous reviewers for providing helpful comments and editing on earlier drafts of this chapter.

References

Acosta-Pérez, V., Zuria, I., Castellanos, I., and Moreno, C. E. 2013. Características de las cavidades y los sustratos de anidación utilizados por el Carpintero Mexicano (*Picoides scalaris*) en dos localidades del centro de México. *Ornitol. Neotrop.* 24:107–11.

Aitken, K. E. H., and Martin, K. 2004. Nest cavity availability and selection in aspen-conifer groves in a grassland landscape. *Canadian Journal of Forest Research* 34:2099–2109.

———. 2007. The importance of excavators in hole-nesting communities: Availability and use of natural tree holes in old mixed forests of western Canada. *Journal of Ornithology* 148:S425–S434.

———. 2008. Resource selection plasticity and community responses to experimental reduction of a critical resource. *Ecology* 89:971–80.

———. 2012. Experimental test of nest-site limitation in mature mixed forests of central British Columbia, Canada. *Journal of Wildlife Management* 76: 557–65.

Arango, X., Rozzi, R., Massardo, F., Anderson, C. B., and Ibarra, J. T. 2007. Discovery and implementation of Magellanic woodpecker (*Campephilus magellanicus*) as an emblematic species: A biocultural approach for conservation in Cape Horn Biosphere Reserve. *Magallania* 35:71–88.

Aubry, K. B., and Raley, C. M. 2002. The pileated woodpecker as a keystone habitat modifier in the Pacific Northwest. USDA Forest Service Gen. Tech. Rep. PSW-GTR-181:257–74.

———. 2006. Ecological characteristics of fishers (*Martes pennanti*) in the southern Oregon Cascade Range. UDSA Forest Service, Pacific Northwest Research Station, Olympia, WA.

Balvanera, P., Pfisterer, A. B., Buchmann, N., He, J. S., Nakashizuka, T., Raffaelli, D., and Schmid, B. 2006. Quantifying the evidence for biodiversity effects on ecosystem functioning and services. *Ecology Letters* 9:1146–56.

Bancroft, W. J., Garkaklis, M. J., and Roberts, J. D. 2005. Burrow building in seabird colonies: A soil-forming process in island ecosystems. *Pedobiologia* 49:149–65.

Beals, E. W. 1970. Birds of a euphorbia-acacia woodland in Ethiopia: Habitat and seasonal changes. *Journal of Animal Ecology* 39:277–97.

Bednarz, J. C., Martin, J. H., Benson, T. J., and Varland, D. E. 2013. The efficacy of fungal inoculation of live trees to create wood decay and wildlife-use trees in managed forests of western Washington, USA. *Forest Ecology and Management* 307:186–95.

Bednarz, J. C., Ripper, D., and Radley, P. M. 2004. Emerging concepts and research directions in the study of cavity-nesting birds: Keystone ecological processes. *Condor* 106:1–4.

Bentz, B. J., Régnière, J., Fettig, C. J., Hansen, E. M., Hayes, J. L., Hicke, J. A., Kelsey, R. G., Negrón, J. F., and Seybold, S. J. 2010. Climate change and bark beetles of the western United States and Canada: Direct and indirect effects. *BioScience* 60: 602–13.

Blanc, L. A., and Martin, K. 2012. Identifying suitable woodpecker nest trees using decay selection profiles in trembling aspen (*Populus tremuloides*). *Forest Ecology and Management* 286:192–202.

Blanc, L. A., and Walters, J. R. 2008. Cavity nest-webs in a longleaf pine ecosystem. *Condor* 110:80–92.

Blendinger, P. G. 1999. Facilitation of sap-feeding birds by the white-fronted woodpecker in the Monte Desert, Argentina. *Condor* 101:402–7.

Brightsmith, D. J. 2004. Effects of weather on parrot geophagy in Tambopata, Peru. *Wilson Bulletin* 116:134–45.

———. 2005. Competition, predation and nest niche shifts among tropical cavity nesters: Phylogeny and natural history evolution of parrots (Psittaciformes) and trogons (Trogoniformes). *Journal of Avian Biology* 36:64–73.

Bull, E. L., and J. A. 2011. Pileated woodpecker (*Dryocopus pileatus*). In *The Birds of North America Online*, ed. A. Poole. Ithaca, NY: Cornell Lab of Ornithology. Accessed March 27, 2015. doi:10.2173/bna.148.

Cardiff, S. W., and D. L. Dittmann. 2002. Ash-throated flycatcher (*Myiarchus cinerascens*). In *The Birds of North America Online*, ed. A. Poole. Ithaca, NY: Cornell Lab of Ornithology. DOI:10.2173/bna.664. Accessed March 25, 2015.

Carlson, A. 2000. The effect of habitat loss on a deciduous forest specialist species: The white-backed woodpecker (*Dendrocopos leucotos*). *Forest Ecology and Management* 131:215–21.

Casas-Crivillé A., and Valera F. 2005. The European bee-eater (*Merops apiaster*) as an ecosystem engineer in arid environments. *Journal of Arid Environments* 60:227–38.

Cockle, K. L. and Martin, K. 2015. Temporal dynamics of a commensal network of cavity-nesting vertebrates: Increased diversity during an insect outbreak. *Ecology* 96:1093–1104.

Cockle, K. L., Martin, K., and Drever, M. C. 2010. Supply of tree-holes limits nest density of cavity-nesting birds in primary and logged subtropical. *Atlantic Forest Biological Conservation* 143:2851–57.

Cockle, K. L., Martin, K., Wesolowski, T. 2011. Woodpeckers, decay and the future of cavity-nesting vertebrate communities worldwide. *Frontiers in Ecology and the Environment* 9:377–82.

Cottee-Jones, H. E. W., Whittaker, R. J. 2012. The keystone species concept: A critical appraisal. *Frontiers of Biogeography* 4:117–27.

Cross, S. P., and Huibregtse, W. 1964. Unusual roosting site of *Eptesicus fuscus*. *Journal of Mammalogy* 45:628.

Daily, G. C., Ehrlich, P. R., and Haddad, N. M. 1993. Double keystone bird in a keystone species complex. *Proceedings of the National Academy of Sciences (USA)* 90:592–94.

Davies Z. G., Fuller R. A., Loram A., Irvine K. N., Sims V. A., and Gaston, K. J. 2009. National scale inventory of resource provision for biodiversity within domestic gardens. *Biological Conservation* 142:761–71.

De Groot, R. S., Alkemade, R., Braat, L., Hein, L., and Willemen, L. 2010. Challenges in integrating the concept of ecosystem services and values in landscape planning, management and decision making. *Ecological Complexity* 7:260–72.

De Jong, M. J. 1996. Northern rough-winged swallow (*Stelgidopteryx serripennis*). In *The Birds of North America Online*, ed. A. Poole. Ithaca, NY: Cornell Lab of Ornithology. DOI:10.2173/bna.234. Accessed March 25, 2015.

Dekker, R. W. R. J., and Brom, T. G. 1992. Megapode phylogeny and the interpretation of incubation strategies. *Zoologische Verhandelingen* 278:19–31.

Diamond, J. M. 1999. Evolutionary biology: Dirty eating for healthy living. *Nature* 400:120–21.

Drapeau, P., Nappi, A., Imbeau, L., and Saint-Germain, M. 2009. Standing dead wood for keystone bird species in the eastern boreal forest: Managing for snag dynamics. *Forestry Chronicle* 85:227–34.

Drever, M. C., Aitken, K. E. H., Norris, A. R., and Martin, K. 2008. Woodpeckers as reliable indicators of bird richness, forest health and harvest. *Biological Conservation* 141:624–34.

Drever, M. C., Goheen, J. R., and Martin, K. 2009. Species-energy theory, pulsed resources, and regulation of avian richness during a mountain pine beetle outbreak. *Ecology* 90:1095–1105.

Eberhardt, L. S. 2000. Use and selection of sap trees by yellow-bellied sapsuckers. *Auk* 117:41–51.

Edworthy, A., Drever, M. C., and Martin, K. 2011. Woodpeckers increase in abundance but maintain fecundity in response to an outbreak of mountain pine bark beetles. *Forest Ecology and Management* 261:203–10.

Ehrlich, P. R., and Daily, G. C. 1988. Red-naped sapsuckers feeding at willows: Possible keystone herbivores. *American Birds* 42:357–65.

Farris, K. L., Huss, M. J., and Zack, S. 2004. The role of foraging woodpeckers in the decomposition of ponderosa pine snags. *Condor* 106:50–59.

Fayt, P. 2004. Old-growth boreal forests, three-toed woodpeckers and saproxylic beetles: The importance of landscape management history on local consumer-resource dynamics. *Ecological Bulletins* 51:249–58.

Fayt, P., Machmer, M. M., and Steeger, C. 2005. Regulation of spruce bark beetles by woodpeckers: A literature review. *Forest Ecology and Management* 206: 1–14.

Fischer, J., Lindenmayer, D. B., and Manning, A. D. 2006. Biodiversity, ecosystem function, and resilience: Ten guiding principles for commodity production landscapes. *Frontiers in Ecology and the Environment* 4:80–86.

Fry, C. H., Fry, K., and Harris, A. 1992. *Kingfishers, Bee-eaters and Rollers*. Halfway House, South Africa: Russel Friedman Books.

Gil-Tena, A., Brotons, L., Fortin, M. J., Burel, F., and Saura, S. 2013. Assessing the role of landscape connectivity in recent woodpecker range expansion in Mediterranean Europe: Forest management implications. *European Journal of Forest Research* 132:181–94.

Grove, S. J. 2002. Saproxylic insect ecology and the sustainable management of forests. *Annual Review of Ecology and Systematics* 33:1–23.

Hansen, E. M., and Goheen, E. M. 2000. *Phellinus weirii* and other native root pathogens as determinants of forest structure and process in western North America. *Annual Review of Phytopathology* 38:515–39.

Hardy, P. C., and Morrison, M. L. 2001. Nest site selection by elf owls in the Sonoran Desert. *Wilson Bulletin* 113:23–32.

Horn, D. J., and Johansen, S. M. 2013. A comparison of bird-feeding practices in the United States and Canada. *Wildlife Society Bulletin* 37:293–300.

Horn, S., and Hanula, J. L. 2008. Relationship of coarse woody debris to arthropod availability for red-cockaded woodpeckers and other bark-foraging birds on loblolly pine boles. *Journal of Entomological Science* 43:153–68.

Jackson, J. A. 2006. Ivory-billed woodpecker (*Campephilus principalis*): Hope, and the interfaces of science, conservation, and politics. *Auk* 123:1–15.

Jackson, J. A., and Jackson, B. J. S. 2004. Ecological relationships between fungi and woodpecker cavity sites. *Condor* 106:37–49.

Jennings, J. 2008. Captive management: Family Ramphastidae (toucans). In *Biology, Medicine, and Surgery of South American Wild Animals*, ed. M. E. Fowler and Z. S. Cubas. Ames: Iowa State University Press.

Jolly, J. N. 1989. A field study of the breeding biology of the little spotted kiwi (*Apteryx owenii*) with emphasis on the causes of nest failures. *J. Roy. Soc. NZ* 19:433–48.

Jones, C. G., Lawton, J. H., and Shachak, M. 1994. Organisms as ecosystem engineers. *Oikos* 69:373–86.

Kattan, G. 1988. Food habits and social organization of acorn woodpeckers in Colombia. *Condor* 90:100 –106.

Koenig, W. D., Liebhold, A. M., Bonter, D. N., Hochachaka, W. M., and Dickinson, J. L. 2013. Effects of the emerald ash borer invasion on four species of birds. *Biological Invasions*. DOI 10.1007/s10530-013-0435-x. Accessed September 9, 2013

Koenig, W. D., Walters, E. L., and Liebhold, A. M. 2011. Effects of gypsy moth outbreaks on North American woodpeckers. *Condor* 113:352–61.

Koivula, M. J., and Schmiegelow, F. K. A. 2007. Boreal woodpecker assemblages in recently burned forested landscapes in Alberta, Canada: Effects of post-fire harvesting and burn severity. *Forest Ecology and Management* 242:606–18.

Kozma, J. M. 2010. Characteristics of trees used by white-headed woodpeckers for sap feeding in Washington. *Northwestern Naturalist* 91:81–86.

Kreisel, K. J., and Stein, S. J. 1999. Bird use of burned and unburned coniferous forests during winter. *Wilson Bulletin* 111:243–50.

Kunz, T. H., and Lumsden, L. F. 2003. Ecology of cavity and foliage roosting bats. In *Bat Ecology*, ed. T. H. Kunz and M. B. Fenton. Chicago: University of Chicago Press.

Leonard, D. L. Jr. 2001. American three-toed woodpecker (*Picoides dorsalis*). In *The Birds of North America Online*, ed. A. Poole. Ithaca, NY: Cornell Lab of Ornithology. DOI:10.2173/bna.588. Accessed September 3, 2013.

Lindell, C. A., McCullough, D. G., Cappaert, D., Apostolou, N. M., and Roth, M. B. 2008. Factors influencing woodpecker predation on emerald ash borer. *American Midland Naturalist* 159:434–44.

Lonsdale, D., Pautasso, M., Holdenrieder, O. 2008. Wood-decaying fungi in the forest: Conservation needs and management options. *European Journal of Forest Research* 127:1–22.

Martikainen, P., Kaila, L., and Haila, Y. 1998. Threatened beetles in white-backed woodpecker habitats. *Conservation Biology* 12:293–301.

Martin, K., Aitken, K. E. H., and Wiebe, K. L. 2004. Nest sites and nest webs for cavity-nesting communities in interior British Columbia, Canada: Nest characteristics and niche partitioning. *Condor* 106:5–19.

Martin, K., and Eadie, J. M. 1999. Nest webs: A community wide approach to the management and conservation of cavity nesting birds. *Forest Ecology and Management* 115:243–57.

Martin, K., Ibarra, J. T., and Drever. M. 2015. Avian surrogates in terrestrial ecosystems: Theory and practice. In *Indicators and Surrogates of Biodiversity and Environmental Change*, ed. D. Lindenmeyer, P. Barton, and J. Pierson, 33–44. Canberra, Australia: CSIRO Publishing and CRC Press.

Martin, K., Norris A., and Drever, M. 2006. Effects of bark beetle outbreaks on avian biodiversity in the British Columbia interior: Implications for critical habitat management. *BC Journal of Ecosystems and Management* 7: 10–24.

McKechnie, S. 2006. Biopedturbation by an island ecosystem engineer: Burrowing volumes and litter deposition by sooty shearwaters (*Puffinus griseus*). *New Zealand Journal of Zoology* 33:259–65.

McWhorter, T. J., and López-Calleja, M. V. 2000. The integration of diet, physiology, and ecology of nectar-feeding birds. *Revista Chilena de Historia Natural* 73:451–60.

Messier, C., and Puettmann, K. J. 2011. Forests as complex adaptive systems: Implications for forest management and modelling. *Italian Journal of Forest and Mountain Environments* 66:249–58.

Meysman, F. J., Middelburg, J. J., and Heip, C. H. 2006. Bioturbation: A fresh look at Darwin's last idea. *Trends in Ecology and Evolution* 21:688–95.

Mikusiński, G. 1995. Population trends in black woodpecker in relation to changes and characteristics of European forests. *Ecography* 18:363–69.

———. 2006. Woodpeckers (Picidae): Distribution, conservation and research in a global perspective. *Annales Zoologici Fennici* 43:86–95.

Mikusiński, G., Gromadzki, M., Chylarecki, P. 2001. Woodpeckers as indicators of forest bird diversity. *Conservation Biology* 15:208–17.

Miller, D. A., Arnett, E. B., and Lacki, M. J. 2003. Habitat management for forest-roosting bats of North America: A critical review of habitat studies. *Wildlife Society Bulletin* 31:30–44.

Miller, R. S., and Nero, R. W. 1983. Hummingbird-sapsucker associations in northern climates. *Canadian Journal of Zoology* 61:1540–46.

Millenium Ecosystem Assessment. 2003. *Ecosystems and Human Well-Being: A Framework for Assessment.* Island Press, Washington.

Montellano, M. G. N., Blendinger, P. G., and Macchi, L. 2013. Sap consumption by the white-fronted woodpecker and its role in avian assemblage structure in dry forests. *Condor* 115:93–101.

Mulder, C. P., and Keall, S. N. 2001. Burrowing seabirds and reptiles: Impacts on seeds, seedlings and soils in an island forest in New Zealand. *Oecologia* 127:350–60.

Newton, I. 1994. The role of nest sites in limiting the numbers of hole-nesting birds: A review. *Biological Conservation* 70:265–76.

Odion, D. C., and C. T. Hanson. 2013. Projecting impacts of fire management on a biodiversity indicator in the Sierra Nevada and Cascades, USA: The black-backed woodpecker. *Open Forest Science Journal* 6:14–23.

Polis, G. A., and Hurd, S. D. 1996. Linking marine and terrestrial food webs: Allochthonous input from the ocean supports high secondary productivity on small islands and coastal land communities. *American Naturalist* 147:396–423.

Remm J., and Lõhmus, A. 2011. Tree cavities in forests: The broad distribution pattern of a keystone structure for biodiversity. *Forest Ecology and Management* 262:579–85.

Remm J., Lõhmus, A., and Remm, K. 2006. Tree cavities in riverine forests: What determines their occurrence and use by hole-nesting passerines? *Forest Ecology and Management* 221:267–77.

Rissler, L. J., Karowe, D. N., Cuthbert, F., and Scholtens, B. 1995. The influence of yellow-bellied sapsuckers on local insect community structure. *Wilson Bulletin* 107:746–52.

Roberge, J. M., and Angelstam, P. 2006. Indicator species among resident forest birds: A cross-regional evaluation in northern Europe. *Biological Conservation* 130:134–47.

Roberge, J. M., Mikusinski, G., and Svensson, S. 2008. The white-backed woodpecker: Umbrella species for forest conservation planning? *Biodiversity and Conservation* 17:2479–94.

Robles, H., and Martin, K. 2013. Resource quantity and quality determine the inter-specific associations between ecosystem engineers and resource users in a cavity-nest web. *PLOS ONE* 8:e74694.

Rohweder, M. R., McKetta, C. W., and Riggs, R. A. 2000. Economic and biological compatibility of timber and wildlife production: An illustrative use of production possibilities. *Wildlife Society Bulletin* 28:435–47.

Ruczyński, I., and Bogdanowicz, W. 2005. Roost cavity selection by *Nyctalus noctula* and *N. leisleri* (Vespertilionidae, Chiroptera) in Białowieża Primeval Forest, eastern Poland. *Journal of Mammalogy* 86:921–30.

Saab, V. A., Russell, R. E., and Dudley, J. G. 2007. Nest densities of cavity-nesting birds in relation to postfire salvage logging and time since wildfire. *Condor* 109:97–108.

Schlatter, R., and Vergara, P. 2005. Magellanic woodpecker (*Campephilus magellanicus*) sap feeding and its role in the Tierra del Fuego forest bird assemblage. *Journal of Ornithology* 146:188–90.

Schumann, N., Dann, P., Hoskins, A. J., and Arnould, J. P. Y. 2013. Optimizing survey effort for burrow-nesting seabirds. *Journal of Field Ornithology* 84:69–85.

Şekercioğlu, Ç. H. 2006. Ecological significance of bird populations. In *Handbook of the Birds of the World*, vol. 11, ed. J. del Hoyo, A. Elliott, and D. A. Christie, 15–51. . Barcelona and Cambridge: Lynx Press and BirdLife International.

Short, L. L. 1979. Burdens of the picid hole-excavating habit. *Wilson Bulletin* 91:16–28.

Southwick, E. E., and Southwick, A. K. 1980. Energetics of feeding on tree sap by ubythroated hummingbirds in Michigan. *American Midland Naturalist* 104:329–33.

Sutherland, G. D., Gass, C. L., Thompson, P. A., and Lertzman, K. P. 1982. Feeding territoriality in migrant rufous hummingbirds: Defense of yellow-bellied sapsucker (*Sphyrapicus varius*) feeding sites. *Canadian Journal of Zoology* 60:2046–50.

Tarof, S., and Brown. C. R. 2013. Purple martin (*Progne subis*). In *The Birds of North America Online*, ed. A. Poole. Ithaca, NY: Cornell Lab of Ornithology. DOI:10.2173/bna.287. Accessed March 25, 2015.

Tate, J. 1973. Methods and annual sequence of foraging by the sapsucker. *Auk* 90:840–56.

Tibbitts, T. A., Pate, A., Petryszyn, Y., and Barns, B. 2002. Determining foraging and roosting areas for Underwood's mastiff bat (*Eumops underwoodi*) using radiotelemetry at Organ Pipe Cactus National Monument, Arizona: Final summary report to Organ Pipe Cactus National Monument, Ajo, Arizona.

Tilman, D., Reich, P. B., and Knops, J. M. 2006. Biodiversity and ecosystem stability in a decade-long grassland experiment. *Nature* 441:629–32.

USFWS (US Fish and Wildlife Service). 2012. *2011 National Survey of Fishing, Hunting, and Wildlife-Associated Recreation*. Washington: US Fish and Wildlife Service.

Valdivia-Hoeflich, T., Rivera, J. H. V., and Stoner, K. E. 2005. The Citreoline Trogon as an ecosystem engineer. *Biotropica* 37:465–67.

Virkkala, R. 2006. Why study woodpeckers? The significance of woodpeckers in forest ecosystems. *Annales Zoologici Fennici* 43:82–85.

Vivanco, L. A. 2001. Spectacular quetzals, ecotourism, and environmental futures in Monte Verde, Costa Rica. *Ethnology* 40:79–92.

Walters, E. L., Miller, E. H., and Lowther, P. E. 2002. Yellow-bellied sapsucker (*Sphyrapicus varius*). In *The Birds of North America Online*, ed. A. Poole. DOI:10.2173/bna.662. Ithaca, NY: Cornell Lab of Ornithology. Accessed August 21, 2013.

Wenny, D. G., Devault, T. L., Johnson, M. D., Kelly, D., Şekercioğlu, Ç. H., Tomback, D. F., and Whelan, C. J. 2011. The need to quantify ecosystem services provided by birds. *Auk* 128:1–14.

Wesołowski, T., Czeszczewik, D., and Rowiński, P. 2005. Effects of forest management on three-toed woodpecker *Picoides tridactylus* distribution in the Białowieża Forest (NE Poland): Conservation implications. *Acta Ornithologica* 40:53–60.

Wiebe, K. L., Koenig W. D., and Martin, K. 2007. Costs and benefits of nest reuse versus excavation in cavity-nesting birds *Annales Zoologica Fennici* 44:209–17.

Winkler, D. W., and Sheldon, F. H. 1993. Evolution of nest construction in swallows (Hirundinidae): A molecular phylogenetic perspective. *Proceedings of the National Academy of Sciences (USA)*: 5705–7.

Winkler, H., Christie, D. A., Nurney, D. 1995. *Woodpeckers: Identification Guide to the Woodpeckers of the World*. New York: Houghton Mifflin.

Wood, D. R., Burger, L. W., Bowman, J. L. and Hardy, C. L. 2004. Avian community response to pine-grassland restoration. *Wildlife Society Bulletin* 32:819–29.

Yorio, P., Frere, E., Gandini, P., and Schiavini, A. 2001. Tourism and recreation at seabird breeding sites in Patagonia, Argentina: Current concerns and future prospects. *Bird Conservation International* 11:231–45.

Zwartjes, P. W., and Nordell, S. E. 1998. Patterns of cavity-entrance orientation by gilded flickers (*Colaptes chrysoides*) in cardon cactus. *Auk* 115:119–26.

CHAPTER ELEVEN

Avian Ecological Functions and Ecosystem Services in the Tropics

Çağan H. Şekercioğlu and Evan R. Buechley

Birds contribute many important ecological functions through their roles as predators, pollinators, scavengers, seed dispersers, seed predators, and ecosystem engineers. Many of these ecosystem functions also translate to ecosystem services, which are defined as natural processes that benefit humans (Şekercioğlu 2010). Birds' abilities to fly and migrate enable them to respond to eruptive resources and to connect varying landscapes in ways that other organisms cannot. Further, the impressive diversity of birds (over 10,500 species) is indicative of their vast adaptive variety, which enables them to fill a wide diversity of niches. While there has long been interest in the relationship between birds and agricultural crops, dating from the 19th century in the United States, targeted research in this field largely lay dormant until the latter part of the 20th century (Whelan et al. 2008, 2015). To date, the vast majority of research on avian ecosystem services and agriculture has taken place in temperate climates, with relatively little research being done in the tropics. Nonetheless, over the last decade in particular, there has been a growing body of research in this regard, particularly in the Neotropics.

Although less than 1 percent of the world's bird species primarily prefer agricultural areas, nearly a third use such habitats occasionally (Şekercioğlu et al. 2007), often providing important ecosystem services such as pest control, pollination, seed dispersal, and nutrient deposition (Sodhi et al. 2011). Even though most bird species are found in the tropics, studies of functional change in bird communities are disproportionately focused on European and North American ecosystems (Şekercioğlu 2006b). There is growing interest in avian functional diversity in tropical forests and agro-ecosytems, and especially in tree-dominated agroforestry systems, such as shade coffee and cacao plantations, which harbor higher bird diversity than do open agricultural systems with few or no trees (Thiollay 1995; Greenberg et al. 1997; Greenberg et al. 2000b; Wang and Young 2003; Perfecto et al. 2004; Waltert et al. 2005; Marsden et al. 2006; Clough et al. 2009; Tscharntke et al. 2008; Van Bael et al. 2007; Kellermann et al. 2008). However, recent research has focused disproportionately on Neotropical coffee plantations (Komar 2006), and we need more studies on other types of tropical agroforest systems (Marsden et al. 2006; Round et al. 2006), particularly in Africa (Naidoo 2004; Waltert et al. 2005; Holbech 2009; Buechley et al. 2015) and on Pacific ocean islands (Marsden et al. 2006). There is a need for a global synthesis of these studies in order to understand how bird communities and the proportions of bird functional groups such as granivores, frugivores, insectivores, and nectarivores change from forests to agroforests to open agricultural systems. Not only is this important for a better understanding of the ecology of tropical bird communities and for improvement of tropical bird conservation, but also for estimating the changes in birds' ecosystem services (Wenny et al., 2011) and for calculating the economical contributions of these services to tropical farmers' incomes.

The objectives of this chapter are (1) to review the tropical avian ecology literature in order to quantify the changes in bird functional groups in tropical forests, agroforest, and agricultural areas, and (2) to improve our understanding of the changes in bird ecosystem services and ecological function in tropical agroforests and agricultural areas as a result of the declines or increases in predators, seed dispersers, pollinators, and other avian functional groups. We reviewed studies that compared tropical agroforestry and open agricultural ecosystems to native forests nearby. We used the combination of keywords "bird* AND tropic* AND forest* AND [agriculture OR agroforest]" in the Web of Knowledge database and in Google Scholar to generate a list of peer-reviewed research articles published between 1970 and 2015. Of these, we chose relevant articles that compared tropical forest birds to agroforest birds, open agricultural birds, or both. Under forests, we included natural primary or secondary forests and woodlands, and excluded plantations. Most tropical woodland species also spend time in forests, so they were included in the analyses.

Agroforests are defined as agricultural areas that have significant tree cover, such as cocoa, rubber, or shade coffee plantations.

Avian Community Structure

Tropical forest biodiversity is often highly specialized and reliant on little-disturbed forest (Turner and Corlett 1996). Nonetheless, agroforests are an important habitat for biodiversity conservation in the tropics, particularly when they are less intensively managed and have high canopy cover (Bhagwat et al. 2008). Although the variety of schemes used in the literature for guild classifications makes generalizations difficult (Komar 2006), some important patterns emerge. When agroforest systems are compared to primary forests, the species numbers of large frugivorous and insectivorous birds (especially terrestrial and understory species) are often lower (Tsharntke et al. 2008). In contrast, nectarivores, small to medium insectivores (especially migrants and canopy species), omnivores, and sometimes granivores and small frugivores do better or even thrive in agroforest systems (Petit et al. 1999; Verea and Solozano 2005; Neuschulz et al. 2011; Ruiz-Guerra et al. 2012), frequently by tracking seasonal resources (Greenberg et al. 1997; Johnson and Sherry 2001; Carlo et al. 2003). However, changes in guild species numbers do not necessarily translate to changes in relative abundance (Verea and Solozano 2005; Marsden et al. 2006), biomass, or function (Greenberg et al. 2000b; Perfecto et al. 2004), and more research is needed to quantify these important measures (Beehler et al. 1987; Komar 2006).

Neotropics

Tropical agroforestry systems often vary in their functional diversity patterns, but insectivores often have lower representation than in forests. In Paraguay, in *yerba mate* plantations shaded by forest trees located close to extensive forest, fruit and insect eaters, insectivores, and nectarivores were less abundant than in nearby forest, and two-thirds of carnivorous species were not found in plantations (Cockle et al. 2005). More than 60% of the birds captured in the understory of Venezuelan shaded cacao plantations were hummingbirds, whereas insectivores had reduced abundance and species richness (Verea and Solozano 2005). Shade cacao plantations in southeast Brazil had fewer species of frugivores and understory insectivores, and more species of nectarivores and omnivores, than nearby forest fragments (Faria et al. 2006). Landscape effects were not pronounced, although the proportional representation of frugivorous species was higher, and that of gleaning insectivores was lower, in the less forested landscape. Barlow et al. (2007) documented significantly more species in primary forest than in second-growth or *Eucalyptus* plantations in the Brazilian Amazon. Primary forest and *Eucalyptus* plantations had almost no species in common. Obligate ant-following and dead-leaf-gleaning insectivores were only recorded in primary forest, arboreal omnivores were most abundant in second growth, and there was a low relative abundance of external bark-searching and terrestrial gleaning insectivores in *Eucalyptus*. *Eucalyptus* also had a high relative abundance of nectarivores. In Ecuador, Canaday (1996) studied changes in the insectivorous bird community along a gradient of human impact, finding a significant reduction in the number of insectivorous birds in areas of greater human impact, including petroleum exploration and small-scale agriculture. In Mexican cacao plantations shaded by 60 species of planted native trees, but isolated from other extensive forest patches, forest specialists were scarce and resident insectivorous species were mostly missing, whereas small foliage-gleaning insectivores comprised most of the migrant birds (Greenberg et al. 2000a). However, omnivorous or frugivorous bird species were also few, again suggesting the importance of landscape composition. In Mexican shade coffee plantations, disturbance-sensitive bark insectivores, understory bark insectivores, and large canopy frugivores had fewer species than did native forest, whereas facultative and obligatory insectivores, omnivores, and midstory and understory/undergrowth granivores increased in shade coffee (Leyequien et al., 2010). On the other hand, Mexican tropical dry forests and tree orchards did not differ in their guild composition (Mac Gregor-Fors and Schondube 2011). In cacao farms in Panama, the diversity of birds and the diversity of canopy tree species were strongly positively correlated (Van Bael et al. 2007). In Costa Rica, one study documented higher species richness in forest edge than in coffee farms, active pasture, or fallow fields (Hughes et al. 2002), while another study found bat and bird assemblages in agroforestry systems to be as abundant and diverse as in forest; however, the species assemblages were highly modified and contained less forest specialists (Harvey and González Villalobos 2007). Another study in Costa Rica contrasting cacao plantations and forest patches documented higher avian density and diversity in cacao, but significantly fewer forest specialist species (Reitsma et al. 2001). In a long-term study (1960–99) of the effects of the conversion of lowland tropical rainforest to agricultural habitat in Costa Rica, Sigel et al. (2006) documented insectivore declines while vegetarian and omnivorous species increased.

Afrotropics

In Ethiopia, shade coffee farms had more than double the species richness of nearby primary forest, while there was a much higher relative abundance of forest specialists, understory insectivores, and Afrotropical-resident understory insectivores in primary forest (Buechley et al. 2015). In these traditional, organic shade coffee plantations where coffee was grown in its native habitat under native forest trees, there were some results that contrasted with most global findings: (1) there was no difference in the relative abundance of insectivores between the two habitats, and (2) there was greater relative abundance of granivores in primary forest. In another study in Ethiopia, considerable overlap was found in species assemblages, higher abundances of open and shrubland bird species were documented at agricultural sites, and higher abundances of woodland and forest species were found in forest patches (Gove et al. 2008). In Kenya, bird communities were sampled in agricultural habitats surrounding a forest reserve to evaluate the habitat characteristics that influence bird diversity and abundance (Otieno et al. 2011). The results indicated that hedge volume was the most important factor in vegetation structure in agriculture, which correlated with bird species richness and insectivorous bird density. The bird density was also shown to increase with overall tree density. In another Kenyan study, bird communities were sampled at 20 sites along a habitat gradient from primary forest to intensive agriculture (Mulwa et al. 2012). The bird density and species richness was higher on average in agriculture than in forest habitat; but within forest and agriculture, density and richness increased with vegetation complexity. Importantly, the bird assemblages in forest and agriculture were distinct, with very few forest specialists occurring in agriculture. Insectivores declined in farmland, while carnivores and herbivores increased. Unusually, in Uganda there were no differences between forests and smallholder agricultural areas in the detection rates of insectivores versus noninsectivores, whereas larger, mostly frugivorous birds were more likely to be detected in the agricultural areas (Naidoo 2004). Newmark (1991) showed a decreasing richness of understory species with shrinking forest patch size in the Usambara Mountains of Tanzania. Insect gleaners, frugivores, salliers, and seed eaters were less frequent in forest fragments than in a large forest control site. In Cameroon agroforestry systems with relatively high tree cover surrounded by primary forest, Waltert et al. (2005) observed reduced species richness compared to primary forest, and in some cases abundance of insectivorous species, especially those of the understory. Frugivores and omnivores did not differ, whereas nectarivores and granivores had higher richness in agroforests. In Ghana, Holbech (2009) found the trophic organization of the lower-story birds in luxuriant tree plantations to be similar to that found in native forest, though there were fewer ant-following birds in the plantations (69% of the numbers found in the forest). Cardamom (*Elettaria cardamomum, Amomum costatum* and *Amomum subulatum*) and coffee (*Coffea arabica and Coffea robusta*) plantations were better for forest birds than were cacao (*Theobroma cacao*) plantations. In the plantations, the presence of a canopy per se was more important than the number of species making up the canopy, and the choice of native versus exotic tree species was less important than the presence of a well developed and diverse secondary plant community, especially in the subcanopy layers (Holbech 2009). A similar finding was reported by Najera and Simonetti (2010) in a review of 167 case studies from 32 countries comparing birds in forests and in plantations. Sixty-eight percent of forest bird species were sensitive to "edge-effects" in Madagascar; the canopy insectivores were edge-sensitive, while the sallying insectivores preferred edges (Watson et al. 2004). In an unusual finding, frugivores declined at the forest edge in comparison to the forest interior. In a regional review of conversion of forest to agricultural and human-dominated landscapes in West Africa, Norris et al. (2010) showed a decline in insectivores and large-foliage gleaners in secondary forest as compared to primary forest. In cacao agroforests, ant followers, insectivores, and species with restricted distributions declined, while nectarivores increased; in annual crops the bird species richness was lower, and ant followers, insectivores, and foliage gleaners were replaced by granivores and nectarivores.

Indomalayan and Australasian Tropics

In traditional agroforests in tropical China, there were no consistent differences in bird guilds between economic forests, monsoon evergreen broadleaf forests, and montane rain forests (Wang and Young 2003); but this was not the case for other Asia-Pacific sites studied. Compared to nearby primary forest, the fruit orchards of Thailand were dominated by smaller frugivores, nectarivores, and widespread generalists, whereas understory insectivores were poorly represented (Round et al. 2006). There was a 60% reduction in bird species richness in oil palm and rubber plantations in southern Thailand as compared to forest (Aratrakorn et al. 2006). Insectivores and frugivores were particularly susceptible to declines, while omnivores fared much better. There was little difference in bird community composition between the two plantation types. In Malaysian mixed agricultural habitats consisting of oil palms, rubber, and fruit trees, smaller primary forest frugivores and trunk-feeding insectivores tended to persist, whereas ground and understory birds were likely to disappear (Peh et al. 2005). Malaysian oil palm plantations, rubber tree plantations, and orchard gardens had only a third of the bird species found in the nearby primary forest, but the proportions of insectivores and frugivores did not differ between habitats (Peh et al. 2006). Schulze et al. (2004) showed a decline in bird species richness from forest ecosystems to agricultural ecosystems in Sulawesi, Indonesia, including a significant reduction in the number of insectivorous birds. They showed a positive relationship between the number of tree species and the number of endemic bird species, frugivores, and nectarivores. In another study in Sulawesi, bird species richness decreased from primary and secondary forest to cacao agroforestry (Waltert et al. 2004). The agroforests supported few frugivores and nectarivores when compared to primary and secondary forest. In Sulawesi cacao plantations, frugivores and nectarivores had lower species richness at increasing distances from the forest, while in granivores the opposite trend was found (Clough et al. 2009). Increasing the tree cover in these cacao plantations led to higher species richness in frugivores and insectivores. Bowman et al. (1990) studied bird community structure along a successional gradient of forest and slash-and-burn agriculture in Papua New Guinea. The primary forest supported more specialist feeders, including frugivores, nectarivores, and branch gleaners, while obligate granivores were restricted to open grassy habitats. In southern India, while the bird species richness varied very little across landscapes, there was significant variation in the composition of bird communities in different habitats (Sidhu et al. 2010). Tea and teak plantations were found to harbor fewer rainforest species, while coffee and cardamom plantations with more native shade trees supported more sensitive rainforest species. A study in Sumatra contrasted rubber plantations, rubber agroforest, and forest, finding that the avian species richness was similar between rubber and agroforest, while lower in the plantations, and that the number of forest specialists was lower in agroforest and plantations than in forest (Beukema et al. 2007). In another Sumatran study, larger frugivores, larger insectivores of both canopy and understory, and terrestrial insectivores of the forest interior had mostly disappeared from the agroforests, while small frugivores, smaller foliage-gleaning insectivores, nectarivores, and edge species persisted (Thiollay 1995). Similarly, large frugivores, some insectivores, and ground foragers declined in the small-scale mixed agriculture-agroforestry systems of Papua New Guinea (Marsden et al. 2006). In a meta-analysis of studies from the region, Koh and Wilcove (2008) showed that oil palm plantations in southern peninsular Malaysia and Borneo harbor 77% fewer forest bird species than does primary forest. Worldwide, insectivorous birds are 40% less frequent in tree plantations, whereas the proportion of granivores is more than three times higher (Najera and Simonetti 2010). In a global analysis of 6,100 entirely tropical bird species, Şekercioğlu (2012) found that the species richness of large frugivorous and insectivorous birds (especially terrestrial and understory species) often declines in agroforests in comparison to primary forests. In contrast, nectarivores, small-to-medium insectivores (especially migrants and canopy species), omnivores, and sometimes granivores and small frugivores do better, frequently by tracking seasonal resources.

Avian Ecosystem Services

As demonstrated above, avian richness, abundance, and guild structure is often influenced by habitat modification in agricultural landscapes. Several studies show that arthropod abundance and plant herbivory increase when birds are artificially exclosed from agricultural crops (Van Bael and Brawn 2005; Kellermann et al. 2008; Maas et al. 2015). However, it remains unclear how bird community structure impacts insect control, seed predation, seed dispersal, and other ecosystem services; further research is needed in this regard.

In Jamaica, coffee plants with birds artificially exclosed had significantly higher coffee borer infestation, more borer broods, and greater berry damage than did control plants (Kellermann et al. 2008). Lower infestation on control plants correlated with higher total bird abundance, but not with specific avian insectivore abundance or vegetation complexity. Pest reduction by birds economically benefited coffee farmers in Jamaica by US$310 per hectare (Johnson et al., 2010). Railsback and Johnson (2014) modeled the avian ecosystem services and habitat usage in Jamaican coffee farms, and concluded that when considering both bird conservation and economic production, shade coffee is preferable to splitting the landscape into forest and sun coffee, because shade coffee supports more birds and benefits more from ecosystem services. Bird exclosure experiments in Panama revealed that birds decreased arthropod densities and leaf damage in the forest canopy during the dry seasons but not the wet ones, and that birds had no effect on the arthropod abundance in the forest understory (Van Bael and Brawn 2005). In Costa Rica, bird exclosures led to an increase in herbivorous arthropod abundance, which in turn led to an increase in leaf damage (Karp and Daily 2014). In a tropical forest restoration experiment in Costa Rica, the insect biomass was highest on tree branches where both birds and bats were excluded, and lowest where neither were excluded (Morrison and Lindell 2012). Interestingly, the predation rates on artificial Lepidoptera larvae in Mexico during the dry season were significantly higher in forest fragments than in continuous forest, potentially due to the less diverse yet more dominant avian insectivore community in forest fragments (Ruiz-Guerra et al. 2012). In a study of seed dispersal by birds in Costa Rica, bird abundance, not richness, best predicted the richness of bird-dispersed seeds (Pejchar et al. 2008). In Brazil, a study of seed dispersal by frugivorous birds showed that isolated trees attracted a greater and more distinct bird assemblage than did trees in forest fragments, and that the seeds of isolated trees were more likely to be dispersed to the largest variety of surrounding habitats (Pizo and dos Santos 2011). A few bird species were particularly important for the long-distance dispersal of seeds, making them valuable links connecting forest fragments. In Tanzania, when birds and bats were excluded from coffee shrubs with nets, there was a significant reduction of fruit set and fruit retention (Classen et al. 2014). Surprisingly, though, there was no difference in ecosystem services along a gradient of land-use intensity. In Kenya, frugivore richness and density declined with forest disturbance in three different rain forest study sites, thus suggesting a regional trend of forest disturbance leading to a decline of frugivores and their valuable seed dispersal services, particularly for large-seeded tree species and trees with small fruits (Kirika et al. 2008). In a study of three frugivore species in the Taita Hills of Kenya, differences in mobility and habitat use caused significant differences in seed dispersal (Lehouck et al. 2009). The most sedentary and forest-dependent species contributed to short-distance dispersal, often within the same forest patch, while the two more mobile species dispersed seeds further away from parent trees, and often into different forest patches or exotic plantations. This suggests that seed dispersal by different species can be complementary, contributing to dispersal into a range of different habitats over varying distances. Retaining frugivore diversity may be integral to maintaining dispersal function. In Ghana, a study of dispersal of the large seeds of *Antiaris toxicaria,* an important timber species, concluded that mammals were responsible for 76.3% of seed dispersal, while birds were responsible for 23.7% (Kankam and Oduro 2009). The authors note, however, that dispersal by birds and fruit bats may be more effective because they are more mobile foragers. They conclude that a population reduction of seed dispersers can affect recruitment of tropical trees, and they suggest conservation of frugivores in order to promote the sustainable management of *A. toxicaria.*

In Borneo, bird exclusion significantly increased herbivory damage to oil palms—up to 28% foliage damage (Koh 2008). The author suggests that this may lead to a fruit yield loss of 9 to 26%. Oil palm is an important agricultural crop, suggesting that insectivorous birds provide important services to farmers in the form of insect control. In the Mariana Islands, a loss of bird diversity caused by the invasive brown treesnake *Boiga irregularis* has led to reduced recruitment in several Mariana Island tree species, many of which are dependent on birds for pollination and dispersal (Mortensen et al. 2008). In Hawaii, seed dispersal by native and introduced bird species was studied in dry forests (Chimera and Drake 2010). The authors found that although trees covered only 15.2% of the study area, 96.9% of the bird-dispersed seeds were deposited beneath them. The invasive bird *Zosterops japonicas* was the leading seed disperser, and of the bird-dispersed seeds, 75% were of the invasive tree *Boccania frutescens*, while the invasive shrub *Lantana camara* accounted for an additional 17%. Exotic bird species were found to rarely disperse the seeds of native tree species, and less than 8% of all bird-dispersed seeds were from native trees. This suggests that avian seed dispersal can operate both as a service and a disservice, depending on which species of seeds the birds are dispersing, particularly when the birds in question are exotic species. This study concludes that current dispersal patterns are likely to contribute to the replacement of native flora by exotic plants in Hawaiian dry forests.

Discussion

The results of the findings of field studies in Neotropical (Leyequien et al., 2010), Afrotropical (Waltert et al. 2005), Indomalayan (Peh et al. 2006), and Australasian (Marsden et al. 2006) regions suggest that the replacement of forests with agricultural areas results in a shift towards less specialized bird communities comprising more widespread and relatively common species, and with altered proportions of functional groups (Şekercioğlu 2012). Insectivores and other invertebrate predators often make up a smaller proportion of bird communities outside forests, whereas seed-dispersing frugivores and pollinating nectarivores are higher in agroforests, especially as compared to open agricultural areas that can experience substantial increases in avian seed predators in comparison to forests and agroforests. Given that there are considerable differences in functional distribution, specialization, global range, population size, mobility, and conservation status between forest, agroforest, and agricultural bird communities, there is an urgent need for detailed field studies that compare bird community ecology and avian function in these habitats.

Agroforest birds are likely to be the primary seed dispersers in agricultural areas, a pattern also observed in the field (e.g., Pizo 2004). Integrating agroforests with open agricultural areas may result in a spillover of nectarivores and partially make up for the decline in avian pollinators among open agricultural species. The high primary productivity of agroforestry in the tropics is likely to attract more fruit- and nectar-eating birds than birds of other groups. Since trophic cascades are more likely in more productive ecosystems (Van Bael et al. 2003), reductions in insectivorous bird species in simplified agricultural systems may lead to increases in insect outbreaks (Mellink 1991). The use of pesticides to control insect pests in agricultural areas may offer a poor prey base for insectivorous birds. This is counterproductive for agriculture, as insectivorous birds can be important by removing and controlling insect pests (Greenberg et al. 2000b; Perfecto et al. 2004; Johnson et al. 2010). The low richness and low numbers of insectivorous forest birds in agricultural areas may also be due to their poor dispersal abilities (Şekercioğlu et al. 2002).

Higher mobility and better dispersal capacity often improve the ability of birds to adapt to land-use change and can reduce the likelihood of extinction, as is indicated by the fact that long-distance migratory bird species are 2.6 times less likely to be threatened or near threatened with extinction than are sedentary species (Şekercioğlu 2007). Field studies on the intensification of coffee management on local bird species also indicate that resident sedentary birds are more sensitive than long-distance migrants to alteration of their habitat (Mas and Dietsch 2004). Even though resident birds decline in response to the conversion of native habitats to coffee plantations, Wunderle (1999) found no effect of plantation size on migrant bird populations. Many migrants make extensive use of habitats with intermediate disturbance, such as shade coffee and other agroforests. Consequently, there may be some conservation trade-offs if the migrant birds reach higher numbers at lower levels of shade (tree cover), and some drop-off in numbers as tree cover increases to levels that benefit the residents more. Such potential conservation trade-offs in different groups needs more study.

Many published studies on agroforest avian communities have focused on Neotropical coffee (Komar 2006) and, to a lesser extent, cacao plantations. We need more research in other agro-ecosystems (particularly traditional mixed agroforests) and in different parts of the world. This is especially important since some studies in other regions and on different agroforest types have found no differences among avian guilds and have observed patterns contrary to the general trends revealed in this review and in global analyses (e.g., Buechley et al. 2015). The research on shade-grown coffee and cacao provides a sound foundation, but it can be improved by incorporating distinctive aspects of regional ecology with more targeted research. There is also a need for studies focused on raptors and seed eaters in agroforest systems, since these groups can be important pest and seed predators respectively, but remain understudied in agroforestry systems (Komar 2006). Furthermore, we know next to nothing about the impact on tropical forests and agroforests of the global declines in avian scavengers (Buechley and Şekercioğlu 2016a, 2016b).

Agro-ecosystems frequently comprise the matrix in which forest fragments, protected areas, and other native habitat remnants are embedded. Tropical agro-ecosystems often have substantial amounts of arboreal vegetation in the form of remnant trees (Fischer and Lindenmayer 2002), living fences (Harvey et al. 2005), riparian strips (Martin et al. 2006), and agroforestry plots (Schroth et al. 2004), all of which often have conservation values disproportionate to their land cover. These trees can provide connectivity (Graham 2001), dietary resources (Şekercioğlu et al. 2007), nesting opportunities (Manning et al. 2004; Şekercioğlu et al. 2007), and microclimatic refugia (Şekercioğlu et al. 2007) to many forest species, and can mediate the effects of forest fragmentation (Kupfer et al. 2006).

Despite the ecological importance of shifts in avian function in agroforests and agro-ecosystems, it is surprising that many studies comparing bird communities in forest, agroforest, and agricultural ecosystems do not report on the ecologically important changes in the proportions of avian functional groups, and that such studies report on relative abundance or estimated biomass even less frequently. Furthermore, as Komar (2006) points out, most coffee studies have failed to sufficiently quantify observer bias or detectability differences between habitats, thus making abundance estimates problematic. Komar also notes that none of these studies has carefully quantified the effects on bird abundance of plantation distance from forest. These criticisms also apply to most studies in other agro-ecosystems. Studies that rigorously compare and manipulate the relative abundance and biomass of avian functional groups in tropical agro-ecosystems, while incorporating landscape effects, comprise a critical frontier in ecology and will help illuminate the ecological causes and consequences of bird community changes in these rapidly expanding, human-dominated landscapes.

This overview has shown that the replacement of forests and agroforests with simplified agricultural systems results in a shift towards less specialized bird communities with altered proportions of functional groups. There is a strong relationship between specialization and extinction risk, and specialized birds are significantly more threatened with extinction (Şekercioğlu 2011). These ecological shifts can affect the ecological functions of and ecosystem services provided by birds in agroforests and other agricultural landscapes. The proportions of insectivores are lower among agroforest and agricultural birds, while the proportions of frugivores and nectarivores, which act as important seed dispersers and pollinators respectively, increase among birds with agroforest habitat preferences, especially in comparison to the bird communities of open agricultural areas. The increased presence of grasses in open agricultural areas contributes to the higher number of granivorous birds, which can become major seed predators and agricultural pests, especially when noncrop species are not producing seeds.

Nevertheless, reduced or increased species richness does not necessarily mean that there will be parallel changes in abundance, biomass, or function (Greenberg et al. 2000b; Perfecto et al. 2004). For example, although insectivorous bird diversity in tropical agroforestry systems is positively related to the magnitude of predator effects, Van Bael et al. (2008) observed no differences between agroforestry systems and forests in the magnitude of bird effects on plant pests, even though agroforest communities have fewer insectivorous bird species, simplified habitat structure, and less plant diversity (Van Bael et al. 2008). Given these uncertainties, there is an urgent need for detailed field studies comparing avian function and functional diversity between forests, agroforests, and simplified agricultural systems, ideally in landscapes that vary in their forest cover and composition.

High biodiversity can be successfully combined with high yields in tropical agroforests (Clough et al. 2011), and Perfecto and Vandermeer (2008) concluded that "diverse, low-input agroecosystems using agroecological principles are probably the best option for a high-quality matrix." Diverse, low-input agro-ecosystems include traditional shade coffee and cacao plantations (e.g., Buechley et al. 2015) and a number of other agroforestry types. The findings of these global analyses and reviews indicate that such agroforestry systems with some native cover also maintain a significantly larger proportion of important avian guilds such as frugivores and nectarivores, and a larger proportion of their respective services. These agroforestry systems also harbor substantially lower numbers and proportions of granivorous birds, some of which are major seed and crop predators. Agroforests and other agricultural habitats rich in tree cover are essential for connecting isolated protected areas and their metapopulations. Maintaining diverse, low-input, and preferably traditional agroforestry systems interspersed with tropical forest remnants not only will sustain more native biodiversity in tropical agricultural areas, but will also support higher proportions of avian seed dispersers, pollinators, insect predators, and their valuable ecosystem services. Nonetheless, most forest specialist species are lost from agroforest and open agricultural habitats, so preserving intact forest habitat is likely necessary to conserve much of the diversity of tropical bird communities.

Acknowledgments

This chapter is an update of Ç. H. Şekercioğlu (2012), Bird functional diversity in tropical forests, agroforests and open agricultural areas, *Journal of Ornithology* 153:S153–61, Dt. Ornithologen-Gesellschaft e.V. 2012, with kind permission of Springer Science + Business Media.

References

Aratrakorn, S., Thunhikorn, S. and Donald, P. F. 2006. Changes in bird communities following conversion of lowland forest to oil palm and rubber plantations in southern Thailand. *Bird Conservation International* 16:71–82.

Beehler, B. M., Raju, K., and Ali, S. 1987. Avian use of man-disturbed forest habitats in the Eastern Ghats, India. *Ibis* 129:197–211.

Beukema, H., Danielsen, F., Vincent, G., Hardiwinoto, S., and Van Andel, J. 2007. Plant and bird diversity in rubber agroforests in the lowlands of Sumatra, Indonesia. *Agroforestry Systems* 70:217–42.

Bhagwat, S. A., Willis, K. J., Birks, H. J. B., and Whittaker, R. J. 2008. Agroforestry: A refuge for tropical biodiversity? *Trends in Ecology & Evolution* 23:261–67.

Bowman, D. M., Woinarski, J., Sands, D., Wells, A., and McShane, V. J. 1990. Slash-and-burn agriculture in the wet coastal lowlands of Papua New Guinea: Response of birds, butterflies and reptiles. *Journal of Biogeography* 17:227–39.

Buechley, E. R., and Şekercioğlu, Ç. H. 2016a. The avian scavenger crisis: Looming extinctions, trophic cascades, and loss of critical ecosystem functions. *Biological Conservation.* In press.

———. 2016b. Vanishing vultures: The collapse of critical scavengers. *Current Biology.* In press.

Buechley, E. R., Şekercioğlu, Ç. H., Atickem, A., Gebremichael, G., Ndungu, J. K., Mahamued, B. A., Beyene, T., Mekonnen, T., and Lens, L. 2015. Importance of Ethiopian shade coffee farms for forest bird conservation. *Biological Conservation* 188:50–60.

Carlo, T. A., Collazo, J. A., and Groom, M. J. 2003. Avian fruit preferences across a Puerto Rican forested landscape: Pattern consistency and implications for seed removal. *Oecologia* 134:119–31.

Chimera, C. G., and Drake, D. R. 2010. Patterns of seed dispersal and dispersal failure in a Hawaiian dry forest having only introduced birds. *Biotropica* 42:493–502.

Classen, A., Peters, M. K., Ferger, S. W., Helbig-Bonitz, M., Schmack, J. M., Maassen, G., Schleuning, M., Kalko, E. K., Böhning-Gaese, K., and Steffan-Dewenter, I. 2014. Complementary ecosystem services provided by pest predators and pollinators increase quantity and quality of coffee yields. *Proceedings of the Royal Society of London B: Biological Sciences* 281:20133148.

Clough, Y., et al., 2011. Combining high biodiversity with high yields in tropical agroforests. *Proceedings of the National Academy of Sciences USA* 108:8311–16.

Clough, Y., Dwi Putra, D., Pitopang, R., Tscharntke, T. 2009. Local and landscape factors determine functional bird diversity in Indonesian cacao agroforestry. *Biological Conservation* 142:1032–41.

Cockle, K. L., Leonard, M. L., and Bodrati, A. A. 2005. Presence and abundance of birds in an Atlantic forest reserve and adjacent plantation of shade-grown yerba mate, in Paraguay. *Biodiversity and Conservation* 14:3265–88.

Faria, D., Laps, R. R., Baumgarten, J., and Cetra, M. 2006. Bat and bird assemblages from forests and shade cacao plantations in two contrasting landscapes in the Atlantic Forest of southern Bahia, Brazil. *Biodiversity and Conservation* 15:587–612.

Fischer, J., and Lindenmayer, D. B. 2002. The conservation value of paddock trees for birds in a variegated landscape in southern New South Wales. 2. Paddock trees as stepping stones. *Biodiversity and Conservation* 11:833–49.

Gove, A. D., Hylander, K., Nemomisa, S., and Shimelis, A. 2008. Ethiopian coffee cultivation-implications for bird conservation and environmental certification. *Conservation Letters* 1:208–16.

Graham, C. H. 2001. Factors influencing movement patterns of keel-billed toucans in a fragmented tropical landscape in southern Mexico. *Conservation Biology* 15:1789–98.

Greenberg, R., Bichier, P., and Angón, A. C. 2000a. The conservation value for birds of cacao plantations with diverse planted shade in Tabasco, Mexico. *Animal Conservation* 3:105–12.

Greenberg, R., Bichier, P., Angón, A. C., MacVean, C., Perez, R., and Cano, E. 2000b. The impact of avian insectivory on arthropods and leaf damage in some Guatemalan coffee plantations. *Ecology* 81:1750–55.

Greenberg, R., Bichier, P., and Sterling, J. 1997. Bird populations in rustic and planted shade coffee plantations of eastern Chiapas, Mexico. *Biotropica* 29:501–14.

Harvey, C. A., et al. 2005. Contribution of live fences to the ecological integrity of agricultural landscapes. *Agriculture, Ecosystems & Environment* 111:200–230.

Harvey, C. A., and González Villalobos, J. A. 2007. Agroforestry systems conserve species-rich but modified assemblages of tropical birds and bats. *Biodiversity and Conservation* 16:2257–92.

Holbech, L. H. 2009. The conservation importance of luxuriant tree plantations for lower storey forest birds in south-west Ghana. *Bird Conservation International* 19:287–308.

Johnson, M. D., Kellermann, J. L., Stercho, A. M. 2010. Pest reduction services by birds in shade and sun coffee in Jamaica. *Animal Conservation* 13:140–47.

Johnson, M. D., Sherry, T. W. 2001. Effects of food availability on the distribution of migratory warblers among habitats in Jamaica. *Journal of Animal Ecology* 70:546–60.

Kankam, B. O., and Oduro, W. 2009. Frugivores and fruit removal of Antiaris toxicaria at Bia Biosphere Reserve, Ghana. *Journal of Tropical Ecology* 25:201–4.

Karp, D. S., and Daily, G. C. 2014. Cascading effects of insectivorous birds and bats in tropical coffee plantations. *Ecology* 95:1065–74.

Kellermann, J. L., Johnson, M. D., Stercho, A. M., and Hackett, S. C. 2008. Ecological and economic services provided by birds on Jamaican Blue Mountain coffee farms. *Conservation Biology* 22:1177–85.

Kirika, J. M., Farwig, N., and Böhning-Gaese, K. 2008. Effects of local disturbance of tropical forests on frugivores and seed removal of a small-seeded Afrotropical tree. *Conservation Biology* 22:318–28.

Koh, L. P. 2008. Birds defend oil palms from herbivorous insects. *Ecological Applications* 18:821–25.

Koh, L. P., and Wilcove, D. S. 2008. Is oil palm agriculture really destroying tropical biodiversity? *Conservation Letters* 1:60–64.

Komar, O. 2006. Ecology and conservation of birds in coffee plantations: A critical review. *Bird Conservation International* 16:1–23.

Kupfer, J. A., Malanson, G. P., and Franklin, S. B. 2006. Not seeing the ocean for the islands: The mediating influence of matrix-based processes on forest fragmentation effects. *Global Ecology and Biogeography* 15:8–20.

Lehouck, V., Spanhove, T., Demeter, S., Groot, N. E., and Lens, L. 2009. Complementary seed dispersal by three avian frugivores in a fragmented Afromontane forest. *Journal of Vegetation Science* 20:1110–20.

Leyequién, E., De Boer, W. F., and Toledo, V. M. 2010. Bird community composition in a shaded coffee agro-ecological matrix in Puebla, Mexico: The effects of landscape heterogeneity at multiple spatial scales. *Biotropica* 42:236–45.

Luck, G. W., and Daily, G. C. 2003. Tropical countryside bird assemblages: Richness, composition, and foraging differ by landscape context. *Ecological Applications* 13:235–47.

Manning, A. D., Lindenmayer, D. B., and Barry, S. C. 2004. The conservation implications of bird reproduction in the agricultural matrix: A case study of the vulnerable superb parrot of south-eastern Australia. *Biological Conservation* 120:363–74.

Marsden, S. J., Symes, C.T., Mack, A. L. 2006. The response of a New Guinean avifauna to conversion of forest to small-scale agriculture. *Ibis* 148:629–40.

Martin, T. G., McIntyre, S., Catterall, C. P., Possingham, H. P., Tscharntke, T. 2006. Is landscape context important for riparian conservation? Birds in grassy woodland. *Biological Conservation* 127:201–14.

Maas, B., Karp, D. S., Bumrungsri, S., Darras, K., Gonthier, D., Huang, C.-C., Lindell, C. A., Maine, J. J., Mestre, L., Michel, N. L., Morrison, E. B., Perfecto, I., Philpott, S. M., Şekercioğlu, Ç. H., Silva, R. M., Taylor, P., Tscharntke, T., Van Bael, S. A., Whelan, C. J., and Williams-Guillén, K. 2016. Bird and bat predation services in tropical forests and agroforestry landscapes. *Biological Reviews.* In press. DOI: 10.1111/brv.12211.

Maine, J. J., Mestre, L., Michel, N. L., Morrison, E. B., Perfecto, I., Philpott, S. M., Şekercioğlu, Ç. H., Silva, R. M., Taylor, P., Tscharntke, T., Van Bael, S. A., Whelan, C. J., Williams-Guillén, K. 2015. Bird and bat predation services in tropical forests and agroforestry landscapes. *Biological Reviews.* Article first published online 23 July 2015. DOI: 10.1111/brv.12211.

Mas, A. H., and Dietsch, T. V. 2004. Linking shade coffee certification programs to biodiversity conservation: Butterflies and birds in Chiapas, Mexico. *Ecological Applications* 14:642–54.

Mellink, E. 1991. Bird communities associated with three traditional agroecosystems in the San Luis Potosi Plateau, Mexico. *Agriculture, Ecosystems & Environment* 36:37–50.

Morrison, E. B., and Lindell, C. A. 2012. Birds and bats reduce insect biomass and leaf damage in tropical forest restoration sites. *Ecological Applications* 22:1526–34.

Mortensen, H. S., Dupont, Y. L., and Olesen, J. M. 2008. A snake in paradise: Disturbance of plant reproduction following extirpation of bird flower-visitors on Guam. *Biological Conservation* 14:2146–54.

Mulwa, R. K., Böhning-Gaese, K., and Schleuning, M. 2012. High bird species diversity in structurally heterogeneous farmland in western Kenya. *Biotropica* 44:801–9.

Naidoo, R. 2004. Species richness and community composition of songbirds in a tropical forest-agricultural landscape. *Animal Conservation* 7:93–105.

Najera A., and Simonetti, J. A. 2010. Enhancing avifauna in commercial plantations. *Conservation Biology* 24:319–24.

Neuschulz, E. L., Botzat, A., and Farwig, N. 2011. Effects of forest modification on bird community composition and seed removal in a heterogeneous landscape in South Africa. *Oikos* 120:1371–79.

Norris, K., Asase, A., Collen, B., Gockowksi, J., Mason, J., Phalan, B., and Wade, A. 2010. Biodiversity in a forest-agriculture mosaic: The changing face of West African rainforests. *Biological Conservation* 143:2341–50.

Otieno, N. E, Gichuki, N., Farwig, N., and Kiboi, S. 2011. The role of farm structure on bird assemblages around a Kenyan tropical rainforest. *African Journal of Ecology* 49:410–17.

Peh, K. S. H., de Jong J., Sodhi, N. S., Lim, S. L. H., Yap, C. A. M. 2005. Lowland rainforest avifauna and human disturbance: persistence of primary forest birds in selectively logged forests and mixed-rural habitats of southern Peninsular Malaysia. *Biological Conservation* 123:489–505.

Peh, K. S. H., Sodhi, N. S., de Jong, J., Şekercioğlu, Ç. H., Yap, C. A. M., and Lim, S. L. H. 2006. Conservation value of degraded habitats for forest birds in southern Peninsular Malaysia. *Diversity and Distributions* 12:572–81.

Pejchar, L., Pringle, R. M., Ranganathan, J., Zook, J. R., Duran, G., Oviedo, F., and Daily, G. C. 2008. Birds as agents of seed dispersal in a human-dominated landscape in southern Costa Rica. *Biological Conservation* 141:536–44.

Perfecto, I., and Vandermeer, J. 2008. Biodiversity conservation in tropical agroecosystems: A new conservation paradigm. *Annals of the New York Academy of Sciences* 1134:173–200.

Perfecto, I., Vandermeer, J. H., Bautista, G. L., Nunez, G. I., Greenberg, R., Bichier, P., and Langridge S. 2004. Greater predation in shaded coffee farms: The role of resident Neotropical birds. *Ecology* 85:2677–81.

Petit, L. J., Petit, D. R., Christian, D. G., Powell, H. D. W. 1999. Bird communities of natural and modified habitats in Panama. *Ecography* 22:292–304.

Pizo, M. A. 2004. Frugivory and habitat use by fruit-eating birds in a fragmented landscape of southeast Brazil. *Ornitologia Neotropical* 15:117–26.

Pizo, M. A., and dos Santos, B. T. 2011. Frugivory, post-feeding flights of frugivorous birds and the movement of seeds in a Brazilian fragmented landscape. *Biotropica* 43:335–42.

Railsback, S. F., and Johnson, M. D. 2014. Effects of land use on bird populations and pest control services on coffee farms. *Proceedings of the National Academy of Sciences* 111:6109–14.

Reitsma, R., Parrish, J. D., and McLarney, W. 2001. The role of cacao plantations in maintaining forest avian diversity in southeastern Costa Rica. *Agroforestry Systems* 53:185–93.

Round, P. D., Gale, G. A., and Brockelman, W. Y. 2006. A comparison of bird communities in mixed fruit orchards and natural forest at Khao Luang, southern Thailand. *Biodiversity and Conservation* 15:2873–91.

Ruiz-Guerra, B., Renton, K., and Dirzo, R. 2012. Consequences of fragmentation of tropical moist forest for birds and their role in predation of herbivorous insects. *Biotropica* 44:228–36.

Schroth, G., Da Fonseca, G. A. B., Harvey, C. A., Gascon, C., Lasconcelos, H. L., and Izac, A. M. N. ed. 2004. *Agroforestry and Biodiversity Conservation in Tropical Landscapes*. Washington: Island Press.

Schulze, C. H., Waltert, M., Kessler, P. J., Pitopang, R., Veddeler, D., Mühlenberg, M., Gradstein, S., Leuschner, C., Steffan-Dewenter, I., and Tscharntke, T. 2004. Biodiversity indicator groups of tropical land-use systems: Comparing plants, birds, and insects. *Ecological Applications* 14:1321–33.

Şekercioğlu, Ç. H. 2006a. Increasing awareness of avian ecological function. *Trends in Ecology and Evolution* 21:464–71.

———. 2006b. Ecological significance of bird populations. In *Handbook of the Birds of the World*, vol. 11, ed. J. Del Hoyo, A. Elliott, and D. A. Christie, 15–51. Barcelona and Cambridge: Lynx Press and BirdLife International.

———. 2007. Conservation ecology: Area trumps mobility in fragment bird extinctions. *Current Biology* 17:R283–R286.

———. 2010. Ecosystem function and services. In *Conservation Biology for All*, ed. N. S. Sodhi and P. R. Ehrlich, 45–72. Oxford: Oxford University Press.

———. 2011. Functional extinctions of bird pollinators cause plant declines. *Science* 331:1019–20.

———. 2012. Bird functional diversity in tropical forests, agroforests and open agricultural areas. *Journal of Ornithology* 153:S153–61.

Şekercioğlu, Ç. H., Ehrlich, P. R., Daily, G. C., Aygen, D., Goehring, D., and Sandi, R. 2002. Disappearance of insectivorous birds from tropical forest fragments. *Proceedings of the National Academy of Sciences USA*. 99:263–67.

Şekercioğlu, Ç. H., Loarie, S. R., Oviedo Brenes, F., Ehrlich, P. R., Daily, G. C. 2007. Persistence of forest birds in the Costa Rican agricultural countryside. *Conservation Biology* 21:482–94.

Sidhu, S., Shankar Raman, T. R., and Goodale, E. 2010. Effects of plantations and home-gardens on tropical forest bird communities and mixed-species bird flocks in the Southern Western Ghats. *Journal of the Bombay Natural History Society* 107:91–108.

Sodhi, N. S., Şekercioğlu, Ç. H., Robinson, S., and Barlow, J. 2011. *Conservation of Tropical Birds*. Wiley-Blackwell., Oxford.

Thiollay, J. M. 1995. The role of traditional agroforests in the conservation of rain forest bird diversity in Sumatra. *Conservation Biology* 9:335–53.

Tscharntke, T., Şekercioğlu, Ç. H., Dietsch, T. V., Sodhi, N. S., Hoehn, P., Tylianakis, J. M. 2008. Landscape constraints on functional diversity of birds and insects in tropical agro-ecosystems. *Ecology* 89:944–51.

Turner, I. M., and Corlett, R. T. 1996. The conservation value of small, isolated fragments of lowland tropical rain forest. *Trends in Ecology and Evolution* 11:330–33.

Van Bael, S. A., Bichier, P., Ochoa, I., and Greenberg, R. 2007. Bird diversity in cacao farms and forest fragments of western Panama. *Biodiversity and Conservation* 16:2245–56.

Van Bael, S. A., and Brawn, J. D. 2005. The direct and indirect effects of insectivory by birds in two contrasting neotropical forests. *Oecologia* 143:106–16.

Van Bael, S. A., Brawn, J. D., Robinson, S. K. 2003. Birds defend trees from herbivores in a Neotropical forest canopy. *Proceedings of the National Academy of Sciences USA*. 100:8304–7.

Van Bael, S. A., Philpott, S. M., Greenberg, R., Bichier, P., Barber, N. A., Mooney, K. A., Gruner, D. S. 2008. Birds as predators in tropical agroforestry systems. *Ecology* 89:928–34.

Verea, C., and Solozano, A. 2005. Avifauna associated with a cacao plantation understory in northern Venezuela. *Ornitologia Neotropical* 16:1–14.

Waltert, M., Bobo, K. S., Sainge N. M., Fermon, H., and Muhlenberg, M. 2005. From forest to farmland: habitat effects on afrotropical forest bird diversity. *Ecological Applications* 15:1351–66.

Waltert, M., Mardiastuti, A., and Mühlenberg, M. 2004. Effects of land use on bird species richness in Sulawesi, Indonesia. *Conservation Biology* 18:1339–46.

Wang, Z. J., Young, S. S. 2003. Differences in bird diversity between two swidden agricultural sites in mountainous terrain, Xishuangbanna, Yunnan, China. *Biological Conservation* 110:231–43.

Watson, J. E., Whittaker, R. J., and Dawson, T. P. 2004. Habitat structure and proximity to forest edge affect the abundance and distribution of forest-dependent birds in tropical coastal forests of southeastern Madagascar. *Biological Conservation* 120:311–27.

Wenny, D. G., DeVault, T. L., Johnson, M. D., Kelly, D. Şekercioğlu, Ç. H., Tomback, D. F., and Whelan, C. J. 2011. The need to quantify ecosystem services provided by birds. *Auk* 128:1–14.

Whelan, C. J., Wenny, D. G., and Marquis, R. J. 2008. Ecosystem services provided by birds. *Annals of the New York Academy of Sciences* 1134:25–60.

Whelan, C.J., Şekercioğlu, Ç. H., Wenny, D.G. 2015. Why birds matter: from economic ornithology to ecosystem services. *Journal of Ornithology*. In press.

Wotton, D. M., and Kelly, D. 2012. Do larger frugivores move seeds further? Body size, seed dispersal distance, and a case study of a large, sedentary pigeon. *Journal of Biogeography* 39:1973–83.

Wunderle J. M. 1999. Avian distribution in Dominican shade coffee plantations: Area and habitat relationships. *Journal of Field Ornithology* 70:58–70.

CHAPTER TWELVE

Why Birds Matter

Bird Ecosystem Services Promote Biodiversity and Support Human Well-Being

Çağan H. Şekercioğlu, Daniel G. Wenny, Christopher J. Whelan, and Chris Floyd

Birds and humans have been interconnected for thousands of years. Birds inspire, entertain, feed, and clothe humans. Throughout the evolution of modern humans (Finlayson et al. 2012; Hardy and Moncel 2011) and the cultural development of our societies (Cocker and Tipling 2013; Mynott 2009; Podulka et al. 2004), birds have mattered.

Ecosystem services, as recognized by the Millennium Ecosystem Assessment (MEA 2005), fall into two primary categories. Cultural and provisioning services accrue directly; these services are themselves products. Bird art and bird eggs, for instance, are commodities that can be bought and sold. Regulating and supporting services, in contrast, accrue indirectly; they are not themselves commodities, but instead they help maintain other components of the world's ecosystems upon which humans depend for both goods (food, shelter) and other services (disease management; pest control). These indirect services facilitate other ecosystem services, and therefore promote biodiversity. Pollination and seed dispersal, for example, are important steps in plant reproductive cycles. Without birds' pollination or seed dispersal services, many plant species experience much lower reproductive success, which leads ultimately to changes in plant community composition (chapters 4, 5, 6, and 7). Declines in avian pollinators and seed dispersers may indirectly affect many other human uses of plants and the habitats in which they occur.

Most bird ecosystem services fall into the indirect categories of regulating and supporting services. These arise primarily through bird behavior—in particular, foraging behavior. An important exception comes from the ecosystem engineering of birds that excavate cavities or burrows for nesting and roosting (chapter 10). Accurately quantifying indirect ecosystem services proves challenging. Ascribing an economic value to them is even more challenging. Nevertheless, the preceding chapters show that the ecological roles of birds translate into vital ecosystem services that benefit humans in many ways. Greater understanding of the contribution of indirect services to economic vitality will help avoid policies that attempt to increase the delivery of one ecosystem service but inadvertently decrease another (Bennett et al. 2009).

In addition to a prevalence of indirect regulating and supporting services, art, literature, music, and folklore also abound with bird imagery or inspiration (Campbell and Lack 1985; Cocker and Tipling 2013; Podulka et al. 2004). Birds also provide provisioning ecosystem services (direct uses) such as meat, feathers, and recreational opportunities. Even putting aside the global chicken production of at least 20 billion animals, which makes this domestic bird the most numerous bird species on the planet (Cocker and Tipling 2013), hundreds of bird species provide provisioning ecosystem services that help the subsistence of millions of people around the world.

The valuation or commodification of ecosystem services has proven problematic. In fact, it may not be possible, or even desirable, to ascribe monetary value to specific ecosystem services (chapter 2). Important goals of describing ecosystem services are to inform and improve natural resource policies that maintain biodiversity and the ecosystem services they provide (Bateman et al. 2013; Hauck et al. 2013; Turner et al. 2007). A key part of this effort is to increase awareness of ecosystem services and how they are important in supporting the global economy. In some cases, however, the apparent mismatch between the ecological importance of a service and its perceived value would make such valuation insufficient to preserve the service (Garcia-Llorente et al. 2011).

A valuable example of such a mismatch is the near extinction of vultures in India and its consequences for increased human mortality from rabies (Markandya et al. 2008). Even though most people perceive vultures as organisms with little value, vultures provide the essential ecosystem service of scavenging, and thereby regulating disease (chapter 8). In India, which the highest rate of human rabies infection in the world, more than 20,000 people die from rabies annually. Populations of four species of *Gyps* vultures collapsed due to their consumption of cattle treated with the veterinary drug diclofenac (Green et al. 2004; Oaks et al. 2004). Populations of feral dogs that had competed with vultures for carcasses exploded following the vultures' decline (Prakash 1999). Markandya et al. (2008) estimated that the vulture population crash in India resulted in an additional 48,000 rabies deaths at a cost of US$34 billion during the 14 years from 1993 to 2006. The vulture populations have not yet recovered (Prakash et al. 2012), and costs continue to escalate. Nonetheless, most people are not aware of these staggering consequences of the decline in vulture ecosystem services (Buechley and Şekercioğlu, Ç. H. 2016a, 2016b).

The assertion that most people are well informed about ecosystem services, and that market economies therefore already value them appropriately (Sagoff 2011), is doubtful on at least two counts. First, externalities— unaccounted consequences of the actions of individuals or groups—distort the value of natural resources (Dasgupta and Ehrlich 2013). Second, perverse subsidies (Myers 1998) skew markets and lead to continued declines in ecosystem service delivery (MEA 2005; Raudsepp-Hearne et al. 2010). Furthermore, the "self-correcting market" approach suggests tacit approval of, or lack of understanding or concern about, increasing extinction rates and declining ecosystem service delivery, despite widespread public support for biodiversity conservation (MEA 2005; Swift 2014). Clearly, we need better data on how ecosystem services are linked with human well-being and how declines in ecosystem services will affect it (Kareiva et al. 2007; Raudsepp-Hearne et al. 2010; Schröter et al. 2005).

Ecosystem "Disservices"

A difficulty with assessing ecosystem services is the tendency to ignore, intentionally or not, "disservices" –those bird species or bird behaviors that impose costs or harm directly or indirectly on humans, human activities (e.g., agriculture), or human structures. Bird disservices include crop damage, spread of disease and invasive species, nutrient loading (eutrophication), collisions with aircraft, and simple nuisance (e.g., large communal roosts). As we argue below, the ecosystem services provided by birds far outweigh their disservices. Nonetheless, a thorough accounting and understanding of disservices is needed for a complete picture, and will help identify solutions that can balance the costs and benefits of integrated management (Triplett et al. 2012) to limit their impact.

Crop damage by birds is arguably the most recognized ecosystem disservice, but the economic losses resulting from birds are not well quantified (chapter 1). The perceived extent of crop loss or damage, as indicated by opinion surveys, is often much greater than the actual loss measured in the field (Gebhardt et al. 2011). Similarly, the perception that the presence of birds in an agricultural field indicates crop damage by those species is often incorrect (Greene et al. 2010; Lindell et al. 2012). Overall, crop losses due to birds are probably fairly low, especially when compared with the damage caused by rodents and insects. A database search on Web of Science with the terms "crop damage and bird" yielded 808 articles published between January 2000 and August 2013, whereas the terms "crop damage and rodent" yielded 2,171 articles, and "crop damage and insect" yielded 27,797.

The main crops affected (in reality or as perceived) are grains such as corn, rice, barley, and oats, but they also include sunflowers and fruit crops (grapes [*Vitis*], blueberries [*Vaccinium*], and cherries, [*Prunus*]). For grain crops, the most frequently cited bird pests are red-billed quelea (*Quelea quelea*), blackbirds (Icteridae), pheasants (*Phasianus colchicus*), crows (*Corvus* spp.), rose-ringed parakeets (*Psittacula krameri*), and house sparrows (*Passer domesticus*). For fruit crops, American robins (*Turdus migratorius*), cedar waxwings (*Bombycilla cedrorum*), European starlings (*Sturnus vulgaris*), Eurasian bullfinch (*Pyrrhula pyrrhula*), mynahs (Sturnidae), some crows (*Corvus* spp.), and various parrots (Psittaciformes) cause some crop loses.

Corn or maize (*Zea mays*) is a globally important crop and can illustrate the level of bird damage to crops (chapter 1). In North America a wide variety of birds feed on corn, but mostly on waste (spilled) grain after harvest. Damage to the crop resulting in reduced yield can be to the seeds shortly after planting, to the seedlings after germination, or to the nearly developed kernels before harvest. Red-winged blackbirds (*Agelaius phoeniceus*) feed on preharvest kernels for a short period in the fall, causing some crop loss (Dolbeer 1990; McNicol et al. 1982). Crop damage was less than 1%, far below government estimates of about 20% (Weatherhead et al. 1982), and was primarily in fields near wetlands and blackbird roosting sites in fall. Most important, blackbirds eat substantial amounts of invertebrates, especially during the breeding season when corn kernels are developing. These invertebrates include insect pests of corn that cause far more crop damage than do blackbirds (Bendell et al. 1981; Tremblay et al. 2001). In the 30 years since these studies, corn varieties less palatable to birds have been developed that likely reduce blackbird damage to corn even further, and anthraquinone-based repellent has reduced blackbird damage to corn by 28% (Carlson et al. 2013). A recent study in North Dakota showed that a mean of 0.2% of the corn crop in randomly selected plots was damaged by blackbirds in 2009–10 (Klosterman et al. 2013). Similar results have been found for other crops. For example, blackbirds had no effect on stink bugs on rice; and though they caused damage themselves, it did not result in any meaningful reduction in crop yield (Borkhataria et al. 2012). Nevertheless, some parts of fields can have more than 10% damage (Conover 1984), and in some cases, even slight damage can prevent corn from being sold for human consumption (Dolbeer 1990).

The most notorious example of an avian seed predator is the red-billed quelea, the world's most numerous wild bird species, with one to three billion individuals (Elliott and Lenton 1989), and the predominant avian pest in Africa. The red-billed quelea may be a bigger problem for crops in Africa than the red-winged blackbird is in North America, but efforts to control it may be more harmful than the damage it causes (Meinzingen et al. 1989).

Detailed studies indicate that while local damage may be high, the bird's impact on food production across Africa is negligible, with losses to cereal crops amounting to less than 1% of the production (Elliott and Lenton 1989). This is in the region of losses caused by bird pests in other parts of the world (Elliott and Lenton 1989; Weatherhead et al. 1982). Also, considering the important ecological roles played by this species as a predator of insects, including pest species; as a provider of nutrients that fertilize fields and orchards; and as an important food source for many birds, mammals, and people (Elliott and Lenton 1989), the extensive environmental damage and nontarget deaths caused by explosives, fire bombs, and especially the aerially sprayed pesticide fenthion used against this bird species (Meinzingen et al. 1989) cannot be justified. Fenthion has especially severe effects on aquatic species found in water bodies near quelea roosting sites, on predatory and scavenging birds (McWilliam and Cheke 2004), and on other nontarget species (Cheke et al. 2012). It can persist in the soil for weeks or months, and may affect soil biota (Cheke et al. 2013). Furthermore, many Africans collect and consume queleas killed by avicides, and are thus routinely exposed to dangerous chemicals (Jaeger and Elliott 1989).

Ecosystem Services from "Pest" Species

The beneficial services of granivores are often understudied and overlooked. Many granivorous bird species consume the seeds of agricultural weeds, potentially reducing the competitive effects of these weeds on crop plants (Ndang'ang'a et al. 2013). This is becoming increasingly important as exotic grasses become established in large parts of the globe. For example, in southern Costa Rican sun coffee plantations, native *Sporophila* seedeaters primarily eat grass seeds, as they are too small to harm the coffee beans substantially (Şekercioğlu, personal observation). Especially during the breeding season, in fact, most granivorous bird species feed their young large quantities of invertebrates, and may thus help reduce invertebrate pest populations (i.e., contribute pest control service) in agricultural plantations. The magnitude of this potential ecosystem service of seasonal granivorous species is a critical knowledge gap that awaits empirical quantification.

Similarly, some frugivorous bird species are considered pests of fruit crops including grapes, blueberries, and strawberries (*Fragaria*), among others (De Grazio 1978; Greig-Smith 1987). Other frugivorous bird species disperse invasive plant species (chapter 5). Like granivorous species, however, many frugivore species also eat invertebrates during the breeding season. The relative costs and benefits of such diet switching have not been evaluated for any species in an ecosystem services context.

A critical research need is to examine in detail the entire life cycle of key "avian pest" species, quantifying both the negative and positive effects of their habit use and foraging behavior on agricultural crops. Such studies need also to account for the economic and ecological costs of avian and nonavian pest control measures. Such detailed analyses may well demonstrate that the cost of avian pests is typically less than what farmers perceive (e.g., Basili and Temple 1999), and that avian "pests" in fact provide benefits to the farmer, owing to their consumption of invertebrate pests, such as corn rootworm (*Diabrotica* spp.), Japanese beetles (*Popillia japonica*), cutworms (Noctuidae), corn earworms (*Helicoverpa zea*), soybean aphids (*Aphis glycines*), and bean leaf beetles (*Cerotoma trifurcate*). Furthermore, the invertebrates likely cause far greater damage to crops than do birds, so that attempts to control the perceived avian pests may backfire and lead to greater crop damage by invertebrates (Becker 1996).

Spread of Plant Diseases

A poorly known bird disservice is the transmission of plant viruses by certain bird species (Broadbent 1965; Peters et al. 2012). In these cases, the birds move infected plant material in the process of nest construction, or spread the virus during contact with the plants while foraging. Peters et al. (2012) state that bird transmission of plant viruses is likely not a highly effective mechanism of virus spread, though it may be an effective mechanism of introduction into previously uninfected areas. They contend that once the viruses are present, other mechanisms will be more responsible for their spread throughout the newly infected population. On the other hand, birds also appear to spread viruses of certain invertebrate herbivores (Reilly and Hajek 2012). To what extent bird transmission of viruses that attack herbivores may counter the negative effect of transmitting plant viruses has not, to our knowledge, been investigated. The fungal pathogens of plants may also be spread by birds, especially by species that excavate cavities in trees, but little is known about this. The little information that exists suggests that insects are the primary vectors for such fungal pathogens (Paoli et al. 2001).

Damage from Bird Ecosystem Engineers

Notwithstanding the well-documented ecological importance of woodpeckers and their popularity among many birders, woodpeckers and a few other tree-nesting bird species such as monk parakeets (*Myopsitta monachus*) can also be notable pests. Wooden utility poles are frequent targets of woodpeckers, especially pileated woodpeckers. In central British Columbia, for example, woodpecker damage was found in 14% of the poles along one transmission line, and more than two-thirds ($n = 60$) of the damaged poles had nest cavities, some of which were used by secondary cavity nesters such as the tree swallow and the mountain bluebird (Parker et al. 2008). Because of safety concerns and the cost and logistical difficulty of replacing poles, utility companies are forced to combat woodpecker damage by patching holes and applying deterrents such as mesh barriers (Harness and Walters 2004). The nuisance activity of woodpeckers most familiar to the public is the birds' drilling of holes into wooden structures on houses. For example, Harding et al. (2009) found woodpecker damage in 33% of homes surveyed ($n = 1,185$) in Ithaca, New York. Based on a mail survey with respondents from 21 US states, mean annual repair costs from woodpecker damage in the United States was estimated at US$300 per home (with some repair costs exceeding US$5,000) and millions of dollars per year (Craven 1984).

By drilling sap wells into living tissue, sapsuckers can inflict highly visible injury to trees and shrubs. In the Rocky Mountains, stems of willow

(*Salix* sp.) shrubs with sap wells of red-naped sapsuckers usually die within a year or two of being drilled (Ehrlich and Daily 1988; Walters 1996; C. Floyd, personal observation). However, the extent to which these effects influence willow populations and associated riparian communities is unknown. The yellow-bellied sapsucker is purportedly responsible for much of the externally visible injuries to deciduous trees in the eastern United States (Smiley et al. 2009). If sapsuckers damage trees that are of value to humans, that damage constitutes an ecosystem disservice. To deter attacks by sapsuckers, growers of sugar maple (*Acer saccharum*) use trunk wraps and chemical repellents (Smiley et al. 2009); they formerly used lethal methods, such as treating sap wells with strychnine (McAtee 1913). Despite the often highly visible nature of sap wells and scars, and the considerable attention that sapsucker damage gets in unpublished literature (e.g., arboriculture pamphlets), there is very little published evidence that hole excavation by sapsuckers increases tree mortality (Erdmann and Oberg 1974; Smiley et al. 2009).

Originally from South America, cold-tolerant monk parakeets have become established in a number of temperate countries, including the United States, the United Kingdom, Spain, and Belgium. In addition to being agricultural pests, as the only parrots that build stick nests, highly social monk parakeets do significant damage to utility structures and cause power outages with their communal nests, which can reach the size of small cars (Avery et al., 2002). The economic damage caused by monk parakeets includes power outages, costs of electrical equipment repair and restoration, costs to the customers who lose power, and the costs of removing the nests from utility towers (Avery et al. 2002). The total estimated cost associated with monk parakeet-caused power outages in Florida in 2001 alone was $585,000 (A. Hodges and C. Newman, unpubl. data in Avery et al. 2002). The cost to remove both a nest and the birds inhabiting it was $1,500 in 2002, with an estimated total of $1,665,000 for just the Florida nests in 2001 (Avery et al., 2002).

The Role of Humans in Bird Disservices

An overlooked aspect of bird "disservices" is that many of them have their roots in human causes and usually have their highest impact in ecosystems that have been most modified by humans. Invasive species, crop damage, and nuisance species are often caused by human-induced habitat loss or modification. These habitat changes can result in conditions suitable for a few species that rapidly take advantage of the new foraging or nesting opportunities. Many avian "disservices" arise from a few adaptable bird species that flourish in human-dominated landscapes, including cities, agricultural fields, suburban residential areas, and golf courses. For example, Canada geese (*Branta canadensis*) thrive in the extensive grass, water traps, and predator-free habitats provided by golf courses and office parks. Many granivorous and gregarious bird species congregate and forage within extensive grain crop fields, occasionally causing economic losses. The proportion of granivorous species in tropical bird communities, for instance, is four to five times higher in open agricultural ecosystems than in tropical forest or agroforest ecosystems (chapter 11).

Rock pigeons (*Columba livia*) have adapted very well to cities, where there is abundant food provided by people, and where many buildings resemble their nesting cliffs in the wild. The perception of avian nuisance species varies among and within cultures. In many countries, rock pigeons are dismissed as "winged rats," and millions of dollars are spent every year reducing them and their droppings. In Turkey, however, being hit by a pigeon dropping is considered good luck, and the feeding of pigeons is viewed as a good religious deed. As a result, pigeons in that country provide an economic service both to the vendors of lottery tickets, which some people purchase after getting hit by a dropping, and to pigeon feed vendors around some mosques (Şekercioğlu 2006a).

Perhaps the most worrying bird disservice is that of birds striking aircraft, which ends badly for both people and birds, causes hundreds of millions of damage every year and can result in plane crashes and human deaths. Again, this often results from people building airports that replace natural bird habitats such as grasslands and wetlands, which provide the large and flat expanses needed for airport runways. In 2004 alone, the US Air Force logged 4,318 aircraft-wildlife collisions, mostly with birds, and this number was estimated at 30,000 for civilian aircraft (Kelley 2005). The most famous example is the "Miracle on the Hudson" flight in 2009, which flew into a flock of Canada geese and subsequently lost engine power. Captain Chesley Sullenberger was able to glide the massive plane with 155 people on board onto the Hudson River without any fatalities or major injuries, bringing the plane-bird collision problem into the global spotlight.

Recognizing that airports can inadvertently create ideal habitat for certain bird species is the first step in finding a solution (Blackwell et al. 2013). For example, the US Air Force has paid $200,000 per year for trained peregrine falcons (*Falco peregrinus*) to drive away European starlings (*Sturnus vulgaris*), Canada geese (*Branta canadensis*), and other birds that congregate around the airfield of McGuire Air Force Base (Kelley 2005). The Israeli Air Force receives daily "bird migration reports" to stay away from certain areas during migration (Pearce 2004). Clearly, more work is needed to devise nonlethal methods to discourage birds from congregating near airports. This is especially true for large, uncommon, or charismatic species, as illustrated by the recent controversies over shooting snowy owls (*Nyctea scandiaca*) at various airports in the eastern United States during the winter of 2013–14 (http://www.nycaudubon.org/issues-of-concern/protecting-raptors/snowy-owls).

Similarly, more information is needed on wildlife-aircraft collisions; the species involved, geographic locations, type of aircraft, and season. A recent analysis of wildlife strikes by US military rotary-wing aircraft found that passerines, shorebirds, seabirds, and bats were most frequently struck, but that collisions with raptors and vultures, waterbirds, pigeons, and gulls caused the most damage (Washburn et al. 2014). Strikes were most common in fall (September to November), and least common in winter (December to February). More detailed data may yield strategies to minimize wildlife-aircraft collisions (Washburn et al. 2014).

Bird Diversity and Ecosystem Services

The chapters in this volume have highlighted important ecosystem services provided by birds, the means by which birds provide each ecosystem service, and the conservation status of the birds involved. A more thorough understanding of the relationship between bird diversity and ecosystem services is a critical research need. The expectation that biodiversity improves ecosystem services is based on investigations of both experimental and natural plant communities. These studies typically measure ecosystem function (usually as plant biomass) from plant communities with different numbers and combinations of herbaceous plant species. The experimental studies face the limitations of relatively low diversity and small scale. Some, nonetheless, are long-term studies replicated at sites around the globe (Diaz et al. 2005). These plant studies often show that some measures of ecosystem service delivery increase with plant species richness (the number of plant species) or with the functional diversity of plants (the different growth strategies). In other cases, ecosystem service delivery depends upon the presence of particularly important species (Chapin et al. 2000; Diaz et al. 2006; Reich et al. 2012).

The plant biodiversity experiments demonstrate that diversity enhances ecosystem service delivery in multiple ways (Isbell et al. 2011). First, different species contribute ecosystem function in some years but not in others, or in some areas but not in others, or only under some environmental circumstances. Second, some species provide a single ecosystem function in multiple years, while other species contribute multiple functions within a single year. These authors concluded that "although species may appear functionally redundant when one function is considered under one set of environmental conditions, many species are needed to maintain multiple functions at multiple times and places in a changing world" (Isbell et al. 2011, p. 200).

Similar findings have been reported from a meta-analysis of functional redundancy in plots representing 18 land-use intensity gradients from five biomes and nine countries, and including over 2,800 plant species (Laliberte et al. 2010). Increased land-use intensification significantly reduced plant species' functional redundancy, though, not surprisingly, specific relationships exhibited considerable variation among the different land uses. To the extent that functional redundancy contributes to the delivery of ecosystem services (de Bello et al. 2010), the findings of Isbell et al. (2011) and Laliberte et al. (2010) suggest that decreased species diversity erodes ecosystem service delivery.

Furthermore, ecosystem feedbacks and interspecific complementarity that depend upon plant species diversity appear to accumulate over time (Reich et al. 2012). Thus, species combinations that appeared functionally redundant during the early years of the Cedar Creek biodiversity experiment became progressively more functionally unique over time (more than 13 years). This implies that the loss of ecosystem functioning due to compositional simplification has likely been underestimated through analysis of short-term experiments.

Still unknown is to what extent these findings from experimental and natural plant communities apply to other groups. For instance, is the delivery of ecosystem services by birds dependent upon species richness, species diversity, some measure of bird functional diversity, or some combination of these things? There may be no simple or universal answer. This important question needs investigation—not only for birds, but also for other animal taxa.

Numerous studies suggest that, as with plant communities, greater diversity of bird species ensures greater ecosystem function and delivery of ecosystem services. For instance, fruit seed dispersal services along a landscape-scale gradient of anthropogenic forest loss were positively related to avian frugivore abundance and richness (Garcia and Martinez 2012). In this study, the richness of frugivorous bird species was the sole characteristic of the assemblage that was related to seed dispersal into deforested parts of the landscape. These results echo and bolster the results of earlier studies. Cordeiro and Howe (2003) demonstrated that decreased avian disperser assemblage diversity due to forest fragmentation in Tanzania reduced fruit consumption, seed dispersal, and seedling establishment. Markl et al. (2012) found through a meta-analysis that human disturbance (forest fragmentation, hunting, and selective logging) often leads to a rapid decline in large frugivores, with the consequence of disproportionate loss of seed dispersal services for large-seeded plant species (see also Kelly et al. 2010). Chapters 5, 6, and 7 provide many other examples of the role of bird species as seed dispersal agents.

As with seed dispersal services, the delivery of both pest control (chapter 3; Barbaro et al. 2014; Philpott et al. 2009; Van Bael et al. 2008) and pollination services (chapter 4; Kelly et al. 2010; Luck and Daily 2003) is enhanced by a greater diversity of bird species. Hence, anthropogenic modifications of habitat that drive the loss of some species inevitably cause a decline in the effectiveness of these ecosystem services.

Moreover, in some circumstances, pest control services delivered by birds interact synergistically with pollination services delivered by insects (Classen et al. 2014), thus leading to increased quantity and quality of coffee (*Coffea arabica*) yields. Similarly, diurnal pest control services provided by birds complement the nocturnal pest control services provided by bats (Maas et al. 2013, 2015).

Conservation

A link between conservation of avian functional guilds and the ecosystem services they provide has attracted increasing attention in the past decade (Şekercioğlu 2006a; Şekercioğlu 2006b; Şekercioğlu et al. 2004; Wenny et al. 2011; Whelan et al. 2008; Sodhi et al. 2011). Some bird functions, such as scavenging by vultures, nutrient deposition by seabirds, and vertebrate predation by birds of prey, are declining especially rapidly due to the increasing endangerment of bird species in these groups (Şekercioğlu et al. 2004; fig. 12.1), especially the specialists (Şekercioğlu 2011). In a single decade, during which the proportion of extinction-prone (globally extinct, threatened,

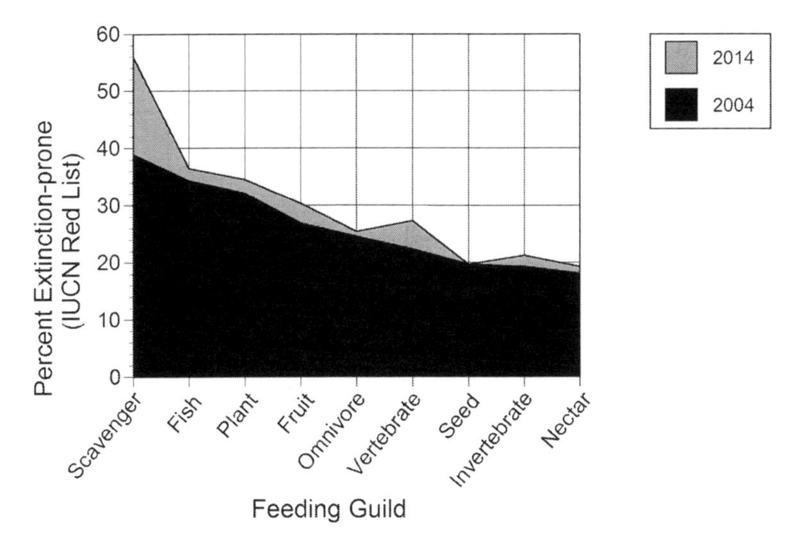

FIGURE 12.1. Changes in the percentages of threatened, near threatened, and extinct species in avian guilds between 2004 and 2014. Note the particularly sharp increase in the threat status of avian scavengers. Reprinted with permission from Buechley and Şekercioğlu (2016a).

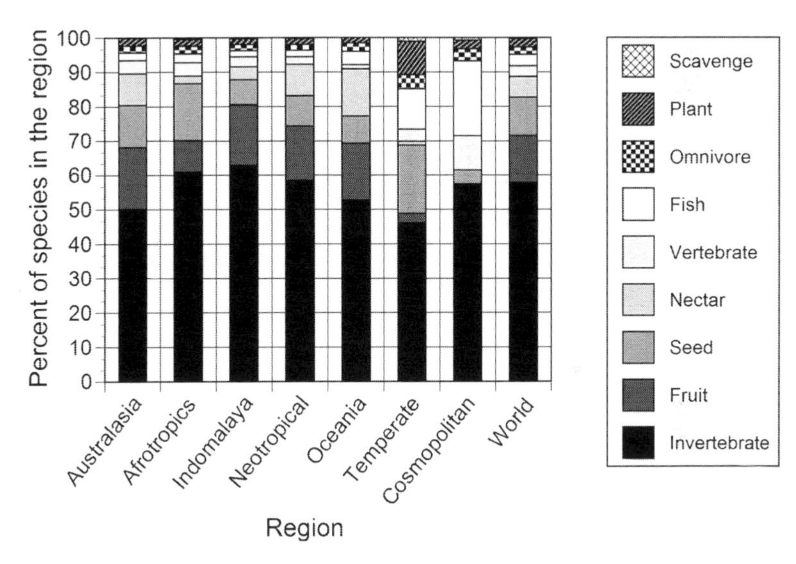

FIGURE 12.2. Proportionate distribution of avian feeding guilds (functional groups) in different bird communities.

or near threatened) bird species increased from 21 to 23%, the same proportion of specialized avian scavengers (36 species, mainly vultures) increased from 38% (Şekercioğlu et al. 2004) to 53% (fig. 12.1; Buechley and Şekercioğlu 2016a). For fish eaters, most of which are seabirds, the increase was from 34 to 37%. The status of birds of prey also declined rapidly, and the percentage of extinction-prone species increased from 22 to 26% between 2004 and 2013. These declines are largely a result of declines in the status of seabirds in general, and of African scavengers and birds of prey in particular (http://www.birdlife.org/datazone/sowb/casestudy/564).

Most avian functional groups reach their peak richness in the Neotropics (fig. 12.1; Kissling et al. 2012). However, there is global variation in the proportionate representation of avian dietary guilds (fig. 12.2). Insectivores and frugivores have their highest representation in the tropics, with frugivores being proportionately lower in the Afrotropics, and insectivores in Australasia. Seedeaters are particularly well represented in drier parts of the world, especially Australasia, the Afrotropics, and the temperate regions. Nectarivores, on the other hand, reach their highest proportions in the Neotropics (home to the hummingbird radiation), the Pacific Ocean islands, and Australia. Scavengers reach their highest species richness in the savannas of eastern Africa, home to the largest remaining congregations of mammalian megafauna left on the planet. In the temperate regions, piscivores (fish eaters) and herbivores are better represented than in the tropics. As expected, wide-ranging carnivores (birds of prey) and piscivores reach their highest proportions among wide-ranging, cosmopolitan species (fig. 12.2).

Perpetuation of the numerous and important services provided by birds requires perpetuation of the species providing them. From the perspective of policy and advocacy, therefore, we suggest that avian conservation should emphasize not only the need to preserve individual species, but the ways in which their preservation can help maintain their contribution of ecosystem services.

Future Directions

As demonstrated in chapter 2, we have made considerable progress in building conceptual frameworks for understanding the economics of ecosystem services. Important considerations include whether a bird species responsible for a service is unique or part of a redundant network (Power et al. 1996). We emphasized the importance of species richness for delivery of ecosystem services, implying that even with redundancies, each species counts. Also critical is the temporal and spatial scale over which any potential service is generated (Chee 2004). Ecosystem service accounting needs to consider the intrinsic variability inherent in the production of any ecosystem service. Ecosystem processes naturally vary over time. Understanding natural variability is thus critical to the proper accounting of ecosystem services (MEA 2005).

The chapters in this volume illustrate the great strides avian ecologists have made in understanding and documenting the many ecosystem services provided by birds. Gaps in knowledge persist and need filling; the science of ecosystem services is in its infancy, and much remains to be learned. For instance, we know considerably more about how birds function as consumers of arthropods and fruits, and hence as control agents of herbivorous insects and as seed dispersers, than we know about birds as pollinators. Additional research regarding which plant species are pollinated by which bird species, and about the effectiveness of birds relative to other pollinators, is critical. More research on avian granivory in agroecosystems is also warranted, particularly with respect to the seasonal switching of diet between seeds and arthropods.

When birds contribute multiple ecosystem services (e.g., reduction of weed seed in nonbreeding season, and insect pest control in breeding season), we must determine the relative demographic impact of each service. The same holds when different co-occurring species contribute such different ecosystem functions and services. Rarely, if ever, do we know the demographic contribution of birds in any one of these roles, let alone their synergistic or antagonistic effects. Cost-benefit analysis could be applied when alternative ecosystem "components" might be available (e.g., for the use of pesticides versus encouragement of bird predation).

Experimental exclusion of birds is a powerful tool for assessing ecosystem function. However, such experiments are not always feasible, and an inherent problem in their interpretation is in the question of how best to extrapolate from the scale of the experiment to that of an ecosystem. In light of such difficulties, ecologists should be poised to take advantage of "natural" experiments that may arise, for instance, from geographically local declines of certain species or groups of species, or from epizootics like that of West Nile Virus (e.g., see Caffrey et al. 2005). Such declines may allow efficient detection of a concomitant decline of important bird services. A case in point is the work of investigating the ecological consequences of the virtual elimination of land birds following the accidental introduction of the brown tree snake (*Boiga irregularis*) on the Pacific island of Guam (Caves et al. 2013; Rogers et al. 2012; see also HilleRis-Lambers et al. 2013).

Conservation: How Much Will It Cost, and Who Pays?

Humans have a proven track record of identifying conservation concerns and needs, and of prioritizing conservation strategies. We have a woeful track record of committing the resources needed to implement conservation policies.

Governments around the world have committed themselves to the goals of conserving biological diversity and promoting sustainable development. They have yet to commit the resources needed to accomplish these goals. Under the auspices of the United Nations Conference on Environment and Development (UNCED), 167 nations signed the Convention on Biological Diversity treaty in 1992. Targets of the treaty include halting extinctions and protecting areas of globally important biological diversity by 2020. The estimated costs of such conservation measures were poorly known until a study by McCarthy et al. (2012) placed the cost of meeting conservation targets for all globally threatened bird species at approximately $4 billion annually. Extrapolating costs for site preservation and including additional taxa raised the estimate to over $70 billion. The authors conclude that their estimates indicate "the need to increase investment in biodiversity by at least an order of magnitude" (McCarthy et al. 2012, p. 949) over current expenditures. Unfortunately, much of the available funds are often squandered as a result of corruption and inefficiency (Barrett et al. 2001; Chan et al. 2007; Smith and Walpole 2005). Furthermore, a particular challenge is posed by the disparity between the greater resources available in richer countries and the higher potential conservation gains in financially poor, biodiversity-rich countries.

In the United States, a useful model for generating funds to support bird and other wildlife conservation and land preservation is provided by the Pittman-Robertson Federal Aid in Wildlife Restoration Act, signed into law by President Franklin D. Roosevelt on September 2, 1937. This legislation, whose passage was urged by organized sportsmen, state wildlife agencies, and the firearms and ammunition industries, provides an excise tax on ammunition, firearms, handguns, and archery equipment to support conservation, restoration, and research on wildlife species and their critical habitat. A similar sort of excise tax, applied to gear related to wildlife recreation (e.g., binoculars, spotting scopes, bird feed, and feeders) could be implemented to generate funds to be used specifically to acquire lands for nongame wildlife habitat, and to promote management and restoration of lands for nongame wildlife conservation and research (Whelan et al. 2010). Bird-watchers in the United States should be encouraged to participate in the Federal Migratory Bird Hunting and Conservation Stamp program, better known as Federal Duck Stamps. Professional and amateur birders should lobby for the development of a similar program of collectible stamps for nongame species and upland habitats to complement the wetland preservation funded by the duck stamps. Citizens elsewhere around the globe, including scientists, conservationists, and amateurs interested in nature, should similarly lobby their respective governments and conservation agencies for the implementation of programs to increase funding for conservation measures that ensure the preservation of birds and their ecosystem services.

The financial costs of conservation appear great, perhaps even insurmountable. But even at the estimate of $80 billion annually, those costs are small compared to the value of biodiversity and ecosystem services. As stated by McCarthy et al. (2012): "More prosaically, the total required is less than 20% of annual global consumer spending on soft drinks." Clearly, the capital to fund conservation exists. It is up to those of us who value birds, and the rest of nature, to urge governments and citizens of the world to find the will.

References

Avery, M. L., Greiner, E. C., Lindsay, J. R., Newman, J. R., Pruett-Jones, S. 2002. Monk parakeet management at electric utility facilities in south Florida. USDA National Wildlife Research Center, Staff Publications. Paper 458.

Barbaro, L., Giffard, B., Charbonnier, Y., van Halder, I., and Brockerhoff, E. G. 2014. Bird functional diversity enhances insectivory at forest edges: A transcontinental experiment. *Diversity and Distributions* 20:149–59.

Barrett, C. B., Brandon, K., Gibson, C., and Gjertsen, H. 2001. Conserving tropical biodiversity amid weak institutions. *Bioscience* 51:497–502.

Basili, G. D., and Temple, S. A. 1999. Dickcissels and crop damage in Venezuela: Defining the problem with ecological models. *Ecology* 9:732–39.

Bateman, I. J., Harwood, A. R., Mace, G. M., Watson, R. T., Abson, D. J., Andrews, B., Binner, A., Crowe, A., Day, B. H., Dugdale, S., Fezzi, C., Foden, J., Hadley, D., Haines-Young, R., Hulme, M., Kontoleon, A., Lovett, A. A., Munday, P., Pascual, U., Paterson, J., Perino, G., Sen, A., Siriwardena, G., van

Soest, D., and Termansen, M. 2013. Bringing ecosystem services into economic decision-making: Land use in the United Kingdom. *Science* 341:45–50.

Becker, J. 1996. *Hungry Ghosts: Mao's Secret Famine*. New York, The Free Press.

Bendell, B. E., Weatherhead, P. J., and Stewart, R. K. 1981. The impact of predation by red-winged blackbirds on European corn-borer populations. *Canadian Journal of Zoology-Revue Canadienne De Zoologie* 59:1535–38.

Bennett, E. M., Peterson, G. D., and Gordon, L. J. 2009. Understanding relationships among multiple ecosystem services. *Ecology Letters* 12:1394–1404.

Birdlife International. 2013. *State of the World's Birds: Indicators for Our Changing World*. Cambridge: BirdLife International.

Borkhataria, R. R., Nuessly, G. S., Pearlstine, E., and Cherry, R. H. 2012. Effects of blackbirds (*Agelaius phoenicius*) on stinkbug (Hemiptera: Pentatomidae) populations, damage, and yield in Florida rice. *Florida Entomologist* 95:143–49.

Broadbent, L. 1965. Epidemiology of tomato mosaic virus. 9. Transmission of TMV by birds. *Annals of Applied Biology* 55:67–70.

Buechley, E. R., Şekercioğlu, Ç. H. 2016a. The avian scavenger crisis: Looming extinctions, trophic cascades, and loss of critical ecosystem functions. *Biological Conservation*. In press.

Buechley, E. R., Şekercioğlu, Ç. H. 2016b. Vanishing vultures: The collapse of critical scavengers. *Current Biology*. In press.

Caffrey, C., Smith, S. C. R., and Weston, T. J. 2005. West Nile virus devastates an American crow population. *Condor* 107:128–32.

Campbell, B. D., and Lack, E., eds. 1985. *A Dictionary of Birds*. Vermillion, SD: Buteo Books.

Carlson, J. C., Tupper, S. K., Werner, S. J., Pettit, S. E., Santer, M. M., and Linz, G. M. 2013. Laboratory efficacy of an anthraquinone-based repellent for reducing bird damage to ripening corn. *Applied Animal Behaviour Science* 145:26–31.

Caves, E. M., Jennings, S. B., HilleRisLambers, J., Tewksbury, J. J., and Rogers, H. S. 2013. Natural experiment demonstrates that bird loss leads to cessation of dispersal of native seeds from intact to degraded forests. *PLoS ONE* 8.e65618.

Chan, K. M. A., Pringle, R. M., Ranganathan, J., Boggs, C. L., Chan, Y. L., Ehrlich, P. R., Haff, P. K., Heller, N. E., Al-Krafaji, K., and Macmynowski, D. P. 2007. When agendas collide: Human welfare and biological conservation. *Conservation Biology* 21:59–68.

Chapin, F. S., Zavaleta, E. S., Eviner, V. T., Naylor, R. L., Vitousek, P. M., Reynolds, H. L., Hooper, D. U., Lavorel, S., Sala, O. E., Hobbie, S. E., Mack, M. C., and Diaz, S. 2000. Consequences of changing biodiversity. *Nature* 405:234–42.

Chee, Y. E. 2004. An ecological perspective on the valuation of ecosystem services. *Biological Conservation* 120:549–65.

Cheke, R. A., Adranyi, E., Cox, J. R., Farman, D. I., Magoma, R. N., Mbereki, C., McWilliam, A. N., Mtobesya, B. N., and van der Walte, E. 2013. Soil contamination and persistence of pollutants following organophosphate sprays and explosions to control red-billed quelea (*Quelea quelea*). *Pest Management Science* 69:386–96.

Cheke, R. A., McWilliam, A. N., Mbereki, C., van der Walt, E., Mtobesya, B., Magoma, R. N., Young, S., and Eberly, J. P. 2012. Effects of the organophosphate fenthion for control of the red-billed quelea *Quelea quelea* on cholinesterase and haemoglobin concentrations in the blood of target and non-target birds. *Ecotoxicology* 21:1761–70.

Classen, A., Peters, M. K., Ferger, S. W., Helbig-Bonitz, M., Schmack, J. M., Maassen, G., Schleuning, M., Kalko, E. K. V., Bohning-Gaese, K., and Steffan-Dewenter, I. 2014. Complementary ecosystem services provided by pest predators and pollinators increase quantity and quality of coffee yields. *Proceedings of the Royal Society B-Biological Sciences* 281:20133148.

Cocker, M., and Tipling, D. 2013. *Birds and People*. London: Jonathan Cape.

Conover, M. R. 1984. Response of birds to different types of food repellents. *Journal of Applied Ecology* 21:437–43.

Cordeiro, N. J., and Howe, H. F. 2003. Forest fragmentation severs mutualism between seed dispersers and an endemic African tree. *Proceedings of the National Academy of Sciences* 100:14052–56.

Craven, S. R. 1984. Woodpeckers: A serious suburban problem. *Proceedings of the Vertebrate Pest Conference* 11:204–10.

Dasgupta, P. S., and Ehrlich, P. R. 2013. Pervasive externalities at the population, consumption, and environment nexus. *Science* 340:324–28.

De Bello, F., Lavorel, S., Diaz, S., Harrington, R., Cornelissen, J. H. C., Bardgett, R. D., Berg, M. P., Cipriotti, P., Feld, C. K., Hering, D., da Silva, P. M., Potts, S. G., Sandin, L., Sousa, J. P., Storkey, J., Wardle, D. A., and Harrison, P. A. 2010. Towards an assessment of multiple ecosystem processes and services via functional traits. *Biodiversity and Conservation* 19:2873–93.

De Grazio, J. W. 1978. World bird damage problems. Proceedings of the 8th Vertebrate Pest Conference: 9–24. http://digitalcommons.unl.edu/vpc8/13.

Diaz, S., Fargione, J., Chapin, F. S., and Tilman, D. 2006. Biodiversity loss threatens human well-being. *PloS Biology* 4:1300–1305.

Diaz, S., Tilman, D., Fargione, J., Chapin, F. S., Dirzo, R., Kitzberger, T., Gemmill, B., Zobel, M., Vila, M., Mitchell, C., Wilby, A., Daily, G., Galetti, M., Laurance, W. F., Pretty, J., Naylor, R., Power, A., and Harvell, D. 2005. Biodiversity regulation of ecosystem services. In *Ecosystems and Human Well-Being: Current State and Trends*, ed. R. Hassan, R. Scholes, and N. Ash, 297–329. Washington: Island Press.

Dolbeer, R. A. 1990. Ornithology and integrated pest management: Red-winged blackbirds (*Agelaius phoeniceus*) and corn. *Ibis* 132:309–2.

Ehrlich, P. R., Daily, G.C. 1988. Red-naped sapsuckers feeding at willows: Possible keystone herbivores. *American Birds* 42:357–65.

Elliott, C. C. H., and Lenton, G. M. 1989. The pest status of the quelea. In *Quelea Quelea: Africa's Bird Pest*. ed. R. L. Bruggers and C. C. H. Elliott, 17–34. Oxford: Oxford University Press.

Erdmann, G. G., Oberg, R. R. 1974. Sapsucker feeding damages crown-released yellow birch trees. *Journal of Forestry* 72:760–63.

Finlayson, C., Brown, K., Blasco, R., Rosell, J., and Negro, J. J. 2012. Birds of a feather: Neanderthal exploitation of raptors and corvids. *PLoS ONE* 7:e45927.

Garcia, D., and Martinez, D. 2012. Species richness matters for the quality of ecosystem services: A test using seed dispersal by frugivorous birds. *Proceedings of the Royal Society B-Biological Sciences* 279:3106–13.

Garcia-Llorente, M., Martin-Lopez, B., Diaz, S., and Montes, C. 2011. Can ecosystem properties be fully translated into service values? An economic valuation of aquatic plant services. *Ecological Applications* 21:3083–3103.

Gebhardt, K., Anderson, A. M., Kirkpatrick, K. N., and Shwiff, S. A. 2011. A review and synthesis of bird and rodent damage estimates to select California crops. *Crop Protection* 30:1109–16.

Green, R. E., Newton, I., Shultz, S., Cunningham, A. A., Gilbert, M., Pain, D. J., and Prakash, V. 2004. Diclofenac poisoning as a cause of vulture population declines across the Indian subcontinent. *Journal of Applied Ecology* 41:793–800.

Greene, C. D., Nielsen, C. K., Woolf, A., Delahunt, K. S., and Nawrot, J. R. 2010. Wild turkeys cause little damage to row crops in Illinois. *Transactions of the Illinois State Academy of Science* 103:145–52.

Greig-Smith, P. W. 1987. Bud-feeding by bullfinches: Methods for spreading damage evenly within orchards. *Journal of Applied Ecology* 24:49–62.

Harding, E. G., Vehrencamp, S. L., Curtis, P. D. 2009. External characteristics of houses prone to woodpecker damage. *Human-Wildlife Conflicts* 3:136–44.

Hardy, B. L., and Moncel, M.-H. 2011. Neanderthal use of fish, mammals, birds, starchy plants, and wood 125–250,000 years ago. *PLoS ONE* 6:e23768.

Harness, R. E., Walters, E. L. 2004. Woodpeckers and utility pole damage. Paper presented at the Rural Electric Power Conference. Paper 04 B3, Scottsdale, AZ, May 23–25, 2004.

Hauck, J., Gorg, C., Varjopuro, R., Ratamaki, O., and Jax, K. 2013. Benefits and and limitations of the ecosystem services concept in environmental policy and decision making: Some stakeholder perspectives. *Environmental Science & Policy* 25:13–21.

HilleRisLambers, J., Ettinger, A. K., Ford, K. R., Haak, D. C., Horwith, M., Miner, B. E., Rogers, H. S., Sheldon, K. S., Tewksbury, J. J., Waters, S. M., and Yang, S. 2013. Accidental experiments: Ecological and evolutionary insights and opportunities derived from global change. *Oikos* 122:1649–61.

Isbell, F., Calcagno, V., Hector, A., Connolly, J., Harpole, W. S., Reich, P. B., Scherer-Lorenzen, M., Schmid, B., Tilman, D., van Ruijven, J., Weigelt, A., Wilsey, B. J., Zavaleta, E. S., and Loreau, M. 2011. High plant diversity is needed to maintain ecosystem services. *Nature* 477: 199–202.

Jaeger, M. E., and Elliott, C. C. H. 1989. Quelea as a resource. In *Quelea Quelea: Africa's Bird Pest*, ed. R. L. Bruggers and C. C. H. Elliott, 327–38. Oxford: Oxford University Press.

Kareiva, P., Watts, S., McDonald, R., and Boucher, T. 2007. Domesticated nature: Shaping landscapes and ecosystems for human welfare. *Science* 316:1866–69.

Kelley, T. 2005. Clearing the runway nature's way. *New York Times* February 25, B1.

Kelly, D., Ladley, J. J., Robertson, A. W., Anderson, S. H., Wotton, D. M., and Wiser, S. K. 2010. Mutualisms with the wreckage of an avifauna: The status of bird pollination and fruit-dispersal in New Zealand. *New Zealand Journal of Ecology* 34:66–85.

Kissling, W. D., Şekercioğlu, Ç. H., and Jetz, W. 2012. Bird dietary guild richness across latitudes, environments and biogeographic regions. *Global Ecology and Biogeography* 21:328–40.

Klosterman, M. E., Linz, G. M., Slowik, A. A., and Homan, H. J. 2013. Comparisons between blackbird damage to corn and sunflower in North Dakota. *Crop Protection* 53:1–5.

Laliberte, E., Wells, J. A., DeClerck, F., Metcalfe, D. J., Catterall, C. P., Queiroz, C., Aubin, I., Bonser, S. P., Ding, Y., Fraterrigo, J. M., McNamara, S., Morgan, J. W., Merlos, D. S., Vesk, P. A., and Mayfield, M. M. 2010. Land-use intensification reduces functional redundancy and response diversity in plant communities. *Ecology Letters* 13:76–86.

Lindell, C. A., Eaton, R. A., Lizotte, E. M., and Rothwell, N. L. 2012. Bird consumption of sweet and tart cherries. *Human-Wildlife Interactions* 6:283–90.

Luck, G. W., Daily, G. C. 2003. Tropical countryside bird assemblages: Richness, composition, and foraging differ by landscape context. *Ecological Applications* 13:235–47.

Maas, B., Clough, Y. and Tscharntke, T. 2013. Bats and birds increase crop yield in tropical agroforestry landscapes. *Ecology Letters* 16:1480–87.

Maas, B., Karp, D.S., Bumrungsri, S., Darras, K., Gonthier, D., Huang, C.-C., Lindell, C.A., Maine, J.J., Mestre, L., Michel, N.L., Morrison, E.B., Perfecto, I., Philpott, S.M., Şekercioğlu, Ç. H., Silva, R.M., Taylor, P., Tscharntke, T., Van Bael, S.A., Whelan, C.J., Williams-Guillén, K. 2015. Bird and bat predation services in tropical forests and agroforestry landscapes. *Biological Reviews.* Article first published online: 23 JUL 2015. DOI: 10.1111/brv.12211.

Markandya, A., Taylor, T., Longo, A., Murty, M. N., Murty, S., and Dhavala, K. 2008. Counting the cost of vulture decline: An appraisal of the human health and other benefits of vultures in India. *Ecological Economics* 67:194–204.

Markl, J. S., Schleuning, M., Forget, P. M., Jordano, P., Lambert, J. E., Traveset, A., Wright, S. J., and Bohning-Gaese, K. 2012. Meta-analysis of the effects of human disturbance on seed dispersal by animals. *Conservation Biology* 26: 1072–81.

McAtee, W. L. 1913. Destruction of sapsuckers. *Auk* 30:154–57.

McCarthy, D. P., Donald, P. F., Scharlemann, J. P. W., Buchanan, G. M., Balmford, A., Green, J. M. H., Bennun, L. A., Burgess, N. D., Fishpool, L. D. C., Garnett, S. T., Leonard, D. L., Maloney, R. F., Morling, P., Schaefer, H. M., Symes, A., Wiedenfeld, D. A. and Butchart, S. H. M. 2012. Financial costs of meeting global biodiversity conservation targets: Current spending and unmet needs. *Science* 338:946–49.

McNicol, D. K., Robertson, R. J., and Weatherhead, P. J. 1982. Seasonal, habitat, and sex-specific food habits of red-winged blackbirds: Implications for agriculture. *Canadian Journal of Zoology-Revue Canadienne De Zoologie* 60:3282–89.

McWilliam, A. N., and Cheke. R. A. 2004. A review of the impacts of control operations against the red-billed quelea (*Quelea quelea*) on non-target organisms. *Environmental Conservation* 31:130–37.

MEA. 2005. *Millenium Ecosystem Assessment: Ecosystems and Human Well-Being: Synthesis*. Washington: Island Press.

Meinzingen, W. W., Bashir, E. S. A., Parker, J. D., Heckel, J.-U., and Elliott, C. C. H. 1989. Lethal control of quelea. In *Quelea Quelea: Africa's Bird Pest*, ed. C. C. H. Elliott, and R. G. Allan, 293–316. Oxford: Oxford University Press.

Myers, N. 1998. Lifting the veil on perverse subsidies. *Nature* 392:327–28.

Mynott, J. 2009. *Birdscapes: Birds in Our Imagination and Experience*. Princeton, NJ: Princeton University Press.

Ndang'ang'a, P. K., Njoroge, J. B. M., Ngamau, K., Kariuki, W., Atkinson, P. W., and Vickery, J. 2013. Avian foraging behaviour in relation to provision of ecosystem services in a highland East African agroecosystem. *Bird Study* 60:156–68.

Oaks, J. L., Gilbert, M., Virani, M. Z., Watson, R. T., Meteyer, C. U., Rideout, B. A., Shivaprasad, H. L., Ahmed, S., Chaudhry, M. J. I., Arshad, M., Mahmoud, S., All, A., and Khan, A. A. 2004. Diclofenac residues as the cause of vulture population decline in Pakistan. *Nature* 427:630–33.

Paoli, G. D., Peart, D. R., Leighton, M., and Samsoedin, I. 2001. An ecological and economic assessment of the nontimber forest product gaharu wood in Gunung Palung National Park, West Kalimantan, Indonesia. *Conservation Biology* 15:1721–32.

Parker, K. R., Kasahara, M., Taylor, P. A., Child, K. N., Morgan, H. W., Rasmussen, G. G., Zellman, R. K. 2008. Mitigation of woodpecker damage to power poles in British Columbia, Canada. In *Environment Concerns in Rights-of-Way Management 8th International Symposium*, ed. J. W. Goodrich-Mahoney, L. Abrahamson, J. Ballard, and S. Tikalsky. New York: Elsevier Science.

Pearce, F. 2004. Bird traffic controller. *New Scientist* 184:48–51.

Peters, D., Engels, C., and Sarra, S. 2012. Natural spread of plant viruses by birds. *Journal of Phytopathology* 160:591–94.

Philpott, S. M., Soong, O., Lowenstein, J. H., Pulido, A. L., Lopez, D. T., Flynn, D. F., and DeClerck, F. 2009. Functional richness and ecosystem services: Bird predation on arthropods in tropical agroecosystems. *Ecological Applications* 19:1858–67.

Podulka, S., Eckhardt, M. and Otis, D. 2004. Birds and humans: A historical perspective. In *Handbook of Bird Biology*, ed. A. Podulka, R. Rohrbugh Jr., and R. Bonney, 1–42. Ithaca, NY: Cornell Lab of Ornithology.

Power, M. E., Tilman, D., Estes, J. A., Menge, B. A., Bond, W. J., Mills, L. S., Daily, J. C. Castilla, J. Lubchenco, and Paine, R. T. 1996. Challenges in the quest for keystones. *Bioscience* 46:609–20.

Prakash, V. 1999. Status of vultures in Keoladeo National Park, Bharatpur, Rajasthan, with special reference to population crash in *Gyps* species. *Journal of the Bombay Natural History Society* 96:365–78.

Prakash, V., Prakash, V., Bishwakarma, M. C., Chaudhary, A., Cuthbert, R., Dave, R., Kulkarni, M., Kumar, K. Paudel, S. Ranade, R. Shringarpure, and Green, R. E. 2012. The population decline of *Gyps* vultures in India and Nepal has slowed since veterinary use of diclofenac was banned. *PLoS ONE* 7:e49118.

Raudsepp-Hearne, C., Peterson, G. D., Tengö, M., Bennett, E. M., Holland, T., Benessaiah, K., MacDonald, and Pfeifer, L. 2010. Untangling the environmentalist's paradox: Why is human well-being increasing as ecosystem services degrade? *Bioscience* 60:576–89.

Reich, P. B., Tilman, D., Isbell, F., Mueller, K., Hobbie, S. E., Flynn, D. F., and Eisenhauer, N. 2012. Impacts of biodiversity loss escalate through time as redundancy fades. *Science* 336:589–92.

Reilly, J. R., and Hajek, A. E. 2012. Prey-processing by avian predators enhances virus transmission in the gypsy moth. *Oikos* 121:1311–16.

Rogers, H., Lambers, J. H. R., Miller, R., and Tewksbury, J. J. 2012. "Natural experiment" demonstrates top-down control of spiders by birds on a landscape level. *PLoS ONE* 7:e43446.

Sagoff, M. 2011. The quantification and valuation of ecosystem services. *Ecological Economics* 70:497–502.

Schröter, D., W. Cramer, R. Leemans, I. C. Prentice, M. B. Araújo, N. W. Arnell, A. Bondeau, H. Bugmann, T. R. Carter, C. A. Gracia, A. C. de la Vega-Leinert, M. Erhard, F. Ewert, M. Glendining, J. I. House, S. Kankaanpää, R. J. T. Klein, S. Lavorel, M. Lindner, M. J. Metzger, J. Meyer, T. D. Mitchell, I. Reginster, M. Rounsevell, S. Sabaté, S. Sitch, B. Smith, J. Smith, P. Smith, M. T. Sykes, K. Thonicke, W. Thuiller, G. Tuck, S. Zaehle, and B. Zierl. 2005. Ecosystem service supply and vulnerability to global change in Europe. *Science* 310:1333–37.

Şekercioğlu, Ç. H. 2006a. Ecological significance of bird populations. In *Handbook of the Birds of the World*, ed. J. del Hoyo, A. Elliott, and D. Christie, 15–51. Barcelona: Lynx Edicions.

———. 2006b. Increasing awareness of avian ecological function. *Trends in Ecology & Evolution* 21:464–71.

———. 2011, Functional extinctions of bird polinators cause plant declines. *Science* 331:1019–20.

Şekercioğlu, Ç. H., Daily, G. C., and Ehrlich, P. R. 2004. Ecosystem consequences of bird declines. *Proceedings of the National Academy of Sciences* 101:18042–47.

Smiley, E. T., Booth, D. C., Wilkinson, L. 2009. Sprays ineffective for preventing sapsucker damage on sugar maple (*Acer saccharum*). *Arboriculture and Urban Forestry* 35:20–22.

Smith, R. J., and Walpole, M. J. 2005. Should conservationists pay more attention to corruption? *Oryx* 39:251–56.

Sodhi, N. S., Şekercioğlu, Ç. H., Robinson, S., Barlow, J. 2011, *Conservation of Tropical Birds*. Oxford: Wiley-Blackwell.

Swift, A. 2014. Americans again pick environment over economic growth: Gallup. Accessed March 23, 2014. http://www.gallup.com/poll/168017/americans-again-pick-environment-economic-growth.aspx.

Tremblay, A., Mineau, P., and Stewart, R. K. 2001. Effects of bird predation on some pest insect populations in corn. *Agriculture Ecosystems & Environment* 83:143–52.

Triplett, S., Luck, G. W., and Spooner, P. 2012. The importance of managing the costs and benefits of bird activity for agricultural sustainability. *International Journal of Agricultural Sustainability* 10:268–88.

Turner, W. R., Brandon, K., Brooks, T. M., Costanza, R., Da Fonseca, G. A., and Portela, R. 2007. Global conservation of biodiversity and ecosystem services. *Bioscience* 57:868–73.

US Department of the Interior, US Fish and Wildlife Service, US Department of Commerce, and USC Bureau. 2014. 2011 National Survey of Fishing, Hunting, and Wildlife-Associated Recreation.

Van Bael, S. A., Philpott, S. M., Greenberg, R., Bichier, P., Barber, N. A., Mooney, K. A., and Gruner, D. S. 2008. Birds as predators in tropical agroforestry systems. *Ecology* 89:928–34.

Walters, E. L. 1996. Habitat and space use of the red-naped sapsucker, *Sphyrapicus nuchalis*, in the Hat Creek valley, south-central British Columbia. Phd dissertation, University of Victoria.

Washburn, B. E., Cisar, P. J., and Devault, T. L. 2014. Wildlife strikes with military rotary-wing aircraft during flight operations within the United States. *Wildlife Society Bulletin* 38:311–20. DOI: 10.1002/wsb.409.

Weatherhead, P. J., Tinker, S., and Greenwood, H. 1982. Indirect assessment of avian damage to agriculture. *Journal of Applied Ecology* 19:773–82.

Wenny, D. G., Devault, T. L., Johnson, M. D., Kelly, D., Şekercioğlu, Ç. H., Tomback, D. F., and Whelan, C. J. 2011. The need to quantify ecosystem services provided by birds. *Auk* 128:1–14. DOI: 10.1525/auk.2011.10248.

Whelan, C. J., Wenny, D. G., and Marquis, R. J. 2008. Ecosystem services provided by birds. *Annals of the New York Academy of Sciences* 1134:25–60.

———. 2010. Policy implications of ecosystem services provided by birds. *Synesis* 1:11–20.

Contributors

SANDRA H. ANDERSON
School of Environment
University of Auckland

JAMES C. BEASLEY
Savannah River Ecology Laboratory
University of Georgia

ANNE-LAURE BROCHET
Centre de Recherche de la Tour du Valat
And Office National de la Chasse et de la Faune Sauvage
CNERA Avifaune Migratrice

EVAN R. BUECHLEY
Department of Biology
University of Utah

MARTINA CARRETE
Department of Physical, Chemical, and Natural Systems
Universidad Pablo Olavide
and Estación Biológica de Doñana (CSIC)

NORBERT J. CORDEIRO
Department of Biology
Roosevelt University
and Field Museum of Natural History

TRAVIS L. DEVAULT
US Department of Agriculture
National Wildlife Research Center

CHRIS FLOYD
Department of Biology
University of Wisconsin–Eau Claire

MOTOKO S. FUJITA
Center for Southeast Asian Studies

JEFF GORDON
American Birding Association

ANDY J. GREEN
Department of Wetland Ecology
Estación Biológica de Doñana
Consejo Superior de Investigaciones Científicas

STEVEN HACKETT
Department of Economics

MATTHEW D. JOHNSON
Department of Wildlife
Humboldt State University

KAYOKO O. KAMEDA
Lake Biwa Museum

DAVE KELLY
School of Biological Sciences
University of Canterbury

ERIK KLEYHEEG
Ecology and Biodiversity Group
Institute of Environmental Biology
Utrecht University

JENNY J. LADLEY
School of Biological Sciences
University of Canterbury

ANTONI MARGALIDA
Department of Animal Production
(Division of Wildlife)
University of Lleida
and Division of Conservation Biology
University of Bern

KATHY MARTIN
Department of Forest and
Conservation Sciences
University of British Columbia

MARCOS MOLEÓN
Department of Applied Biology
Universidad Miguel Hernández
and School of Animal, Plant, and
Environmental Sciences
University of the Witwatersrand

ZACHARY H. OLSON
Department of Psychology
University of New England

ALASTAIR W. ROBERTSON
Institute of Agriculture and Environment
Massey University

HALDRE S. ROGERS
Department of BioSciences
Rice University

JOSÉ ANTONIO SÁNCHEZ-ZAPATA
Department of Applied Biology
Universidad Miguel Hernández

ÇAĞAN H. ŞEKERCIOĞLU
Department of Biology
University of Utah
Faculty of Sciences
Koç University

MEREL SOONS
Ecology and Biodiversity Group,
Institute of Environmental Biology
Utrecht University
and Department of Animal Ecology
Netherlands Institute of Ecology

DIANA F. TOMBACK
Department of Integrative Biology
University of Colorado Denver

DANIEL G. WENNY
Museum of Vertebrate Zoology
University of California, Berkeley

CHRISTOPHER J. WHELAN
Department of Biological Sciences
University of Illinois at Chicago

Index

Page numbers in **bold** refer to figures.